Advances in 21st Century Human Settlements

Series editor

Bharat Dahiya, Bangkok, Thailand

More information about this series at http://www.springer.com/series/13196

T.M. Vinod Kumar
Editor

E-Governance for Smart Cities

Springer

Editor
T.M. Vinod Kumar
School of Planning and Architecture,
New Delhi (SPA-D)
Calicut, Kerala
India

ISSN 2198-2546 ISSN 2198-2554 (electronic)
Advances in 21st Century Human Settlements
ISBN 978-981-10-1241-9 ISBN 978-981-287-287-6 (eBook)
DOI 10.1007/978-981-287-287-6

Springer Singapore Heidelberg New York Dordrecht London

Softcover reprint of the hardcover 1st edition 2015

Printed on acid-free paper

Springer Science+Business Media Singapore Pte Ltd. is part of Springer Science+Business Media (www.springer.com)

Foreword I

Around the world there are about 3.5 billion people living in cities, but their areal extension covers only 2 % of the Earth's land surface. This indicates that a large number of people live in a small area, resulting in overcrowding. In such a situation, the challenge lies in the management of safety and facilitation of needs for people to live comfortably. Towards this direction, there is continuous improvement in transportation, energy distribution, construction, technology development, provision of facilities and utilities. In recent years, the negative impacts of rapid urbanization are recognized, as many problems are being faced by cities. For example, traffic congestion, poor services of facilities, high cost of housing, etc.

The good news is that due to advancement and innovation in technology, especially Information Technology (IT), it is not difficult to manage cities. However, one needs to be smart in the application of IT for people. Such smart application of new technology for better living for people in cities is 'smart cities'. Smart cities are well planned with innovations and knowledge that develop extraordinary capabilities. They work all the time, i.e. 24 h and 7 days a week. Such a facility makes city citizens to live and enjoy in a satisfactory way of life with easy public services, comfortable mobility, conserving energy, clean environment and better utilization of natural resources. It helps to withstand even hardships due to natural calamities, including economic downturns. The key concern here is governance. This is because, even with all the facilities available in cities, if governance is bad, then smart cities are of no use. Thus good governance is of utmost importance. To achieve good governance, the smart way is to follow e-governance.

E-governance or electronic governance implies the application of Information and Communication Technology (ICT) for all services and facilities for easy operation and management. Day by day, with more and more innovations and new technological development, there is continuous improvement in operations and maintenance of facilities. In such a scenario, developed countries are quick to implement the new technology for betterment of the people, while developing and underdeveloped countries are slow and far behind. The time has come now to make all the cities of the world to be smart cities, which makes the people in cities live more comfortably.

Thus, this book on E-governance of smart cities is timely and is a must read for all decision-makers and practitioners to follow. The present volume is a landmark among other books of similar nature as it covers all the important elements of smart cities. Earlier series of similar nature from the same editor, Sri Vinod Kumar, on—"Geographic Information System for Smart Cities"—with 23 international authors have already been published. This book forms part of a series of "Advances in 21st Century Human Settlements", which Springer will be publishing soon. The chapters in the book are written by eminent personalities and contains mainly four parts: Part I is on *E-Smart City Governance-State of the Art Studies*, Part II on *E-Smart City Governance-Domain Studies*, Part III on *E-Smart City Governance- Tools and Issues*, and Part IV on *E-Smart City Governance-Futures*. Surprisingly, only three Indian cities—Hyderabad, Bangalore and Jaipur—are listed among the 8,000 identified smart cities in the world by the Intelligent Community Organisation recently. This indicates that much more needs to be done in other cities of India in a smart way. In fact, the Government of India has already planned to come out with 100 smart cities soon.

The concept of e-governance for cities is a continuous process that improves every year with new ideas and philosophies with local cultural values. The articles written by eminent scholars in this book, I am sure can be transmitted to the younger generation to enrich the world of knowledge.

August 2014 T. Harinarayana

Foreword II

There is a new type of convergence that is slowly happening; a convergence of ideas, issues and action with speed and efficiency. There are many reasons for this, of which three factors contribute critically. They are individual empowerment, efficient and quick use and access to resources and processes, and lastly records and documentation to be transparently generated for such initiatives. Welcome to the world of Smart cities; Involving Smart people, Smart processes and Smart gadgets to begin with. It is in this context that this book by Prof. Vinod Kumar and group can be seen.

There is both an urgent need and efficient process that will 'liberate' common citizens and protect our environment and climate from reaching catastrophic levels and at the same time provide for efficient governance. One clear way ahead is the 'Smart Way'. Despite the increased automation and ICT deployed worldwide in the last decade relating to issues of governance, it is still way below what can and needs to be done. Almost everywhere, governance is failing, and this is not limited to poor or developing countries. Recent examples from the Middle East and Europe endorse this. Thus the 'Smart' approach to various concerns that directly and indirectly affect citizens is to be looked at in a positive way. Also, given the fact that urbanization is a global phenomenon, it makes sense to address this at the city level, which also has a positive spillover effect on making things easy at the remote and rural level due to increased quality and efficiency of governance in the city.

Keeping these issues in mind the team of authors under Prof. Vinod Kumar have come up with a well-researched book that addresses a wide range of issues in various domains. It is essential to remark that currently, even the best practices of governance are neither efficient in terms of resource use (water, electricity, etc.) nor in terms of capital. Thus a better e-governance approach is much wanted universally, more so at urban levels. A city like Bengaluru in Karnataka, India, consumes almost 75 % of all resources for just 20 % of the state population, creating great inequity and inefficiency in resource use with the remaining population. Perhaps New Delhi could be worse, which only motivates one to initiate Smart approaches to E-Governance to urban issue management everywhere.

It is my privilege to write the foreword to this book. The book is timely and covers topics that can be termed "For Urgent" consideration. I am confident this book will be valuable to academicians, researchers, NGOs, administrators and urban planners. I look forward to the wide use and feedback on this book.

Mysore, July 2014 Ashok Rao

Preface

At a time when 100 new smart cities seem to have captured the attention of the nation, it becomes all the more important to have a close look at where our existing cities stand with regard to e-governance and how they are moving in the direction of becoming smart cities. Smart city as seen here is an integrated urban system of Smart Urban Economy, Smart People, Smart Governance, Smart Mobility, Smart Environment and Smart Living.

One of the 23 reforms which form part of the reform agenda of the Jawaharlal Nehru National Urban Renewal Mission (JNNURM) was the introduction of a system of e-governance using IT applications like GIS and MIS for various services provided by Urban Local Bodies (ULBs)/para statals. All the 65 mission cities and all the small towns that obtained projects approved under the UIG or UIDSSMT component of the Mission had to undertake implementation of this reform as well during the 7-year mission period of 2005–2012, subsequently extended by two more years. This goes along with the government of India's National E-Governance Plan (NeGP). NeGP intends to institute and enable mechanisms to improve the system of governance and thus provide better services to citizens by effective use of ICT. E-governance in municipalities is one of the Mission Mode projects under NeGP, which is expected to result in improved service delivery by local governments for the citizen. The broad aims of implementing e-governance in municipalities as per the Handbook on Primers include,

- Focus on identified citizen services that would be coveted with clearly laid down service levels and outcomes to be achieved,
- improve efficiency and effectiveness in interaction between local government and its citizens and other stakeholders,
- improve quality of internal local government operations and management information systems to support and stimulate good governance,
- bring out transparency and accountability in urban local body operations,
- help improve the delivery of services to citizens.

In line with this, the services to be covered under the reform agenda of the Mission are:

Birth and death registration and health programmes (under basic citizen services), property tax and licences (revenue earning services), water supply and other utilities and building plan approval (development services), procurement and monitoring of projects (efficiency improvement services), accounting and professional management system (back office improvements), citizen grievance redressal (monitoring).

The rationale for this particular reform is that it will benefit citizens by providing single window access to various services, simplification of procedures, and opportunity for greater participation in decision-making and so on. For urban bodies, a common information base becomes available across departments, creation of an effective MIS becomes possible, there will be improvement in revenue collection and efficient citizen grievance redressal becomes possible. For the management like the Mayor, Commissioner, Standing committee, availability of standardized and meaningful MIS on timely basis and ability to monitor and track progress, services and revenue effectively will be the impact.

This effort goes along with the Service Level Benchmark (SLB) concept for e-governance in municipalities which eventually involves having a set of Service level Key Performance Indicators (KPI) defined for e-governance services at the national level. The SLBs for birth/death registration includes number of birth/death registered as against applied for registration and number of birth/death registered as against actual number of birth/death as per KPIs. Similar KPIs have to be laid down for calculation and payment of property tax, payment and management of utility bills, grievance handling, building plan approval, e-procurement and project/ward works, solid waste management system, licences, accounting system and personnel management system.

Where do the Mission cities stand in regard to implementation of the e-governance reform at the end of the 9-year period of the Mission? In the e-governance set-up, the following 27 cities could achieve full implementation:

Ahmedabad, Ajmer, Bangalore, Bhopal, Chennai, Coimbatore, Delhi, Greater Mumbai, Hyderabad, Indore, Jabalpur, Jaipur, Kanpur, Kochi, Lucknow, Ludhiana, Madurai, Mysore, Nagpur, Nanded, Nashik, Pune, Rajkot, Surat, Thiruvananthapuram, Vizag.

It is difficult to believe that four mission cities figure in the category of 'no implementation'.

This review shows how uneven the implementation of such a key reform is, that too one which has direct relevance for the ordinary city dweller. What could be the reason for such poor performance? Is it lack of sufficient support and guidance from the concerned state government? Is any one single authority or level held responsible for drawing up the reform road map and ensure completion? Is it absence of capacity at the city level and poor infrastructure support to take up implementation?

As regards the National Mission Mode Project on E-governance, which further facilitates the smart city agenda, a total of 29 projects were approved as a supplement to the e-governance agenda during the period February 2009 and May 2013 for as many mission cities. The total cost of the projects amounts to Rs. 315.89 crore. Although a full assessment of this additional injection of funds has not yet taken place because total projects completion is still pending, what is clearly visible is that wherever the e-governance reform has been properly implemented, the change is visible. In Hyderabad, the Offsite Real-Time Monitoring System is a unique easy-to-use mobile phone mechanism that generates real-time offsite images of municipal activities by citizens. It makes real-time data available to more than 7.4 million citizens living in over 600 sq km. Building plan approval, which used to take 2–3 months now takes only 7–10 days. In Surat, first copy and subsequent duplicate copies of birth and death certificates are issued within 15 min after filling up the form and providing necessary proof.

While some cities are steadily moving towards what could be called Smart governance, the fact remains that this is only a small fraction of the 4,041 urban local bodies we have. This highlights the need for making this reform an essential part of the next phase of the Urban Renewal Mission and insisting on time bound compliance, a must for achieving the Smart city objective, thereby providing citizens the full advantage of technology.

While the full elements of what a Smart City is still needs to be detailed by the government, varying degrees of action are visible in some select cities at least. The new Greenfield townships coming up along the Delhi–Mumbai Industrial Corridor are planned to be Smart cities. GIFT city coming up next to Ahmedabad is also planned to be a Smart city. Where are the proposed 100 new Smart cities to come up? These are expected to be along the Delhi–Mumbai Industrial Corridor, the proposed Bangalore–Chennai, Chennai–Vizag, Kolkata–Delhi Corridors plus the Yamuna Expressway. The new capital of the new state of Seemandhra can also be planned as a Smart city. It is possible that new private township initiatives like the case of Lavasa would plan to be Smart cities. However, what is important at this stage of evolving of the Smart city concept in India is that a policy pronouncement should lay down the elements of a Smart city and how the process of setting up such cities would be encouraged, promoted and supported. There would also be the important question of integration with other departments and entities which would need to be properly outlined. This is a big challenge and unless we as a nation gear ourselves to rise to the occasion, with the huge IT strength that we have, we would be denying our city residents the innumerable advantages of smart living.

This book, I believe, will play its role in conveying this message loud and clear so that the system readies itself to undertake the big challenge of being Smart.

M. Ramachandran

Contents

E-Governance for Smart Cities 1
T.M. Vinod Kumar

Part I E-Smart City Governance-State of the Art Studies

E-Governance State-of-the-Art Survey: Stuttgart, Germany 47
Satyendra Singh

E-Governance and Smart Cities: Cases of Ahmedabad and Hyderabad 65
Vinita Yadav

Part II E-Smart City Governance-Domain Studies

E-Governance and Its Role in Infrastructure Services of UAE, Case Study—Dubai 81
Ashmita Karmakar

E-Governance for Public Realm: Around Panniyankara Monorail Station, Kozhikode, Kerala 99
T.M. Vinod Kumar and P. Bimal

Smart Systems and Smart Grids for Effective Governance of Electricity Supply in India 159
Veena Aggarwal and Parimita Mohanty

E-Governance for Solar Photo Voltaic Powergrid: Solar City Gandhinagar, Gujarat, India 177
Jignesh G. Bhatt and Omkar K. Jani

Part III E-Smart City Governance-Tools and Issues

Can Smart City Be an Inclusive City?—Spatial Targeting (ST) and Spatial Data Infrastructure (SDI) 233
N. Sridharan

M-Governance: Smartphone Applications for Smarter Cities—Tapping GPS and NFC Technologies 245
Ummer Sahib

Participatory E-Budgeting Using GIS-Based Spatial Decision Support System: Kozhikode Municipal Corporation 307
M.A. Naseer, P. Bimal and T.M. Vinod Kumar

Part IV E-Smart City Governance-Futures

Smart City E-Governance: Issues and Future 353
T.M. Vinod Kumar

Authors Biography

Veena Aggarwal is a Fellow at the Energy and Resource Institute. She has worked on governance and regulatory issues in the electricity and energy sector for over a decade. She has worked with regulatory commissions on regulation and tariff-related assignments in Rajasthan, Punjab, Jammu and Kashmir and Jharkhand. She has also worked on studies in the electricity and energy sector for the Ministry of Power, Rural Electrification, Planning Commision and Central Electricity Regulatory Commission. In the urban space, she has worked as project lead for TERI's work as a 'Centre of Excellence' for the Ministry of Urban Development on urban reforms, rationalising user charges and participatory governance. Her research interests include pricing, governance and reforms in the infrastructure sectors.

Prof. Jignesh G. Bhatt received B.E. (Instrumentation and Control Engineering) from Gujarat University, India, 1997 and M.Tech. (Electrical Engineering, Specialization: Measurement and Instrumentation) from IIT Roorkee, India, 2010. He has post industrial experience on Managerial Positions during 1997–1999, since 2000 he has been serving at Dharmsinh Desai University (DDU), India as assistant professor (Instrumentation and Control Engineering Department). He served as Co-ordinator and Principal Investigator for Government Sponsored Research and Community Development Projects, PG Thesis Supervisor, Invited Expert-Session Chair in Conferences and Journal Reviewer-Editor. His research interests include Automation, e-Governance, e-Learning, Instrumentation, Solar Energy, Solar City, Smart Grid and Wireless Sensor Networks.

P. Bimal holds a master's degree in City Planning from Indian Institute of Technology Kharagpur, and a bachelors' degree in Architecture from Kerala University. He had been working at JUSCO, Jamshedpur, before joining as assistant professor at Department of Architecture, NIT Calicut. He is currently pursuing Ph.D. in urban modeling.

Prof. T. Harinarayana is Director, GERMI and also an Independent Director of GSPC Company. He holds a Ph.D. from Edinburgh University, UK and Ph.D. from Indian School of Mines, Dhanbad. He was a fellow member of IGU, GSI, APAS and elected member of RANS, Moscow. He was also an executive member, AEG and APAS. He is a member of various international associations including American Geophysical Union, SEG, etc., with more than 100 publications. Currently, he is executive member of EMSEV. Professor Harinarayana has over 35 years of experience earlier with NGRI as Scientist "G".

Dr. Omkar K. Jani is currently the Principal Research Scientist (Solar Energy) at Gujarat Energy Research and Management Institute (GERMI). His primary goals include advisory for various solar-related government and private initiatives, developing and undertaking fundamental and applied research, and capacity building. Dr. Jani is a member of the State Advisory Committee of the Gujarat Electricity Regulatory Commission. He is also an advisor to several clean-tech companies and NGOs. Dr. Jani received his Ph.D. in Electrical Engineering from Georgia Institute of Technology with specialization in Solar Photovoltaic Science and Engineering, and his post-doctoral fellowship from the Institute of Energy Conversion, Delaware.

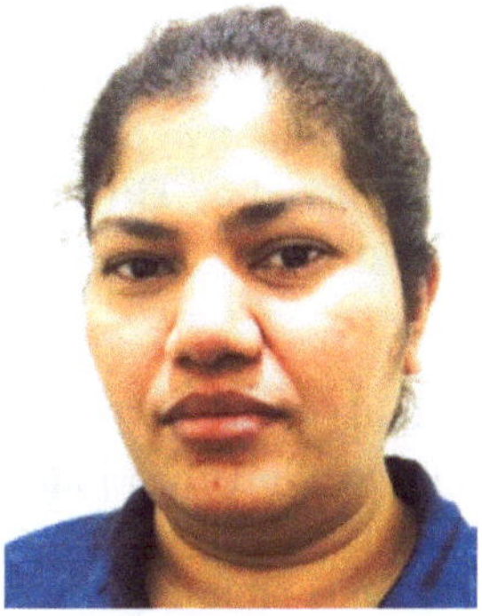

Ashmita Karmakar has graduated in Architecture from LAD College, Nagpur, India in 1998. She completed her master's in Town and Regional Planning from Bengal Engineering College, Howrah, India and Ph.D. in Environmental Planning from School of Planning and Architecture, New Delhi, India in 2012.

She was professionally associated with some of the prestigious organizations in India before moving to Dubai, UAE. Currently she is working in the field of Urban Planning, Master Community planning and Development controls for various signature projects of Dubai, UAE.

In addition, Ashmita karmakar has presented/published some of her research work on urban and environmental planning in the international journals and newspapers.

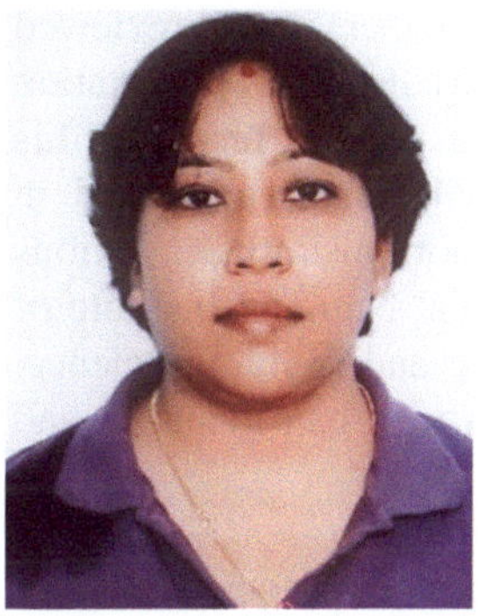

Dr. Parimita Mohanty Fellow and Team Leader Energy Environment Technology Development Division Centre for Distributed Generation, TERI.

She has a Ph.D. degree from IIT, Delhi on control strategies for distributed generation based micro-grids. Currently, over a span of more than one decade, she has vast experience in the areas of smart mini grid and smart grid, solar PV-based product design and customization, battery storage, testing and assessment of various RE products, etc. She has been involved in feasibility study, detailed designing and execution, project management and commissioning as well as performance assessment of various distributed generation-based mini-grid projects. She has more than 16 published papers to her credit, which were published in reputed national and international journals She is also a member of various technical committees for standard formulation and quality assurance aspect of solar photovoltaic.

She has initiated and plays the critical role in establishing various solar PV-based laboratories such as solar lighting lab, battery testing lab, solar hybrid and smart controller lab, etc. She is currently responsible for strategic alliance with various industry partners working in solar lighting, smart grid and smart mini-grid domain. Along with her team, she is currently dealing with more than 20 industry partners nationally as well as internationally. She is also responsible for strategic linkages with academic institutions and Universities for carrying out advanced research on solar lighting and smart mini-grid.

She has handled over 15 projects as team leader or project manager dealing with various aspects of renewable energy and distributed generation such as technology development and customization, product testing and approval, technology adaption and technology transfer, monitoring and impact assessment, project execution both nationally and internationally. She has experience in working in various countries such as India, Nepal, Bhutan, Myanmar, Thailand, Indonesia, Philippines, Australia,

Norway, The Netherlands, etc. She has more than 15 technical papers, published in reputed national and international peer-reviewed journals, more than 15 invited and featured articles in various newspapers and magazines and more than five patents applied for in her credit. She is also the co-author of a book published by TERI on Renewable Energy Technologies titled "Renewable Energy Engineering and Technology—A knowledge Compendium".

Dr. M.A. Naseer possess Doctoral Degree in Transportation Planning and Masters Degree in Urban Planning from Indian Institute of Technology Kharagpur, India, and a graduate degree in Architectural Engineering from Regional Engineering College Calicut, India. He started his teaching career in Architecture at Manipal Institute of Technology Manipal, India and later on joined as a faculty in National Institute of Technology Calicut, India. He has published papers in peer reviewed international and national journals in Transportation Planning and Building Science. He has presented his research works in International Conferences held in USA, Japan and Malaysia. He has served in different administrative positions including that of Head of the Department of Architecture at National Institute of Technology Calicut. In addition to Transportation Planning and Building Science, his research interests include remote sensing application in urban planning and Smart Planning.

Dr. M. Ramachandran has had a distinguished career in the IAS at top levels of policy making and implementation in India. He has been Chief Secretary of the state of Uttarakhand and then Secretary, Urban Development in government of India. A well recognized urban expert and credited with the whole lot of new initiatives taken during his stewardship of the Urban ministry, Dr. Ramachandran has a Master's degree in Economics, M.Phil. in Economic Planning and his Ph.D. is in the area of Project Planning. He has seven books to his credit, the latest, India's Urban Confusion-Challenges and Strategies was released in September 2014.

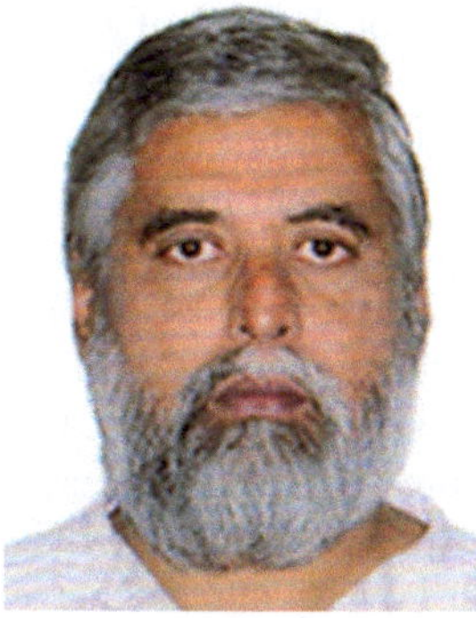

Dr. Ashok Rao holds a B.E., M.E. and Ph.D. all in EE from Mysore University (1982), Indian Institute of Science, Bangalore (1985) and Indian Institute of Technology, Bombay (1991) respectively.

He specializes in Digital Signal Processing area and is known to be one of the leading figures in this area in India. In addition, he teaches and researches in areas of Image Processing, Multimedia, Linear Algebra, Renewable Energy, Bio-Metrics, Appropriate Technology and Sustainable Development, Management, Archeology, Education and Academic Reforms and ICT.

He has over 27 years of Teaching and Research Experience and has successfully guided Ph.Ds both in E and C and Computer Science areas and has over 100 publications. He is also a Solar Photo Voltaic (SPV) specialist having been trained by Siemens Solar California during 1996–1997.

Mr. Ummer Sahib is a City Planner with 29 years experience in geospatial technology and development of innovative solutions for City Planning and Management. He has been responsible for process automation in executing large GIS, LIS and Urban mapping projects in the Middle East, Europe and Africa.

Starting his career in GIS software design for Department of Electronics, Government of India, Mr. Sahib's expertise lies in open source web mapping applications, navigation data production, Geo-analytics and perfecting a virtual addressing system called Geo Reference Locator (GRL) for the UAE. Since 2005, Mr. Sahib is the driving force behind his company's creative innovation on GPS and NFC technology solutions.

Satyendra Singh is a Geospatial Technology Consultant and Environment Planner with over 16 years of experience. With education profile ranging from Architecture, Planning and Geoinformatics, he has developed synergies across these disciplines through professional work and involvement in academic settings in Germany and India. He is based in Stuttgart-Germany and works as Visiting Lecturer at the University of Applied Sciences in Stuttgart. He has also delivered special lectures including CEPT University-Ahmadabad and SPA New Delhi India. He has been managing projects related to urban planning, utility mapping, Energy and infrastructure management in Germany and India involving hardcore technical knowledge of GIS, ICT, Photogrammetry and LiDAR systems.

Prof. Dr. N. Sridharan is Director, School of Planning and Architecture, Vijayawada since 2013. He has more than 30 years experience in the field of urban and regional planning, poverty alleviation, and for the past 10 years in Spatial Data Infrastructure and Urban Governance. He has worked with the Government of India, a Nationalized Bank, School of Planning and Architecture—New Delhi, a National Level NGO, and in the private sector. He is a visiting Professor in several European Universities and resource person for ADB. His research interests include Urban Poverty and Inclusive Governance, Spatial Governance, and Peri-urban development. e-mail: dr.nsridharan@gmail.com

Prof. T.M. Vinod Kumar is an urban planner, with 44 years of experience in teaching, research and consultancies in India, China, Bhutan, Nepal, Indonesia and Malaysia. Academic positions he had held include Dean of Studies, HOD Urban Planning, Head Centre for Analysis and Systems Studies, Head GIS and Remote Sensing Centre; School of Planning and Architecture, New Delhi and Visiting Professor Institute of Technology, Bandung and National Institute of Technology, Calicut. He worked also as Regional Program Coordinator, International Centre for Integrated Mountain Development. He is author of many books and journal articles and his last edited book is "Geographic Information System for Smart Cities" (2014).

Dr. Vinita Yadav has a total work experience of over 13 years in diverse consultancy organizations and in domains ranging from institutional strengthening, urban development, governance, infrastructure development, financial planning and capacity building. She has completed her Master's in Planning (with specialization in Urban Planning) from S.P.A, New Delhi. Later, she obtained Ph.D. from Jawaharlal Nehru University, New Delhi and M.B.A (financial management) from Indira Gandhi National Open University (IGNOU), New Delhi. She started her career as a project manager in SEEDS India in Delhi and later worked on many projects related to urban development as a freelancer. She joined J.P.S Associates (P) Ltd. as a Consultant in October 2004 where she has managed varied projects as Project Coordinator in the field of urban and rural development. She has also managed consultancy portfolio in a non-government organization, i.e. Praxis—Institute for Participatory

Practices, where her responsibilities included project management and business development. Since December 2009, she is working as associate professor in Department of Regional Planning, School of Planning and Architecture, New Delhi where she is teaching subjects such as Poverty and Development, Institutional Analysis and Governance and handling studio assignments in the areas of village, block and district planning. Recently, she published a book titled "Metropolitan Governance: Cases of Ahmedabad and Hyderabad".

E-Governance for Smart Cities

T.M. Vinod Kumar

Abstract This chapter 'E-Governance for Smart City' is intended to give conceptual clarity of the subject matter of the book. The chapter is expected to give broad direction to the book. Until now E-Governance literature is dominated by authors who are computer scientists and IT specialists since there are several technical issues to be solved (like, for example, big data management) but this book takes a departure and will be authored by domain specialists related to city. We are jointly working out the emerging E-Governance System for Smart City from the point of view of urban domain specialists such as specialists in urban development, urban planning, climate change, carbon accounting, water Governance, energy governance, public realm and so on. To start with definition of E-Governance and E-Government is collated from different organizations point of view. Then how consumers of Governance, such as Citizen, Government and Business benefit from E-Governance is enumerated. Further, it traces five distinct phases of E-Governance development such as Phase-1(1996–1999): Basic Web Presence, Phase-2(1997–2000): Interactive web, Phase-3(1998–2003): Transaction web, Phase-4(2000–2005): Integrative and Transformation web and Phase-5(2005+): Smart City Governance web. Prior to 2005 there has been rapid transition in phases but with the emergence of phase-5, there is a trend in stabilization of urban E-Governance. This chapter enumerates new functional specialization of E-Governance in different phases and discusses Smart City E-Governance System and its prime requirements.

Keywords Electronic governance · E-Government · Definitions · Benefits of E-Governance · Evolution · Status of E-Governance in India · E-Governance impact to politician · Administrators and citizen · Smart city E-Governance requirements

---

T.M. Vinod Kumar (✉)
School of Planning and Architecture, New Delhi (SPA-D), Calicut, Kerala, India
e-mail: tmvinod@gmail.com

T.M. Vinod Kumar (ed.), *E-Governance for Smart Cities*, Advances in 21st Century Human Settlements, DOI 10.1007/978-981-287-287-6_1

1 A Voyage to the Unknown Undefined and Untested

None has until now designed and tested an appropriate E-Governance system for a Smart Cities in India. There are no universal or Indian standards of E-Governance available for Smart Cities. Is E-Governance the brain of the City under e-Democracy? Can E-Governance be automated by emerging smart city technologies with no role to humans? Can India with very low ranking of achievement in United Nations E-Government Survey of 2012 ever aspire to have an E-Governance System suited to Smart City? There are many questions, unknowns and undefined. That explains the need for this book. This book had adopted city-specific and domain-specific urban case study approaches to find out what can be E-Governance for a Smart City. Most of the literature on E-Governance is available for States in India with a strong bias towards State and predominantly rural-district level E-Governance, but focus of the book is city level studies. The case studies presented in this book are not limited to Indian cities alone. It is meant to look at many innovations in e-city Governance from world over. Some E-Governance System presented is never attempted and only a suggestion for adoption based on research and development.

This chapter introduces the subject matter of the book. We have in India, like in many countries, a National E-Governance Plan [1] and several E-Governance Mission encompassing different sectors. Under federal constitution of India, many States and Union Territories have also their own E-Governance Plans and E-Governance Missions, which need not necessarily be an exact replica of the National Plan of E-Governance. Many cities in India have their own E-City Governance system. These cities have many layers of constitutional entities with differing in different states, constitutional powers to govern them such as Ward, Municipality or Municipal Corporation and, Metropolitan Planning Committee with their own distinct domain and jurisdiction of Governance. It is not mandatory that there shall be uniformity in E-Governance under one and the same Indian Constitution. E-Governance in India is like Buddhism or Hinduism with many diverse schools of thoughts, all of which live together without violating the Indian Constitution. Hence, E-Governance in India truly represents a unity in diversity, for a multi-racial, multi-cultural and multilingual Indian Society with varying levels of development. Hence, it can be expected that unique city-specific E-Governance for Smart city can come into existence since each city is unique spatially. Many believe if you deploy all available Smart city technologies in any city we have a Smart City, which we do not agree. All components that make a Smart City shall be there along with Smart City technology which includes Smart E-Governance. This is explained below in Sect. 1.11.

This book is being written in 2014 by many authors when '*Mangalyan*', the first Indian mission to planet Mars is very rapidly moving towards its long journey to Mars which may take many more months to reach Mars and I found a striking similarity with this book and its goal which may takes many levels of refinement in existing E-Governance systems like midway corrections on *Manglayan* movement before we reach an acceptable E-Governance for Smart City in the distant future. One *Mangalyan* is not sufficient to reach the goal but many more are required with varying objectives.

However, most of the countries had made some advancement in E-Governance. All over the world there is an attempt to make city more liveable and best performing for the benefit of people and environment. This book from India, second after Geographic Information System for Smart Cities [2] is an attempt to decipher what is largely unknown, E-Governance for Smart City and focus more on India. It can be seen here, that Indian scholars are writing about international experience of E-Governance of smart cities in this book.

Innovations in E-Governance in India mostly reach from Central or Union Government to State Government under its many Central Government Sponsored schemes. This is because more amount of tax is collected by Central Government than State Government and there are governing system to share the tax with states. While sharing the tax many centrally sponsored innovative schemes on e-city Governance are also introduced and it was mandatory to State Government to introduce some elements of E-Governance practice in their state if they accept funds for certain schemes such as Jawaharlal Nehru National Urban Renewal Mission (JNNURM) since 2005.

The term E-Governance and E-Government is understood in many ways emphasizing its different aspects which will be discussed.

1.1 Electronic Governance or E-Governance

The concept of governance has a duality built into that even when it moved from autocracy to oligarchy and then to democracy. In the scheme of this duality, there is one or a group of people who are given legal powers to govern called competent authority and there are another large group who are recipient of governance. They can be, in a democracy for example, the elected politicians and appointed bureaucrats and electorate which need not necessarily 100 % population. This duality gets gradually dissolved to a non-duality in an evolving and progressive e-democracy. It is the constitution of a country that hasten the dissolution of this non-duality to one in which the democracy wants people to be totally involved in all aspects of governance whether you are elected, appointed or electorate who need to act beyond voting once in every 5 years. This is considered a minimum requirement in Smart City. India is striving to reach that non-duality by a series of constitutional amendment such as 74th constitutional amendment and innovative models of public involvement designed by different cities and states at their own initiatives.

Electronic Governance or E-Governance is a movement towards 'government online' to deliver their services and programs, to provide government information and to interact with the citizen. This results in a user friendly relationship with citizen, business and state/government. E-Governance includes the vision, strategies, planning, leadership and resources to organize political and social powers within the framework of National constitution. This may call for a different governance frame for all governmental action than practiced today by dissolving the duality mentioned. This means E-Governance by all.

1.2 Definitions of E-Government

Municipal, State and Union Government uses information technology and internet to support government operations, engage citizens and provide government services. The interaction may be in the form of obtaining information, filings, or making payments and a host of other activities via the World Wide Web [3–5].

World Bank Definition of E-Government: In their website it is given that 'E-Government' refers to the use by government agencies of information technologies (such as Wide Area Networks, the Internet and mobile computing) that have the ability to transform relations with citizens, businesses and other arms of government. These technologies can serve a variety of different ends: better delivery of government services to citizens, improved interactions with business and industry, citizen empowerment through access to information or more efficient government management. The resulting benefits can be less corruption, increased transparency, greater convenience, revenue growth and/or cost reductions.

Traditionally, the interaction between a citizen or business and a government agency took place in a government office. With emerging information and communication technologies (ICTs) it is possible to locate service centres closer to the clients. Such centres may consist of an unattended kiosk in the government agency, a service kiosk located close to the client in the neighbourhood, or the use of a personal computer in the home or office or mobile phone.

Analogous to e-commerce, which allows businesses to transact with each other more efficiently (B2B) and brings customers closer to businesses (B2C); E-Government aims to make the interaction between government and citizens (G2C), government and business enterprises (G2B) and inter-agency relationships (G2G) more friendly, convenient, transparent and inexpensive.

WikiBooks definition of E-Governance: Definitions of E-Government range from 'the use of information technology to free movement of information to overcome the physical bounds of traditional paper-and physical-based systems' to 'the use of technology to enhance the access to and delivery of government services to benefit citizens, business partners and employees'. The common theme behind these definitions is that E-Government involves the automation or computerization of existing paper-based procedures that will prompt new styles of leadership, new ways of debating and deciding strategies, new ways of transacting business, new ways of listening to citizens and communities and new ways of organizing and delivering information.

Ultimately, E-Government aims to enhance access to and delivery of government services to benefit citizens. More important, it aims to help strengthen government's drive towards effective governance and increased transparency to better manage a country's social and economic resources for development. The key to E-Government is the establishment of a long-term, organization-wide strategy to constantly improve operations with the end in view of fulfilling citizen needs by transforming internal operations such as staffing, technology, processes and work flow management. Thus, E-Government should result in the efficient and swift delivery of goods and services to citizens, businesses, government employees and agencies. To citizens and businesses, E-Government would mean the simplification of procedures and streamlining of the approval process. To government employees and agencies, it would mean the facilitation of cross-agency coordination and collaboration to ensure appropriate and timely decision-making.

United Nations definition [6]: 'E-Government is defined as utilizing the Internet and the worldwide web for delivering government information and services to citizens'.

Global Business Dialogue on Electronic Commerce—Gibed [7] definition 'Electronic government (E-Government) refers to a situation in which administrative, legislative and judicial agencies (including both central and local governments) digitize their internal and external operations and utilize networked systems efficiently to realize better quality in the provision of public services'.

Gartner Group's definition: According to Gartner Group e Government is 'the continuous optimization of service delivery, constituency participation and governance by transforming internal and external relationships through technology, the Internet and new media'.

Definition of the Working Group on E-Government in the Developing World [8]: 'E-Government is the use of ICTs to promote more efficient and effective government, facilitate more accessible government services, allow greater public access to information and make government more accountable to citizens. E-Government might involve delivering services via the Internet, telephone, community centres (self-service or facilitated by others), wireless devices or other communications systems'.

1.3 Definitions of E-Governance

'Governance implies the processes and institutions, both formal and informal, that guide and restrain the collective activities of a group. Government is the subset that acts with authority and creates formal obligations. Governance need not necessarily be conducted exclusively by governments. Private firms, associations of firms, non-governmental organizations (NGOs) and associations of NGOs all engage in it, often in association with governmental bodies, to create governance; sometimes without governmental authority [9]'. Hence, constitutional ward committee in a municipality is as important as non-constitutional Residential Welfare Association (RWA) in the process of ongoing governance assuming different roles complementary to each other.

The UNESCO definition (www.unesco.org): 'E-Governance is the public sector's use of ICTs with the aim of improving information and service delivery, encouraging citizen participation in the decision-making process and making government more accountable, transparent and effective'. E-Governance involves new styles of leadership, new ways of debating and deciding policy and investment, new ways of accessing education, new ways of listening to citizens and new ways of organizing and delivering information and services. E-Governance is generally considered as a wider concept than E-Government, since it can bring about a change in the way citizens relate to governments and to each other. E-Governance can bring forth new concepts of citizenship, both in terms of citizen needs and responsibilities. Its objective is to engage, enable and empower the citizen. 'E-democracy builds on E-Governance and focuses on the actions and innovations enabled by ICTs combined with higher levels of democratic motivation and intent' [10].

The concept of electronic governance chosen by the Council of Europe covers the use of electronic technologies in three areas of public action; relations between the public authorities and civil society; functioning of the public authorities at all stages of the democratic process (electronic democracy); and the provision of electronic public services [11]. E-Governance is defined as the, 'application of electronic means in (1) the interaction between government and citizens and government and businesses, as well as (2) in internal government operations to simplify and improve democratic, government and business aspects of Governance' [12]. According to Kati [13], 'Governance" is a way of describing the links between government and its broader environment—political, social and administrative'. The application of electronic links means the interaction between government and citizens and government and businesses, as well as in internal government operations to simplify and improve democratic, government and business aspects of Governance [9]. It is then important to enumerate specifically, its benefits.

1.4 Benefits of E-Government

1.4.1 Benefits to Citizens

E-Government benefits the citizens in a number of ways. Some of them are [14, 15]:

1. 24 × 7(24 h, 7 days a week) round-the-clock government service and not 10–5,
2. Economical and convenient service (no need for physical visit to an office),
3. Fast and efficient service (electronic and broadband bandwidth determined),
4. Transparent (no corruption and so-called speed money),
5. Equitable (any one can access it by the definition of smart people) and
6. Convenience (can be accessed while on move using mobile phones or at home using desk top computer).

1.4.2 Benefits to Business

It benefits businesses:

1. Reduced time in setting up new business (reduced or absence of red tape),
2. Help conducting e-business and e-commerce (online business),
3. Better conformity to government rules and regulations for running business because computer software acts on the basis of rule-based system with no discretion,
4. More convenient and transparent way of doing business with government through e-procurement after e-tendering,
5. Better control over movement of goods through online monitoring of clearances and
6. Conducting monetary transactions online (e-banking, e-payment) avoiding corruption during release of government payment to contractors.

1.4.3 Benefits to Government

The government also benefits in a number of ways:

1. Better policy making and regulatory and development functions as a result of better and up-to date information,
2. Very fast acquisition, storage and retrieval of data leading to better decision making,
3. Better management of government processes,
4. Better dissemination of government rules, regulations and activities,
5. Better performance in regulatory functions like taxation,
6. Better performance in social sectors like education, health and social security and last but not the least,
7. Creates the positive image of modern and progressive government.

These benefit streams has shown distinct pattern of E-Governance evolution.

1.5 Evolution of E-Governance

There is emerging, a definite pattern of evolution of E-Governance in its functional aspect, if one takes its presence in different states of India or world, on a chronological order. The United Nations identifies five stages of E-Government evolution: The distinct phases are

- Phase-1(1996–1999): Emerging with Basic Web Presence
- Phase-2(1997–2000): Enhanced with Interactive web
- Phase-3(1998–2003): Interactive graduating to Transaction web
- Phase-4(2000–2005): Integrative and Transformative web
- Phase-5(2005+): Smart City Governance web or well-connected web.

During 1996–1999, Municipal government for the first time took advantage of internet and made their website. It was nothing but a notice board like, one-way communication for displaying information about a municipal government. Some of them also had two 'push services' by which from time to time emails and SMS were sent to selected consumers if registered. According to the UN Global E-Government Readiness Report 2008, 94 % of countries have websites, 90 % provide information, such as publications, laws, policies; 70 % offer databases of some kind; and 26 % offer true 'single-window' portals. Additionally, 164 governments have contact information online and 125 offer downloadable forms. Information published online can include laws, regulations, policies, budgets, judicial opinions, official publications and reports, forms, executive decisions and a wide range of government advice and information on matters such as health and agriculture. It also can include government directories, organizational structures and contact information for government offices and key officials, including addresses and telephone numbers. Provincial, local and municipal governments are also offering localized and specialized government services. Peru's 'Public Window' system gives citizens in three cities the ability to learn how their local governments are structured, to access information on municipal officials, to see how public funds are spent, and to obtain information on procedures for obtaining a birth certificate, restaurant permit and other official documents.

During the period 1997–2000, E-Governance, showed the emergence of Interactive web sites. An attempt was to make the web two ways usually for uncomplicated type of data collection such as registering comments. This phase is characterized by speedier interactions using electronic channels for some part of a service or transaction. Internet sites provide search capabilities and host forms to download and linkages with other relevant sites. In most instances, this stage enables the public to access critical information online, but requires a visit to a government office in order to complete the task. Citizens and businesses may be able to comment on proposed regulations or file corruption complaints via e-mails, generate downloadable forms, submit forms and information and search various specialized databases. Information and content is regularly updated. Interacting tools include e-mail, web-based forms, chat rooms, web forums, bulletin boards, list-serves and online question and answer (Q&A) sessions with government officials. Some interesting examples of Interaction functionality include Mexico's online forum and a feedback system in Singapore to cut government waste. The development of Interaction services is sometimes combined with the establishment of government-owned or—sponsored access points, such as kiosks, community centres or mobile units. The Bahia provincial government in Brazil has developed citizen assistance service centres, which are based in public places such as shopping malls and offer some 500 services.

During the period of 1998–2003, showed complex transaction features appearing in the web site that may involve intergovernmental work flows and legally binding procedures for example motor vehicle registration. Here citizen and business persons can complete entire tasks on line. For example, when a car is sold print out of registration certificate can be obtained online by car sellers. The 'Transaction' phase of E-Government involved a mutual exchange of information (and sometimes funds) between government and citizens or businesses through step-by-step online self-service processes. Citizens are able to file tax returns, obtain visas, passports, birth and death records, licenses, and permits, pay parking fines and utility bills and apply for government jobs. Electronic identity and digital signatures may be recognized, and secure sites and user pins/passwords are also required. Online transactions make government services available at any time from any Internet-connected computer and more and more frequently via a smart mobile phone, whose cost, accessibility, and wide penetration make it a suitable device to access E-Government services. Traditionally, government services may have required long waits, confrontation with time-consuming bureaucracy, and the occasional bribe. Innovations such as citizen service kiosks located in shopping centres in Brazil, portable government computers that can be carried into rural pockets of India, or medication reminders available through SMS bring E-Government directly to the citizens. Business process reengineering is critical in the streamlining of time-consuming procedures, saving labour costs and increasing productivity in the long run. In addition, governments can help to stem corruption by utilizing new levels of automation and business intelligence to make transparent and/or electronically audit the processes, the transactions and payments. Consequently, online transactions may require significant investments in back office consolidation and harmonization of information and technology systems, as well as changes for the government workforce. The success of these transactions, as with other E-Government applications, will depend on assessing and responding to the needs and capabilities of the intended users. One of the few available surveys of rural users of E-Government found the following services most in demand: personal documents, including birth, marriage, and death certificates, land registry or cadastral services, anti-corruption complaints and other grievances with public services and transportation-related services, including car registration and purchase of bus and rail passes.

During 2000–2005, Integration and transformation progressed and user-centred experience, as well as multiple agency connections appeared in several websites. This fourth level service integrates a wide range of services across a whole government administration as characterized by the many emerging portals. The eCitizen portal developed by the Government of Singapore offers a prime example of this system. This fourth phase is characterized by redefined relationships between government, citizens, businesses, communities and employees delivering seamless experiences and rich levels of engagement derived from new connectivity, interoperations and business models for service, and policy design and development. The integration of information, processes, and channels across multiple government, non-government and private sector organizations enable a user to start and complete an entire task easily, confidently and securely. This is the concept of integrated and transformational service. Users are able to access any service or information in a complete "end-to-end" package where the existing boundary between departments/ministry or organizations do not interfere with or interrupt the service outcome, and where the services are clustered along common "customer" needs. The integration demands are substantial and require:

- Understanding and regular monitoring of customer experiences and expectations.
- Trained and informed staff.
- Interoperability and standardization of information, processes and technologies particularly at interfaces of organizations.
- Multi-channel strategies—ensuring consistent and reliable experiences for users within and across individual channels of service (online, on-call, on-paper and on-site).
- Cross-organizational governance controls—i.e. Memoranda of understanding, contracts, funding, service level agreements.

The opportunities for enabling a stronger dialogue between citizens and government in governance and policy development is also reflected in this phase highlighting the two-way interaction and responsiveness expected of government—a new relationship compared to the traditional 'government-to' approaches. The OECD's paper on Citizen Focus: Public Engagement for Better Policy and Services highlights the importance of having public engagement policy to utilize the technology that supports engagement and dialogue.

After 2005 many nations witnessed the emergence of Smart City Governance with European Union taking a major lead. Its full nature is yet to be specified. Governments transform themselves into a well-connected entity that responds to the needs of its citizens by developing an integrated back-office infrastructure. This is the most sophisticated level of online E-Government initiatives and is characterized by:

- Horizontal connections (between government agencies).
- Vertical connections (between central and local government agencies).
- Infrastructure connections (interoperability issues).
- Connection between governments and citizens.
- Connections between stakeholders (government, private sector, academic institutions, NGOs and civil society).

It is participatory governance of the highest order. It strives to solve the issues of digital divide and inclusion and exclusion of e-democracy and E-Governance. Smart City provides immense opportunity for e-learning and e-medicine. Smart city inhabitants are to be in continuing education life-long. There is full conscious of carbon emission, energy efficiency, environmental efficiency and pollution controls. ICT build in the Smart City technologies were utilized to self-govern these parameters 24 h a day and 7 days a week.

How these phases are evident in many states of India is attempted below.

1.6 Status of E-Governance in India

India has undertaken massive initiatives to introduce E-Governance at the national, state and local levels. In terms of the total number of government websites, India is ranked seventh in the global list [16] during the phase-1 development discussed above. The policy-makers in India tend to justify the adoption and expansion of E-Governance on the grounds that it costs less, reduces waste, promotes transparency, eliminates corruption, generates possibilities to resolve inequality and guarantees a better future for citizens [17–20]. The government tends to portray E-Governance as the panacea for all ranges of problems confronting India. No evidence is there of earlier protest against computerization in administration is evident now. There are critics who, in general, suggest that the whole enterprise of ICT may have created a new class of 'untouchables' living in 'information poverty', compromised equal access to government services and eroded accountability and individual privacy [21, 22]. As the largest democracy, venturing into E-Governance, Indian Government has set the target of delivering at least 25 % of its dealings and services electronically [23, 24]. The major policy measures have been to increase computer density, connectivity, content and cost and cyber laws [25]. Under this overall policy framework, the government has introduced various measures for E-Governance, which can be categorized into national- and state-level initiatives and institutions.

Information Technology Act (2000), which is to regulate cyberspace and define offences and penalties related to information technology such as tampering with computer source documents, breach of confidentiality and privacy, publication of false digital signatures and so on [20, 21]. The Indian Government has also adopted Right to Information Act that requires all public authorities to maintain information and records, and appoint Public Information Officers to assist citizens in gaining access to such information [18, 26]. Every Government website exhibits details of Right to Information Act. Furthermore, it has introduced citizens' charters under which the ministries and departments at both national and state levels are required to adopt charters specifying their respective service provisions, time frames, service standards and channels for redressing grievances. Punitive measures are published for delayed delivery of Government services.

Some of the leading examples of E-Governance include Andhra Pradesh, Gujarat, Karnataka, Kerala, Madhya Pradesh, Maharashtra, New Delhi and Tamil Nadu. Andhra Pradesh Government took the initiative of E-Governance known as the Andhra Pradesh State Wide Area Network, which is a network for data, voice and video communication [27]. For example, through this network, it launched the Twin Cities Network Services to provide various services to citizens in two main cities (Hyderabad and Secunderabad) through one stop Integrated Citizen Services Centres (ICSCs). Through ICSCs, the citizens can access information about state and central governments; pay utility bills and property taxes; purchase certificates and licenses; and receives information regarding building permits, property registration and transport procedures [27, 28].

Madhya Pradesh introduced its own E-Governance with extensive computerization in dealing with payrolls, the budget, accounts, personnel, official communications, land records, public programmes and relief operations. It trains public servants in E-Governance at the State Academy of Administration. Initiatives have also been taken to use Hindi as a popular medium to communicate among governments, provide information to the public, use email services and maintain government web pages [29].

Karnataka state government has begun to computerize most departments. Bangalore is known as an IT hub attracting over 1,500 IT companies from advanced industrial nations; and its Indian Institute of Information Technology has a very advanced infrastructure and IT facilities. Furthermore, under its Department of Information Technology, the government created a "Centre for E-Governance" supporting E-Governance.

The Government of Tamil Nadu is strongly committed to transforming the state into an advanced system of E-Governance by computerizing its major departments and building technical capacity, with the ultimate objective of restoring public confidence and creating an effective relationship between government and citizens [19, 29]. To reinforce its mission of restoring citizens' confidence, the Government has adopted projects to ensure computerization of land records, registration, the education system, transportation and so on. It has also established a Tamil Internet Research Centre to promote the use of Tamil on the internet in order to increase access for citizens.

Kerala uses selected nodal officers in each department to accelerate the application of IT. One unique feature of Kerala is its comprehensive programme aimed at decentralizing E-Governance to the district and Town level—many district cooperative banks and credit societies have been networked and is now having core banking facility. Activities and documents which have been put on the internet may cover tax collection, accounting, welfare schemes, court rulings and government orders and directives. Moreover, the government is using a transliteration technology that allows its web pages to be available in the local language (Malayalam). It launched a project called 'A PC for Every Home through the Kerala Electronics Development Corporation, which aimed to reach the target of 10 personal computers (PCs) per 1,000 people by the end of 2001 and it is still being continued. There are loans available to procure computer. In Kerala, Information Kerala Project was adopted to computerize and network about 1,214 local bodies. In this state, a district panchayat of Ernakulum has adopted a model of E-Governance that provides guidance to three village panchayats in the district to build an information network covering information related to land holdings, age structure, health, tax payments. It gave internet access to cooperative societies and aids to deliver information relating to agriculture, health and education to the people by creating networked 'facilitation booths' in these villages. The work of "Akshaya" centres is well known and it will be discussed in Chap. 5.

According to FICCI—Ernst and Young Report, 'E-Governance 2020' in India (undated); the state of Kerala is also, pioneer in adopting mobile technology to deliver citizen services. M-Governance initiatives in Kerala are initiatives focusing on the utilization of mobile technologies to deliver citizen services and is aimed at making services available to citizens on the mobile platform on an 'anytime anywhere' basis. It was launched as a series of projects in various districts of Kerala on a variety of sectors including healthcare, education, transport and utilities. The following is a summary of the various ongoing initiatives in the state:

- Dr. SMS: This project aims to provide comprehensive information on the healthcare facilities available within a locality to citizens through Short Messaging Service (SMS).
- Transport-related services: These collectively provide information on the schedule of state transport buses as well as details on road tax paid and vehicle registration.
- Utility services: Initiated with the Kerala State Electricity Board and the Kerala Water Authority, solutions in this area are aimed at consumer billing and complaint resolution.
- Education services: These projects aim to provide various examination-related information to students.

The Rajasthan government has taken measures to strengthen E-Governance, proposing the creation of a state wide network to provide information and video communication to both public and private organizations. The Department of Information Technology in Rajasthan has developed such programs as RajSWIFT to facilitate the use of online data and email communication among officials; and RajNIDHI to provide services to citizens in a transparent and responsive manner.

The Government of Gujarat has introduced a state-wide network (Wide Area Network) connecting all office complexes and corporations in the state and has launched E-nagar (E-city) program. SWAGAT is their state-wide attention on grievances by application of technology. It is an on line Four-Tier Grievances Redresser system which got United Nations Award. Geo-Spatial mapping and Geographic Information System is being implemented in most Cities since 2003. Since 2004, there is e-Procure system which is an efficient and transparent procurement process through e-tendering. Since 2010 bar coded ration cards are issued as part of integrated public distribution system. E-Jamin (E-land) since 2010 can access land records in an easy and transparent way. E-nagar concept is centred around Smart and intelligent cities with Citizen service portal, m-Governance, Intelligent transit system with GPS, Self-help Kiosk and Call Centres, Digital Empowerment through SATCOM/Video conferencing, Geo-Spatial Mapping, Wifi-hotspots and last mile connectivity called e-Ward. {Vide The Hindu Kozhikode Edition 26-2-2014}.

The Uttar Pradesh government has taken a special interest in transforming its Allahabad district into a Smart District [30].

Similarly, the Government of Maharashtra is trying to develop Mumbai and Pune into major IT hubs while pursuing the state-wide expansion of E-Governance. It has taken on the responsibility for developing it skills and awareness among employees through training, and to link all district-level offices through the Wide Area Network.

Although these state level initiatives gives some hope on E-Government in India, the position of India in comparison with other countries is very poor and considerable effort is required to progress from this status.

1.7 E-Governance Challenges of Smart Cities in India

E-Governance challenges of emerging Smart Cities are many. 2011 census of India showed that more number of urban populations was added than rural population to Indian population during 2001–2011. As per estimates of Planning Commission, by 2021 nearly 28–40 % of total population of India shall be urban. With rapid economic growth, the next wave of urban development is likely to concentrate in and around 60–70 large cities with a population of one million or more. Some of these million plus cities can be potential Smart Cities. The main area where Smart City E-Governance challenges will be the following:

(a) Water supply and sanitation
(b) Power supply
(c) Urban transport and traffic management
(d) Pollution control and environmental sustainability
(e) Regulation of land use
(f) Management/decongestion of development within crowded zones
(g) Maintenance of civic infrastructure
(h) Policing
(i) Disaster management
(j) Urban poverty

These challenges call for appropriate E-Governance solutions in the above enumerated aspects. There is need to invest in research and development in creating a common E-Governance solution for municipal administration, which may be easily replicated across existing and emerging urban centres. Further, large number of municipal departments are required to come together to deliver citizen services in urban centres, and the need for extensive co-ordination between them. The concept of 'whole-of-the-E-Government' or 'connected government' becomes very relevant for Citizen. Municipal administrations and the supporting agencies should adopt appropriate E-Governance standards and systems, paving way to easy integration and information exchange.

1.8 The E-Municipality Mission Mode Projects

E-Governance projects being designed to address the challenge mentioned above is the e-Municipality Mission Mode (MMP) Project. JNNURM is such MMP for municipalities which is expected to provide a major fillip to India's urban reform agenda. The vision of the National MMP for E-Governance in municipalities is to leverage ICT opportunities for sustained improvement in the efficiency and effectiveness of the delivery of municipal services to citizens. The MMP intends to achieve the following goals:

- Provide single window services to citizens on an anytime, anywhere basis.
- Increase the efficiency and productivity of Urban Local Bodies (ULBs).
- Develop a single and integrated view of ULB information system across all ULBs.
- Provide timely and reliable management information relating to municipal administration for effective decision-making.
- Adopt a standard-based approach to enable integration with other related applications.

Some of the areas of intervention in JNNURM project include:

- Registration of births and deaths.
- Payment of taxes and utility bills.
- Approvals for land use and property development.
- Procurement and monitoring of projects being implemented by the ULB.
- Heath programs and Solid Waste Management.
- Accounting systems of ULB.
- Grievances handling, including the implementation of the Right to Information Act.

A unique feature of e-Municipality solutions will be the significant emphasis on Public–Private Partnerships (PPPs) and augmenting the revenues of municipal administrations so that they can fund their growth and development to a significant level.

1.9 Status of India on United Nations E-Governance Survey 2012

India has fallen by five places in ranking in the UN E-Government survey 2012 (www.unpan.org/egovernment) from 119 in 2010 to 124 in 2012. The survey which was released on March 6, 2012 places the Republic of Korea at the top position followed by The Netherlands. United Kingdom (UK), Denmark and the United States of America (USA) follow in ranking in that order. USA has slipped by three positions and is placed at fifth position as compared to the second position in the last UN survey. As compared to its neighbours, India even lags behind Maldives and Sri Lanka who are placed at 95–115 respectively. The survey acknowledges that with a population of 1.2 billion, India has to deal with many challenges associated with this huge population. It highlights that e-services in India are in the formative stage (Phase-1 discussed).

India is low on E-Government ranking in the crucial area of web-services. There are as many as 5,000 websites of various government agencies supported by the National Informatics Centre (NIC) and there are as many as additional 1,000 websites independently developed by government agencies. However, a majority of these websites are just notice boards conforming to Phase-1 development. Even the vital information is not updated on a regular basis. The last updated counter which is generally provided at the bottom of websites shows a gap of few months. Also, despite India being a multilingual country, a majority of its websites are in English.

However, India ranks high globally as far as legal framework and policy is concerned. Even the IT infrastructure is not that bad. The E-Government index measures online service component, telecommunication infrastructure component and human capital component. India scored 0.5359 on online services; 0.5025 on human capital and 0.1102 on telecommunications infrastructure.

Although India has a moderate score on web presence and human capital, the index is low with regard to infrastructure. The infrastructure index is a composite weighted average index of six primary indices—PCs/1,000 persons; internet users/1,000 persons; telephone lines/1,000 persons; online population; mobile phones/1,000 persons; and TVs/1,000 persons. India despite having high mobile penetration is still low on parameters like computers, internet, fixed line telephones and TVs. Therefore, infrastructure is one key component that India's global ranking remains low.

It is now important to assess the impact of E-Government from whatever India had achieved in E-Governance.

1.10 Impacts of E-Governance on the Relationship of Citizens with Politicians and Administrators

Wadia [18], mentions that in India, E-Governance creates an avenue for its citizens to communicate with top political leaders and local ministers through such tools as video-conferencing, online grievance channels and complaint cells. A major means through which such interaction between citizens and politicians occurs is the parliamentary website, which is supposed to facilitate the top-down flow of information from the legislature to citizens, allow a bottom-up channel for feedback from citizens to the elected members, increase transparency by providing detailed information about legislative procedures and activities, expand the number of avenues for greater public scrutiny of the nature and processes of public policies and thus enhance the accountability of these elected politicians to their constituencies [14].

The emergence of E-Governance has significantly changed the nature of the relationship between citizens and public servants. The E-Governance movement not only promises higher quality and better delivery of services and a greater realization of entitlements, it also claims to offer stronger bonds between public servants and citizens based on transparency and accountability [19, 31, 32]. With regard to this new mode of relationship, emphasizes that E-Governance provides equal access to government and speedy and transparent responses from public servants. For Ghere and Young [33], public agencies now have to justify their decisions based on feedback from the people and conduct their business in public. However, for the critics, instead of a citizen–administration relationship based on equality and accountability, E-Governance may strengthen a top-down bureaucratic process by posting information about the structures and functions of public agencies and reinforcing the existing mode of interaction through documents and reports.

It is important to find out what shall be the impact of Smart City E-Governance which is the subject matter of this book. To start with Smart city concept is explained below and discusses how it opens up new E-Governance opportunities.

1.11 Smart City and Its Use of SCADA for E-Governance

While debating the definition of Smart Cities in the book "GIS for Smart Cities" [2] author group of that book had great difficulty in adopting existing definition. All of them felt that, definition available was like a group of blind men trying to figure out and define an elephant. Many of them came with their own definition. We also analysed the standard definitions available some ten of them. At the same time smart cities was known by several names and it compounded the difficulties. Finally, author collective of that book had accepted Smart cities definition as given below. This book 'GIS for Smart Cities' defines Smart City as a knowledge-based city that develops extra ordinary capabilities to be self-aware, how it functions 24 h and 7 days a week and communicate, selectively, in real-time knowledge to citizen end users for satisfactory way of life with easy public delivery of services, comfortable mobility, conserve energy, environment and other natural resources and create energetic face to face communities and a vibrant urban economy even at a time there is National economic downturns. With this definition it was easy to connect GIS with Smart City. Data recorded is converted into information and these informations are selectively used for ongoing decision making in real-time for solving real urban issues to consumers using analytics.

The primary user of this total smart city information is SCADA. SCADA is acronym for 'Supervisory Control and Data Acquisition', a computer system for gathering and analysing real-time data. SCADA system existed since 1940 but from time to time it absorbed the current technological innovation and ready to solve complex urban problems today irrespective of its complexity and size of population. It can also monitor a self-aware Smart city in totality. In the early stage, SCADA systems are used to monitor and control a plant or equipment in industries such as telecommunications, water and waste control, energy, oil and gas refining, nuclear reactors and transportation. Now it is being used also for large urban systems whether it is a mega city or metropolitan city. For SCADA like Smart City, size does not matter.

With Smart City and SCADA a billion or Trillion dollar business is born and many multinationals such as CISCO, IBM, Microsoft, Mitsubishi, Siemens, Toshiba, Arup and others are part of this emerging business opportunity. However, most of the SCADA in Smart City is not total system but partial or micro system such as in the case of Delhi Metro and Mumbai Traffic system which industry calls micro-SCADA, being small scale. A SCADA system gathers information, such as where a leak on a pipeline has occurred, transfers the information back to a central site, alerting the home station that the leak has occurred, then SCADA carry out necessary analysis and control, such as determining if the leak is critical, and displaying the information in a logical and organized fashion leading to appropriate responses. SCADA systems can be relatively simple, such as one that monitors environmental conditions of a small office building, or incredibly complex, such as a system that monitors all the activity in a nuclear power plant or the activity of a municipal water system, an electric supply system or a total Smart City. SCADA systems were first used in the 1940 and then it was called Supervisory control system (SCS). In 1950, Telemetry instrumentation got added to SCS, and in 1965 integration with IT took place evolving the earlier generation of the present SCADA. In 1970, Automatic Generation Control in terms of Advanced Instrumentation got added to SCADA and then in 1980 Systems Analytics got integrated and in 1990 process-specific management system became a reality and now SCADA is self-adjusting which is an aspect of self E-Governing system. Hence, SCADA is ideal for Smart City. Now SCADA incorporates wireless LAN and uses intelligent electronic devices and have a distributed architecture and adaptable to a scale of metropolitan regions. The most important evolution of SCADA is its self-adaptive nature using efficient communication network.

Further details of Smart Cities are given in the following diagrams. Figure 1 shows the building blocks of Smart City System. Smart City shall be considered as a system with six components and I have given most importance to Smart People followed by Smart economy and the building blocks represented give the relative weights. From Governance point of view Smart City means citizen centred mobile phone-based intelligent Governance with digital infrastructure. Digital infrastructure are network of sensors, cameras, wireless devices, data centres and powerful analytics to enable the Government to provide more efficient services, maintain low-carbon footprints and create an entrepreneurial environment for citizen. Components of Smart Cities are as shown in Fig. 1.

Figure 2 shows Smart People. Smart People require many attributes as given. The most important aspect of smart people is their high Human Development Index. Next to importance is their Graduate Enrolment ratio. The third important factor is the metropolitan regional and the State shall have high percent of urban population and GDP contribution from urban area need to be highest. A soul searching is here to find out whether any existing city has population with the above attributes. Further, one need to assess whether infrastructure and investment required nurturing smart people exists in any such urban centres in India.

Figure 3 describes, Smart City Economies. It is driven by innovation and supported by universities which are not teaching shop but producers of patents.

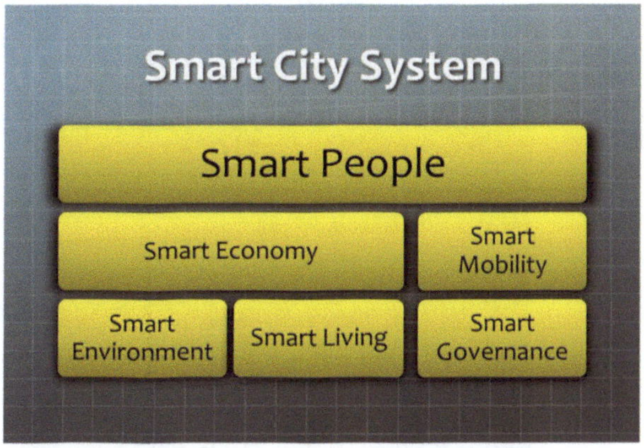

Fig. 1 Smart city system building blocks

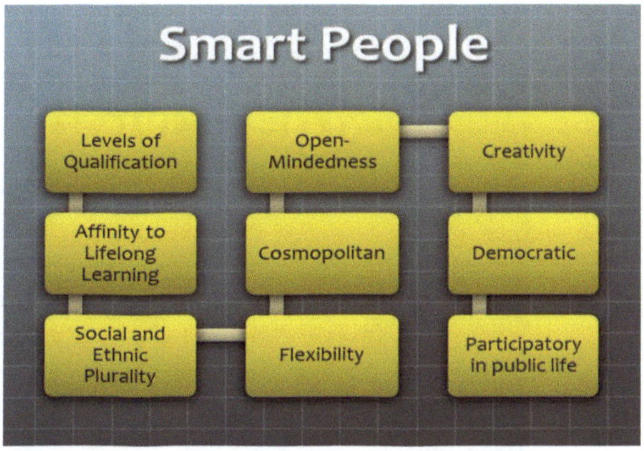

Fig. 2 Smart people

Entrepreneurship will be encouraged by easily available venture capital funds. This will be an economy of high-quality image, high productivity and flexible labour market and not strikes. This economy will have an international branding and generates highly diversified economy.

Figure 4 shows smart mobility. Here international, National and local accessibility are fully taken into account. So also last mile accessibility will be given due importance by urban design and urban compaction. Metros, monorails, intelligent transportation system will be there in such Smart cities.

Figure 4 shows importance given to carbon footprint, pollution, natural sceneries and natural environment and sustainable resource management.

Figure 5 is all about Smart living. Smart living is characterized by access to diverse cultural facilities for minority and majority communities, educational

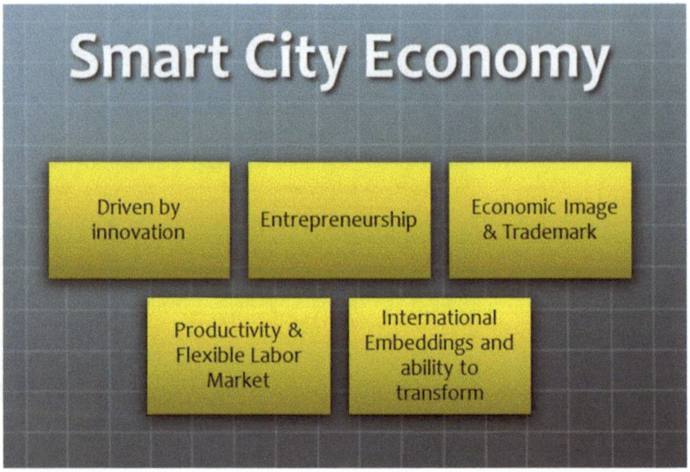

Fig. 3 Smart city economy

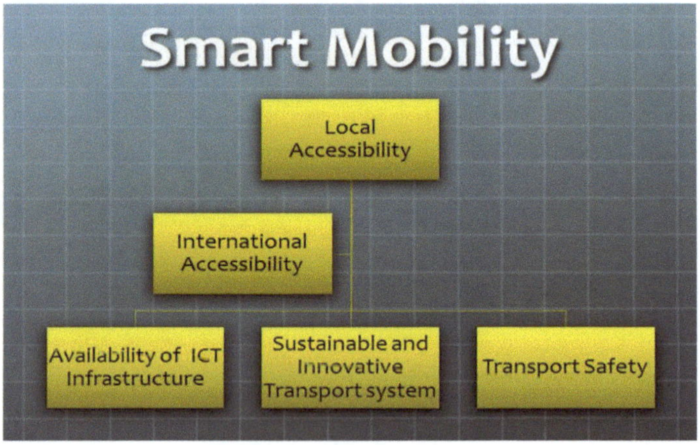

Fig. 4 Smart mobility

facilities of world standards, more attractive tourist attraction, higher health condition and good housing quality and social cohesion (Fig. 6).

This book is all about Smart Governance and we emphasize here participatory decision making, transparent corrupt free Governance, best public and social

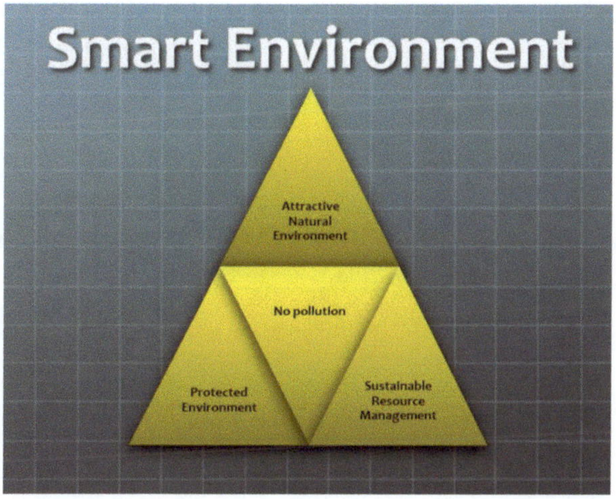

Fig. 5 Smart environments

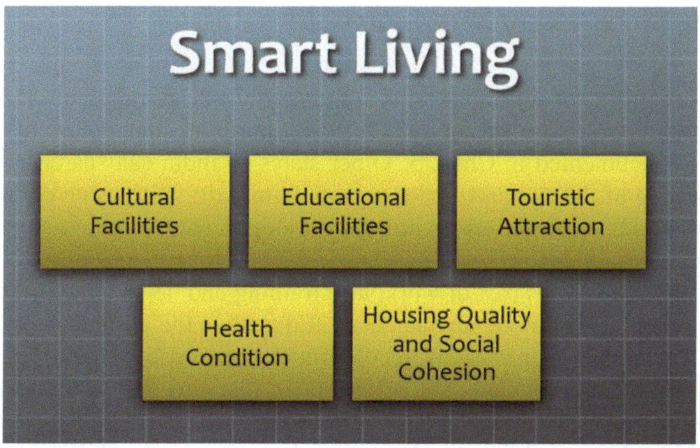

Fig. 6 Smart living

services and above all well thought out political strategies and perspective by the inhabitants (Fig. 7).

Once smart city and SCADA is desired for a particular human settlement, then it became important to identify prime requirements for E-Governance of Smart Cities.

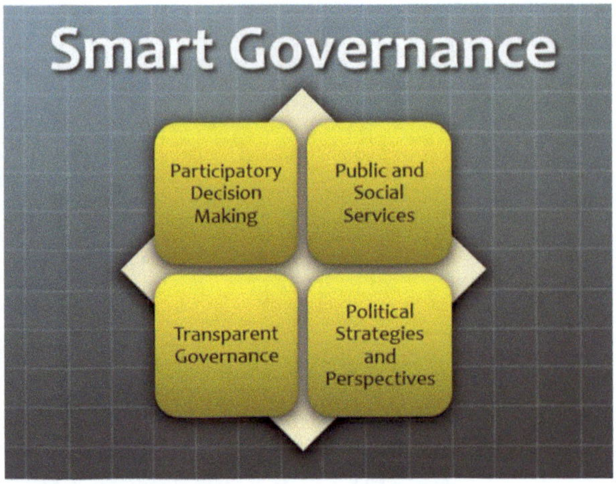

Fig. 7 Smart governance

2 Prime Requirements of E-Governance for Smart City

2.1 Universal E-Literacy Program

The first state in India to become fully e-literate will be Kerala with 33.3 million populations in 2011. The successful e-literacy program undertaken by Akshaya in Kerala is documented in Chap. 5. Other states in India can search other models. Irrespective of the age and sex of person, familiarity with using computer and the Internet can be achieved using multiple strategies such as community training in local government centres such as libraries, local government offices, colleges, school and ICT labs or as part of workplace training and development. Addressing the issue of literacy is more fundamental. 'Education for all' has become a priority in developing countries, as recognized in the Millennium Development Goals and strategies and plans are being put in place to ensure that every child has access to basic education, at least at primary level. Kerala and Manipur had already achieved near full literacy. As part of this commitment, ICTs are gradually being embedded in the educational system and the delivery of a range of subjects. Primary education shall give e-literacy to every child. However, adults who are no longer in the education system can be denied access to the benefits of ICTs due to a range of issues including lack of recognition of distance learning as a legitimate educational mechanism, and the lack of accreditation systems for non-formal education. Information on institutions approved to offer distance learning can be posted on government education portals. Online educational programs can be especially valuable to women who cannot participate in mainstream classroom learning due to family responsibility as housewife. While English is spoken in many developing countries; it is typically a second or third language, used for conducting business, but there are attempt to have selected Indian languages even in operating system but all Indian languages are not covered as of today. Even native language use poses a challenge in countries that have many local or ethnic dialects. Effective E-Government strategies should emphasize content in the national language or languages and should also provide for translation into various ethnic dialects.

Individuals with limited mobility, who may be blind or deaf, are able to benefit greatly from E-Government information and services with sufficient forethought, planning and appropriate modification of the computers. The World Wide Web Consortium develops protocols and guidelines to support accessibility to the Internet for people with disabilities. Those efforts include software solutions, policy initiatives and educational conferences. Governments may use a range of tools to measure the accessibility of their websites.

It is important for governments to encourage high-level participation of women in the ICT sector, including policymaking positions. At present, the ICT field is male-dominated in most developing countries. Most ICT policymakers are men, and ICT programs are more likely to be managed by men than women. As a result of the current imbalance, it is likely that men will reap the majority of the benefits of E-Government, including business opportunities associated with E-Government software development and support services. Programs that address women's participation in the ICT sector could also provide funding and mentoring services on establishing and managing businesses related to E-Government.

2.2 E-Democracy

The term e-democracy refers to the processes and structures that encompass all forms of electronic interaction between the Government (elected representatives) and the citizens (electorate). Online democracy includes access to elected officials by all electorate, availability and use of discussion forums (e-participation), access to meetings and meeting documentation, voter registration and ultimately online voting, also known as e-voting. UK Government hopes to use e-voting in order to bring youth into the democratic process since their participation in elections was regarded as very low [34]. This is also true in urban middle class dominated India where voting percentage has been low in comparison with rural areas. The advent of AAP party in Delhi had pushed up the voting percentage in Delhi in recent election using social media for such mobilization. It implies greater and more active citizen participation and involvement enabled by ICT in the decision-making process. The objective of e-democracy can be grouped in the following two categories:

(a) **Passive access-related objectives**: To provide citizens access to information and knowledge about the political process, about services and about choices available.
(b) **Active access-related objectives**: To make possible the transition from passive information access to active citizen participation by:

- Informing the citizen
- Representing the citizen
- Encouraging the citizen to vote
- Consulting the citizen
- Involving the citizen.

Besides all these potentially positive outcomes of various E-Governance initiatives, particularly e-democracy, some serious caution needs to be exercised. The expectations for the potential of information technology to promote participation in democracy were extremely high. However, information technology will not automatically result in increased democratic participation. There are also examples which demonstrated Government reluctance to engage in any sort of e-democracy.

2.3 E-Participation

E-Participation tools are used to collect and discuss citizens' and businesses' views so their interests and needs are better represented in government programs or processes. The tools include online surveys and polls, electronic newsletters, e-mail, feedback forms and web forums where citizens can express their opinions. They can be used to supplement public forums or meetings. E-Participation applications may have to publish in web site, presenting relevant background information, decisions and other materials to help citizens and businesses understand certain public policy or regulatory issues. New Zealand has aggregated its policies, community of practice, knowledge pool and ways to participate in government in one site, participations.

Feedback or comment forms may support anti-corruption measures. For example, the Philippine Civil Service Commission (CSC) implemented an m-Government system that enables citizens to text complaints or corruption charges on government officials, and mobile phone users can report grievances against the police using SMS.

More advanced e-Participation functionality can include personalization features. For instance, e-mails may be sent to interested citizens or businesses based on their registered interests, alerting them to new decisions, reports, or resources on E-Government sites.

2.4 E-Consultation

E-Consultation takes the process one step further than e-Participation, facilitating online comment on specific policy or regulatory issues, while those issues are under active consideration by the government. E-Consultation resources can provide online access to government proposals and other key documents including wikis, written filings of stakeholders, and audio and visual recordings of public meetings. Resources can be organized to facilitate online consultation with citizen or business stakeholder groups and to receive formal inputs into government policy or regulatory processes. Online consultations can reach a wider range of stakeholders and ensure that consultative processes are transparent.

Canada's government has created a useful site giving citizen's access to all ongoing e-Consultations Singapore also has a site for e-Consultation. The UK, which in recent years has scored highest on the UN's global assessment of e-Democracy, has created resources for local e-Democracy. E-consultation is not present in Indian E-Government practice.

2.5 E-Decision-Making

E-decision-making functionality facilitates intra-government decision-making processes. These applications are generally associated with cabinet-level decision-making or parliamentary procedures and aim to increase the efficiency of decision-making processes within governments. Estonia is among the countries that have implemented e-Decision-making processes. E-Decision-making applications may also be combined with "Publish functions" that provide the public with information about government decision-making.

Integration/Transformation initiatives require established and sophisticated back-end capabilities and infrastructure. In fact, some E-Government projects have components of more than one phase and a government can offer information and services across all phases during the evolution of E-Government in a department, a jurisdiction and a country.

2.6 E-Voting

Europe appears to have big plans for electronic voting (eVoting). There are attempts by European Governments to make the voting process attractive and convenient for citizens. Examples include the British government, which saw the lowest voter turnout in 90 years (59 %) in 2001, and has run many trials in local-council elections, and ballots over the Internet. In Geneva, Switzerland, eVote cast in April 2006 was a first step in a plan to wire the entire country. The European Union has funded a 3.2 million euro, 3-year pilot program to use eVoting in three local elections, one in France and the other two in Germany and in Sweden. More than 1 million Estonians have voted via the Internet in the 2006 local elections. To counter the possibility of coercion, Estonians were able to log back on an unlimited number of times and change their vote anonymously or walk into a polling station and cancel their e-vote.

It may be noted that eVoting is far from fool proof. In this respect, the US (despite the 2000 presidential election incidents) has scrapped plans to introduce eVoting for the armed forces, on the grounds of safety and privacy. The major issues are: "How do you verify that people aren't selling their voter-ID codes, if you can't actually watch them vote at a station?" What seems to be a compromise is the use of voting machines, which started in the US and are also catching on in Europe. India conducts voting using electronic voting machine and found fool proof but there are few who are critical of these voting machines. Election Commission of India had studied the argument and had disapproved the view. The Netherlands and parts of Germany have used them in elections. The United Kingdom has completed two pilot programs using more than 300 machines for local elections. France authorized the first use of electronic polling booths in elections held in 2006. And more than 3.2 million people in Belgium (44 % of all voters) voted electronically in their national elections in May 2005. In Brazil, by as early as 1998, almost two-thirds of the voting population had cast electronic ballots in federal, state and local elections. Some experts argue that even these systems are vulnerable; therefore, an active field of research deals with developing systems to let people know that their votes have been counted. The authors point out that using a simple digital audit trail to re-create what happened on Election Day would mean revealing who voted for whom, thus violating the principle of secret ballots. The most sophisticated systems deliver verifiability without a cumbersome, possibly vulnerable, set of printed-out ballots. With clever cryptographic algorithms and innovative viewing devices, it is possible to envision a process that provides specific proof after the fact that a vote was included in the total, without compromising the privacy of the specific selection. Cryptographers around the world have currently been working on such issues.

Besides technical matters however, political culture does play an important role. E-Voting in the US is a divisive issue; the possibility of it becoming a feature of the electoral process in the foreseeable future there is extremely remote. This is in contrast to Europe, where states such as the UK, Switzerland and Estonia lead the e-Voting experimentation. E-voting in India is conducted in electronic machines but e-Voting from home is a distant possibility and it can drastically increase the voting percentage. Ideally, E-voting is best executed as part of M-Government with fool proof security.

2.7 Other Requirements

Other requirements are briefly tabulated in Table 1.

These requirements shall be dovetailed with the rights and responsibilities of constitutional bodies embedded in the Constitution.

Table 1 Citizen-centric tools and requirements for smart of E-Government

S.no.	Tools	Remarks
1	E-citizen	Develop e-citizen's charter. Only Netherlands is taking the initiative [35]
2	E-inclusion	Set up access points. Common services centres scheme is a good example [36]
3	E-empowerment	Promote use right to information (RTI) act 2005 [12] through official websites [37]
4	Citizen-centric government	Keep citizen at the centre stage of any e-government intervention
5	Single portal	Cradle to the grave services for citizens. Singapore portal is good example [38]
6	Single sign on (SSO)	Only one user id and password. Singapore's SingPass is a good example [39]
7	All public services online	Joined-up, seamless government without need to go elsewhere
8	Efficient onsite search	Set up a specialty search engine. Learn from USA's new search engine [40]
9	Easy site navigation	Set up user-friendly web site so that citizens know where to expect what
10	Participative e-government	Set up uncensored discussion forums on official web sites as a matter of policy
11	Grievance redress	Every web site must have a time-bound grievance redress mechanism
12	Anywhere e-government	Provide e-government through multiple channels including cell phones (m-gov)
13	Anytime (24 × 7) e-government	Provide "always on" e-government including railway reservation's [41]
14	Privacy and security	Assure citizens that their privacy is protected and transactions financially secure
15	Legal support	Make computer-generated documents legally acceptable by legislation
16	Customer relationship	Provide for customer-led [42] customer relationship management (CRM)
17	Quality-marked e-government	Assure a minimum standard of quality. Check [43] and UK's Charter Mark [44]

2.8 Indian Constitution and Governance of Cities

It is the constitution of India that directs and guides city governance based on laws which itself is based on rights of the citizen such as for example right for information, right for owning property, rights for clean drinking water and clean air as well as rights for common property, rights for public realm, eminent domain, etc. Law makers are the elected representatives under constitution, in National, State and Municipal level. They are empowered by constitution to translate principles and processes laid down in the constitutions to make laws and implement laws on several aspects of governance say for example as related to city planning,

Urban management and implementation, provision of communication, transport, water or land for public use and so on given in the 12th schedule of 74th constitutional amendment as adopted by states and union territories. In a federal system like India, there are constitutional bodies such as National Parliament, State Assemblies, Metropolitan Planning Committee, Municipal Council and Ward Committees with distinct legislative, administrative and financial powers and responsibilities. Constitution of India divides the subject matters of Governance as those Parliament is solely responsible to legislate and State Government are responsible jointly (concurrent list) to legislate, State Government are solely responsible, Municipal and state government concurrently or Municipal government solely responsible. Manner in which each of the hierarchical level of constitutional bodies legislates and implement is solely left the respective bodies within their limit of powers and legal process of legislation. Hence, the extremely wide range of issues on which law can be made is divided between Union of India and State in Schedule VII of constitution. While List I and List II in schedule VII delineate the exclusive domains of Parliament and State legislative assemblies respectively. List III is called concurrent list. In concurrent list union of India and state assemblies have joint jurisdiction. As noted earlier, Seventy Fourth Constitutional amendment, spell out in its 12th schedule responsibility of Municipal Government. All those suggested by Indian constitution need not be taken in totally by State Government. State Government decides what Municipal Government shall accept in the 12th schedule based on their assessment of what is most suitable for the respective government in Municipal and State, and leave other responsibility of law making to State Government. Hence, there is no uniform pattern of law-making power among all states and municipality in India however all these diverse methods are firmly rooted in Indian constitution. The amount of decentralization the fundamental basis of Smart City Governance is the prerogative of higher hierarchy of such as Metropolitan Planning Committee and State Government. Central Government have power to override the earlier given decentralized power and so also the State Government by enacting new legislations within the constitutional framework to Metropolitan Planning Committee, and Municipalities to Ward Committees and even RWA. RWA is not a constitutional body like ward. RWA cannot get funds from consolidated funds of Government but Ward can. Hence, it is important to dwell on Tax sharing by constitutional bodies in India.

2.9 Sharing of Tax Collected in India

Taxes in India are collected by all constitutional bodies using the vehicle of annual budget under different heads. However more amounts are collected by union of India, then State and least municipal city government. Finance Commission at Central and State levels are empowered to determine how tax collected can be shared between hierarchies of constitutional bodies at State and Union of India and below metropolitan Planning Committee, Municipality and even ward level. Gadgil formula of yester years and Ranghuram Rajan Committee Report in 2013 which is under consideration enunciates formula by which tax shall be shared between Union Government and State Government based on certain set of indices. State government financial commission also creates such formula to share tax below state government to other constitutional bodies such as village panchayat, district planning committee and Metropolitan planning committee and Municipalities. Generally, indices of underdevelopment and development are used as criteria for a formula to share the tax between central and state government and state and other lower bodies. Indices of underdevelopment are indices of needs but indices of development are used to encourage development initiatives. Since State Government Finance Commission from periodic interval suggests how state tax is shared between cities and villages and with this it is possible to accurately determine how much annual grant will be available at each municipality to make annual plans and implement it using the approved planning process. Powers are also given to take loans and raise capital from open market in India or from foreign sources to Municipality by instrument such as Municipal Bond or others with the approval of Central Government. However, there has not been much achievement in raising money for development other than tax revenues. It is expected that Smart city will excel in raising money for city development.

Having given requirements and constitutional provision, an attempt is made below to enumerate basic and advanced step to set up E-Governance for Smart Cities.

2.10 Basic Steps Towards E-Governance for Smart Cities

There are series of basic steps as well as advance steps to be undertaken by Municipalities or Metropolitan Planning Committee to convert existing governance to E-Governance for Smart City. Basic steps are given below.

2.10.1 Increase City Expenditure on ICT

It is important to increase percent of household connected by internet, and increase existing internet bandwidth available through fourth generation (4G) Broadband for an aspiring smart city. Bharat Sanchar Nigam Limited (BSNL), a public sector undertaking in India has connected all their telephone exchanges by WiMAX towers and with intermediate towers which is a 4G system of broad band in selected states in India such as Kerala. Hence Kerala achieved full coverage of WiMAX towers. However, still there is more number of Adsl 2 users than WiMAX users in such states and percentage of households with 4G broad band connections is still very low. Private internet providers have planned to introduce 4G broad band in India. There are public call offices but public facility for internet connection is only limited number of Internet Café and Akshaya offices in Kerala. One way of measuring increase expenditure in ICT in Municipalities is to find out percentage of ICT expenditure on the total city budget. There is not much to show on this index in Indian Municipalities. It is not even a head of expenditure in Municipal Budget. However, State and Union Government is increasing such expenditure.

2.10.2 Website Availability

Availability of Municipal Website [45] is the second step. Internet is the channel for communication of people with Municipality, business and Citizen and vice versa. Most of the major cities in India have web portal and there are many more functions that can be incorporated to convert this web to a worthy portal for Smart city. Some of them are already discussed above.

2.10.3 Strategic Plan to Promote E-Government and ICT

Based on Strength, Weakness, Opportunities and Threats (SWOT) each municipality that aspire to become a Smart City with E-Governance shall have a strategic plan to promote E-Government and ICT [46]. In the majority of states in India, there is state level and Union level initiatives' but none exists at city level or ward level. Only thing available is some budget allocation to use the state level initiative. The city of Madrid is implementing a roadmap for the development of E-Government and there are plan for implementation of Law on Electronic Access of citizens to public services. Such initiatives are not known in India barring few initiatives based on the requirement of receiving grant for centrally sponsored big ticket JNNURM projects mentioned earlier and introduction of cyber laws.

2.10.4 Increasing Percentage of Services Available on Line

Introduction of ICT in public administration greatly improves the productivity and quality of administrative services of cities. These are achieved by a well designed and technologically superior web design of cities [47]. With the introduction of E-Government, both citizen and municipality had made services 24 h and 7 days a week, efficient, higher quality services, saved cost of providing services and receiving service, reduce waiting time for service delivery, improve transparency and eliminate corruption; for citizen, business and city government. Government of India now have a target only to bring 25 % of Government services through internet. It is important to assess how many percentage of total government services are on line and there shall be a time bound action programme to bring in all administrative services if not majority of services on line. There after a city shall be declared as 100 % online service provider.

2.10.5 Easiness to Lodge Complaints, Claims and Right to Information

City website shall have provision to lodge complaints and claims [48]. Most of the Municipal website host at least one page devoted to Right to Information. These rights are being made more user-friendly. It is ideal to use M-Government protocol for such activity.

2.10.6 Electronic Signature

An electronic signature is legally equivalent to a handwritten signature. It operates using digital certificates embedded in a chip. The electronic or digital certificate includes data that is used to identify the certificate holder, to exchange information with other persons and entities, in a secure manner and to sign the data being sent electronically to verify its integrity and provenance. None of the Indian cities have provision of electronic signature. This books contract is electronically signed with Springer.

2.10.7 Increase in Number of Staff Using Computers Connected to Internet

Statistics of percentage of administrative staff connected to computer having internet connection is a good indicator of advances in E-Governance. This should be published annually by municipality.

2.10.8 Increasing Number of Citizen Using Internet Connected Computers

Statistics of percentage of citizen connected to computer having internet connection is a good indicator of advances in E-Governance

2.10.9 Transparent Governance

Administrative process need to be transparent to citizen and business dealing with city Government. It is only a corruption free city government with maximum percentage of administrative process that is conducted through internet can achieve this. It facilitates easy access to information by individuals and groups.

2.10.10 E-Democracy

ICT can be used to enhance democratic process and increase opportunities for individuals and communities to interact with the government. Internet leads to greater participation with no limitation of geography, and disabilities. E-democracy improves politics, citizen participation, communication and decision making. Indian constitution requires public participation in Governance at local level. E-democracy helps in inclusion of social groups, involvement of citizen taking into account plurality of views and dissents, and gives precedence to general interest as against individual interest. E-democracy results in support for E-Governance. There are many ways e-democracy can be enhanced. They are e-mails, sms, chat/forum, online surveys and social networks.

2.10.11 Electronic Voting

Electronic voting has two parts. They are casting votes electronically and counting votes with no possibility of errors and corruption. All computer-based security that identify voters and security of not hacking on counting of votes shall be there. These make electronic voting cheaper, more efficient, secure and accurate system than traditional voting methods. In a smart city it may be necessary to vote for all policies and plans suggested by all population from time to time and hence, a cheaper e-Voting system need to be developed which is technologically most superior.

2.10.12 Continuing E-Education

Smart city is inhabited by smart people and they are supposed to be provided with all facilities of continuing education [49]. They have also to be educated on emerging ICT technologies they use, such as security of e-voting, e-signature and also shall be able to discuss on emerging policy and planning issues in city level.

2.11 Advanced Steps Towards E-Governance of Smart Cities

E-Governance of Smart Cities can take two roads in its advance steps. An efficient, liveable and sustainable Smart city can use a combination of hardware and software solutions to improve the efficiency of urban operating system for E-Governance of Smart City. Thereafter part of the constitutional responsibility of municipal governance is taken over by this system which cannot be executed by municipal administration alone. It can achieve by adopting Smart City technologies, in the use of software and hardware to govern utilities and services by municipal government using SCADA.

Another approach is to use E-Democracy to enhance E-Governance and involve all interested in various aspects of Governance using web technologies and participatory social engineering. It can be used in SWOT analysis, Strategic Planning, zonal planning and urban land management which is possible easily for city government who had already executed basic steps enumerated above.

2.11.1 E-Governance by Adopting Smart City Technologies

Smart cities, implement a vision of efficient, liveable and sustainable human settlement based on ICT. As a first step to realise these objectives, smart city attempts to be self-aware of all its functioning using sensor technology which is ICT enabled. These sensor technology can be local (for example, traffic flow measurement in real time by sensors under roads) or remote (using remote sensing satellite to find out traffic flow measurement and vacant car park site), a combination of many technologies, and involve heavy use of computer driven mathematical modelling based on past trend data. Informations are generated on real time by, for example, smart water or electricity metres or smart grid and it is used by consumers, utilities and services systems of a city or related institutions, business and citizen to optimise performance and create least cost solution in real time. This can only be achieved by a combination of hardware and software. Hardware of smart city-urban services and infrastructure whether it is social, physical or economic is based on an integration of ICT in the technology used. These technology based on Geographic Information System has got capability to transmit data in real-time to end users. End users use a set of software to make this information for decision making which optimize resources uses such as energy, water, etc., and serve both providers and consumers of service efficiently. Some of the examples of such system as existing today in many smart cities is summarized below.

1. **Beijing, China**: One of the Government responsibilities is to control traffic for the benefit of travellers. Urban and Interurban traffic control system of Beijing uses traffic signals in real time, which includes an enforcement system, and supervision and monitoring system for expressway. It also incorporates CCTV traffic flow detection, weather detection and guidance system all integrated in 10 control centres.
2. **Tiajin, China**: Geographic Information System is used for upgrading power grid, district heating management and tolling system.
3. **Guiyang, China**: Here Smart grid collects and analyses real-time operational data across a distribution network covering 4 million populations.
4. **Dallas, Texas**: In Texas SCADA improves efficiency and resilience to disruption of water distribution system. SCADA is acronym for supervisory control and data acquisition. It is a type of industrial control system (ICS). ICSs are computer-controlled systems that monitor and control industrial processes that exist in the physical world. SCADA systems historically distinguish themselves from other ICS systems by being very large-scale processes that can include multiple sites, and large distances like a city. These processes include industrial, infrastructure and facility-based processes. Dallas also use integrated mobility management platform that improves efficiency of multimodal transportation in major corridors using SCADA.
5. **Houston, Texas**: SCADA improves efficiency of water and electricity distribution system. Houston undertook to retrofit for energy efficiency of 40 municipal buildings delivers that had $3 million yearly warranted energy and water saving.

6. **Paris-Issy-les-Moulineaux, France**: Smart Grid energy management project at district level within city is used for energy efficiency, renewable installations and electric vehicle charging infrastructure.
7. **Lyon and Genoble, France**: Lyon uses Smart grid project at regional scale, including smart metre deployment, low voltage and medium voltage grid management, user automation for energy efficiency and peak shedding.
8. **Madrid, Spain**: Madrid uses weather information system for airport, Storm water management, and Energy management system for buildings, real-time adaptive traffic control system in 739 intersections, integrated platform manages traffic control, CCTV system, city access control and web-based traveller information.
9. **Barcelona, Spain**: Barcelona have a command and control centres for city traffic, critical buildings, metro, parking, CCTV surveillance; quality monitoring; and also metro fare collection.
10. **Abu Dhabi, UAE**: UAE had implemented energy audit that paves way for Building SCADA for 70 buildings, including management of Street lighting energy use.
11. **Dubai, UAE**: Dubai integrated SCADA building management system and security system that serve buildings for 900 residents; SCADA improves water distribution system and CCTV provides access control.
12. **Rio de Janeiro, Brazil**: SCADA improves efficiency of water distribution system, Electricity grid and Gas distribution system through all Metropolitan areas. Traffic management system optimizes mobility throughout the city; all interfaced with Rios's intelligent operation centre (IOC) providing 50 % of its data for holistic view of city system. Smart city is based on holistic real-time view of city.
13. **Sao Paulo, Brazil**: CCTV optimizes surveillance, and SCADA optimizes traffic in tunnels and highways. Two command and control centres manage city areas with real-time adaptive traffic control system in 382 intersections.
14. **Quito, Ecuador**: Quito uses real-time adaptive traffic control system covering 600 crossings, 6,700 traffic lights and 1,500 CCTV cameras. Weather information system for city airport is also used. Integration platform manages traffic control, using CCTV system, city access control and provides web-based traveller information.
15. **Oskarshamm, Sweden**: Performance Contracting optimizes energy in 28 buildings, saving Euro 0.3 million, 350 MWh and 80 tons CO, through active control of heating and ventilation and building management system.
16. **Mumbai, India**: SCADA improves efficiency of water distribution system; real-time adaptive traffic control system and optimizes traffic of 253 crossings. Central traffic management control centre supervises and react to disruptions in real time.

2.11.2 E-Governance for Smart City by Strengthening E-Democracy

Attractive form of e-democracy is one in which citizens could in principle extends their participation in decision making in number of areas of Governance. Clift [7] describes e-democracy as referring to 'how internet can be used to enhance our democratic process and provides increased opportunities for individuals and communities to interact with government and for the government to seek input from community'. E-democracy creates e-participation which is less limited by geography, time and disability. Most of the ICT deployment in Governance is to increase the provision of information to the citizen. There are many key questions for strengthening e-democracy.

1. Are we satisfied with the current level of citizen and business participation in our democracy at city level?
2. How to make the political process of democracy at city level more participatory?
3. How can we make public engagement in governance such as in policy making, planning and monitoring of development more deliberative at city level [50]

In order to answer all these questions Smart City Governance uses ICT. Basic steps discussed above help, in e-democracy at city level irrespective of the fact it is a mega city or million plus city. It could contribute to create a more transparent, interactive government engaged in wide range of dialogue with citizenry. E-democracy participation is achieved in the following way.

1. More informed citizenry is created by internet access to Government website. The capability of web is enhanced to create interactivity with the citizen.
2. Government website can be used as a forum for the free exchange of ideas and ability to share informed debates on issues and policies under consideration.
3. Citizen's inputs can be made online for consultation into decision-making process of government on those issues that directly affect them.
4. Opportunity to vote on line for candidates, policies or plans easily by eVoting.
5. Ability to create network of like-minded people group creating network of community in order to influence politicians, legislators and public officials.

Indian constitution calls for participatory planning and internet can be utilized for ongoing legal processes. They are preparing Master Plans, zonal plans, land management plans and urban design in a city. A structured approach using internet can be utilized for debating preliminary draft of policies and plans for law makers and citizen and planning, managing and monitoring plan implementation.

This discussion leads to the question of searching alternate models of E-Governance for Smart cities.

3 Alternative Models of Smart E-Governance

Smart City uses Smart Governance, an advanced form of E-Governance. It is important to identify first potential Smart cities as indicated in Chap. 15 of 'Geographic Information System for Smart Cities' [2]. This will lead to study nature of such cities whether it is small settlement, million plus metropolitan areas or megacities. Then it is time to decide on appropriate Smart City Governance models. A case to case approach seems to be more appropriate than a state specific or nation specific approach. There are considerable resistance to change in Governance but Smart Governance is not possible with great changes in Governance. Two areas in India, where Government transformation is slow, is in applying changes in constitution for municipal governance and modifying existing pattern of administration both in the national and State levels. In a heterogeneous India, with coalition politics, it is unlikely that there will ever be a single party with absolute majority to rule India who can easily make amendments to constitution. Well-established administrative practice never dies however bad it may be or gets transformed, unless there is far reaching change as dictated by constitutional amendment of Municipal Governance. With this scenario in background, four alternatives model of E-Governance can be visualized.

3.1 Model-1 No Transformation of Government Structure and Processes

In model-1, smart governance is just the governance of 'an aspiring smart city' like any other type of city with *no transformation in existing practice of Governance is envisaged at all.* According to Model 1, smart governance is all about making the right policy choices and implementing these in an effective and efficient manner utilizing the existing Governance practice. This can be done within the existing political and administrative structures. Batty et al. [51] highlight that smart governance is only an attribute that is associated to a governmental management and nothing special that merit structural change. Gil-Gracia [52] highlight that smart governance comprises aspects of political participation, efficient services for citizens as well a well-functioning administration. Here, Government must approve the development of the smart city and it can prioritize some areas of action appropriate to Smart City. Smart governance is about the promotion of smart city initiatives which can be executed with no change in existing governance.

3.2 Model-2 Innovation in Decision-Making Process and Implementation

This model of E-Governance advocates *low level of transformation which is not even apparent* to accommodate capacity for innovative decision making and implementation process. It is not a complete structural change in Governance at all. It is supported by multi-level training programs by institutions such as Kerala Institute of Local Administration (KILA) which trains political and administrative executives in different levels. This is generally based on long research inputs to training. UNESCAP [53] emphasizes that smart governance is 'the process of decision making and the process by which decisions are implemented (or not implemented). Walravens [54] adds that decision making can be innovated by using network technologies like SCADA requiring low level transformation of governance. SCADA can be a tool for Urban Governance. Schuurman et al. [55] define smart governance as the process of collecting all sorts of data and information concerning public management by sensor or sensor networks. New technologies are used to strengthen the rationality of government by using more complete—and more readily available and accessible—information for governmental decision-making processes and the implementation of these decisions.

Decision Support System (DSS) has been used by Business for many decades using information inputs that is available in real time. Strategic decisions are made using DSS to prosper in business based on current reality. Municipal Governance can also use DSS for participatory decision making based on most recent information. Since city is a spatial entity, location-based decision assumes importance. Geographic Information System (GIS) allows for developing Spatial Decision Support System (SDSS) which can be used for variety of tasks in Governance.

Some examples are discussed briefly below. Municipal budget of a million plus agglomeration in India can be around 250–500 Crores of Indian Rupees or more annually depending upon the type of Town. As per Governance procedure given in constitution of India, Ward Committees, Municipality and Metropolitan Planning Committee are involved in making decision annually for spending Rs. 500 Crores in a city based on rational decision made as per the approved Governance mechanism of State Government. Chapter 10 in this book is an attempt to develop such approach for Municipal E-Budgeting System for Municipal Corporation of Kozhikode, Kerala. The trend of hosting web-based Municipal GIS with annually updated database, becomes useful not only to Government for e-annual planning and budgeting but also for business men or women. Not even one interactive Municipal GIS websites are hosted by any Municipal Government in India. If such web sites are available an entrepreneur located several thousands of miles can make investment decision on his business based on SDSS she design based on Municipal GIS.

Imagine a business woman who wants to cultivate flowers and market it in Europe or United states from any metropolises of India. She can do it from this municipal website based on number of requirements for such business, involving cultivating flowers and marketing using airports and decide for selection and purchase land and complete all administrative process for such purpose from her home says in United States using SDSS and E-Governance.

3.3 Model-3 Creation of Smart Administration

This model believes that smart governance is all about *creating smart administration*. Smart Administrator shall be well versed in Indian Constitution and Nitti gritty of innumerable legislations of Union and State Government. He should be in a position to work with people, get participation of elected political executives and local business, using his social engineering skills. He should be having capability for speedy decision making and speedy implementation using modern E-Governance practices and, project management tools like Primavera, real-time responsive systems like SCADA and SDSS. Hence this model incorporates Model 2. He should have very high professional capability in ICT with academic attainments and facility shall be there to renew his ICT capabilities by annual training programs. Big data, spatial data processing and data security shall be his concerns. He should have supporting staff in creation of database which can be spatial or non-spatial. He should have capability to collect revenues as per law and spend budget in a timely fashion using E-Governance tools. He will automate all rule-based administrative decision making through E-Governance leaving only few important decision to be made by him.

Gil-Garcia [52] indicates that a 'smart state' is a new form of electronic governance that use sophisticated information technologies to interconnect and integrate information, processes, institutions and physical infrastructure to better serve citizens and communities. This type of smart governance is at a higher level of transformation in administration since it requires the restructuring of the internal organization of government. Administrations need to be innovated to deal with the requirements for differentiated policies. Batty et al. [51] highlight that 'smart governance is a much stronger intelligence function for coordinating the many different components that comprise the smart city. It is a structure that brings together traditional functions of government and business. It is appropriate smart administration suited to an age of very high deployment of ICT in every household.

Can India aspire for a smart administration in near future? Is it a doubtful preposition? In Smart Administration, there cannot be many low-level administrative staff like peons to carry files since files are replaced by electronic records in computers. Stenographers to type dictation must have disappeared since all Government functionaries have to use internet connected computers with their own digital signature. All file movements in; computers are date and time stamped to determine the efficiency of the Government servant. These self-generated data can be used to determine provision of citizen service conforming to the Right for Government service time norm.

At present in India, there are 5 million Central Government servants/administrators at different levels and smart administration means cutting the number by say half the present size, for example to 2.5 million. In addition to Central Government servants, there are almost equal number of State Government administrators and Municipal Government servants. If you analyse budgets of many Municipalities, a substantial percentage in certain states are used for sustaining administration leaving much less for development work. This is noted by State Government but actions to achieve reduction of administrative expenses have not shown success.

Further, a qualitative change in capabilities of Government administrator is required. This means high professional qualifications and competence for Smart Governance. In the past, Administrative Reforms Commission came out with many recommendations, but 'the tail of the administration dog cannot be straightened'. If we want 'smart administration with straight tail', then we could have followed what France has done. France disbanded, the then existing administrative system and created brand new with no semblance to the old one after it declared itself as a Republic. India under the first Prime Minister, Jawaharlal Nehru, only changed the name from Indian Civil Service to Indian Administrative Service and did not transform it totally, minimally or marginally. The mind behind East India Company (established in 1,600) invested the scholarship of Anthropologists to design a colonial administration with limited manpower highly suited to rule Indian Colony to achieve their limited objectives such as revenue collection and law and order. This system has nothing that can trigger dynamically spatio-economic development. This administration that is followed till now is not suited to Smart Cities. Alternate system can only be E-Governance system in which 90 % Government Tasks are electronically performed. Now the target is only 25 %. So India is now is 125th Rank among 190 countries in E-Governance status by United Nations Report on E-Governance in 2012, while UK is 3rd, who never used the administrative structure designed for Colonial India in UK. Is it possible for India to be top 10 countries of the world in E-Governance? A big restructuring for appropriate administration for smart city is required but seems impossible to achieve now. In coalition politics in India, absolute majority is impossible to dream of now a days and that is required for constitutional change to bring about administrative restructuring and far reaching changes. This system of Governance will have capability to manage big data, use SDSS extensively and will have data security safeguards with them.

3.4 Model-4 Rearranging Governance as Dictated by Smart Cities

The fourth Model of Smart Governance advocates that smart governance is all about *rearranging the position of government within the urban system as dictated by the urban system itself.* This involves the highest level of transformation since it is not only about the transformation of the internal organization by smart governance (vide Model 3), but also of the external organization that connects Smart city with other cities inside or outside the Nation. This establishes a highly responsive urban system of smart city with interacting capabilities with external cities even if it is not part of the same nation. In this model, one can question, how Bangalore, Hyderabad or Jaipur be part of a State for Governance on the basis of linguistic division of State. In the world of Smart Cities, are linguistic states being the rational basis for economic realities that is faced by a smart cites?

A state in India is a contiguous area based on some criteria of homogeneity such as language spoken and then transformed to a programming or administrative region in different spatial hierarchy. In addition to homogeneous regions, there are also polarized regions based on interaction and flow of money, communication and goods. A good example of polarized region is a City region. It is a polarized region based on mutual dependencies of city and rural areas interacting with cities mutually benefiting each other economically. The question is in Smart City scenario, such city region is relevant or not? Since Smart city can also interact with cities outside the National boundary, can Smart City Governance be made as demanded by its international city-based interaction? In fact city regions are planned based on city and rural area interaction and demand for mutual needs. The Smart City interaction can be more than face to face talk in same language but can be telecommunication, SMS or email communications, transliteration and above all flow of money for investment and flow of goods and services much beyond geographical space of one nation, of the yesteryear's city region. It is also possible for smart cities to plan for most economical and high return production system combining a series of cities around it which is most suitable for this purpose, irrespective of the fact many of the cities in this system may be outside one national boundary. Smart city governance shall facilitate such operations.

For example, economically, Singapore, New York, Boston, London, Hong Kong, Mumbai or Chennai are more connected outside their own respective nation, States and cities. These connections are not based on same languages spoken, identical culture or nationality. These cities shall economically progress only based on sensitivity of their policies to these external cities speaking Arabic, Mandarin or Japanese. It has been seen in the past any subtle changes of policies of Singapore can make money flow in or out of Singapore to Hong Kong and vice versa. If this is the reality why not give the freedom of designing government policies and system of international interaction based on these sensitivities to the Smart Cities than dictated by Central Government and not even State Government in India. These smart city foreign affair policies can be well within the constitutional framework. This is the essence of model 4. It cannot be easily implemented in India.

Caragliu and Del Bo [56] state that 'space-specific characteristics could influence on the smart cities development and, therefore, there is a need for geographically differentiated policy actions'. Bătăgan [57] indicates that 'smart governance means collaborating across departments and with communities, helping to promote economic growth and at the most important level making operations and services truly citizen-centric'. It may be noted that smart governance is a widespread adoption of a more community-based model of governance with greater connectivity being facilitated by new technologies. The community links can be outside the nation and state. Schuurman et al. [55] stress that governments in smart cities are called to play a key role in promoting the involvement of all relevant stakeholders which may include foreigners, in order to create an interactive, participatory and information-based urban environment. Finally, Kourtit et al. [58] argues that 'smart governance is the pro-active and open-minded governance structures, with all actors involved, in order to maximize the socio-economic and ecological performance of cities, and to cope with negative externalities and historically grown path dependencies.'

4 Recommended Model

Since each Smart City is unique, it is not possible to recommend any one model to all potential smart cities. Based on many considerations Government shall adopt most suited models. This is to bring out the best output of smart city. Models 1 and 2 are easy to implement than Models 3 and 4. It is likely smart city technologies can spread in many cities which is much easier than implementing all other difficult components that make the smart city system; discussed in this chapter. In that case Model 2 becomes more relevant today. Long experience and critical evaluation of smart cities are required to move to Models 3 and 4.

References

1. Department of Information Technology, Ministry of Communication and Information technology, Government of India (2010) Background papers national e-governance plan, meeting of the national e-governance advisory group, New Delhi, 12 Nov 2010
2. Vinod Kumar TM and Associates (2014) Geographic information system for smart city. Copal Publishing Group, New Delhi
3. Sharma SK, Gupta JND (2003) Building blocks of an e-government—a framework. J Electron Commer Organ 1(4):34–48
4. Sharma SK (2004) Assessing e-government implementations. Electron Gov J 1(2):198–212
5. Sharma SK (2006) An e-government services framework, encyclopaedia of commerce, e-government and mobile commerce, Mehdi Khosrow-Pour, information resources management association, Idea Group Reference, pp 373–378
6. www.unpan.org
7. www.gbde.org
8. www.pacificcouncil.org
9. Keohane RO, Nye JS (2000) Introduction. In: Nye JS, Donahue JD (eds) Governance in a globalization world. Brookings Institution Press, Washington, D.C.
10. Clift S (2003) E-governance to e-democracy: progress in Australia and New Zealand toward information-age democracy. http://www.publicus.net/articles/edempublicnetwork.html
11. http://www.coe.int/T/E/Com/Files/Themes/e-voting/definition.asp
12. Backus M (2001) E-governance and developing countries, introduction and examples, Research Report, No. 3, April 2001
13. Kettl DF (2002) The transformation of governance. John Hopkins University Press, USA
14. Norris P (2001) Digital divide? Civic engagement, information poverty and the internet in democratic societies. Cambridge University Press, Cambridge
15. The Internet and Poverty: Real Help or Real Hype', Briefing No. 28
16. Dev B (1999) E-governance: India in the 21st century. Cyber India Online Ltd. http://www.ciol.com/content/services/egov/
17. Schware R (2000) Information technology and public sector management in developing countries: present status and future prospects. Indian J Publ Adm 46(3):411–416
18. Wadia J (2000) Welcome to digital democracy. Times computing, 22 November. http://www.timescomputing.com/20001122/nws1.html
19. Siliconindia (2001) E-states in India. http://www.siliconindia.com/
20. Ghere RK, Young BA (1998) The cyber-management environment: where technology and ingenuity meet public purpose and accountability. Publ Adm Manage Interact J 3(1). http://www.pamij.com/gypaper.html

21. Hariharan V (1999) Information poverty: India's new challenge. Inf Technol Developing Countries 9(2). http://www.iimahd.ernet.in/egov/ifip/april99.htm
22. Upadhyaya R (2000) E-governance and computer applications. Position paper. International Development Research Center, Ottawa
23. Ministry of Information Technology (MIT) (2001a) Electronic governance—a concept paper. Ministry of Information Technology, India. http://egov.mit.gov.in/
24. MIT (2001b) About Electronic-governance, Ministry of Information Technology, Delhi
25. Vittal N (2000) E-governance and telecommunication. http://www.jansamachar.net/communication/telecom_main1.shtml
26. Global Campaign for Free Expression (2000) Article 19, the Indian freedom of information bill. Global Campaign for Free Expression, London
27. MIT (2001c) EG initiatives in states/ministries/departments. Ministry of Information Technology, India. http://egov.mit.gov.in/
28. Schware R (2000) Information technology and public sector management in developing countries: present status and future prospects. Indian J Public Adm 46(3):411–416
29. PC World (2000) Cyber states in India and e-governance. http://www.pcwindia.com/2000jun/cs.htm
30. Dev B (1999) E-governance: India in the 21st century. Cyber India Online Ltd. http://www.ciol.com/content/services/egov/
31. Heeks R (2001a) Understanding e-governance for development, i-government working paper series, No. 11. Institute for Development Policy and Management, University of Manchester, Manchester
32. Schware R (2000) Information technology and public sector management in developing countries: present status and future prospects. Indian J Public Adm 46(3):411–416
33. Ghere RK, Young BA (1998) The cyber-management environment: where technology and ingenuity meet public purpose and accountability. Public Adm Manage Interact J 3(1). http://www.pamij.com/gypaper.html
34. Riley TB (2003) E-government versus E-governance: examining the difference in a changing public sector climate, the commonwealth secretariat and government telecommunications and information services, Public Works and Government Services, Canada
35. Misra DC (2006) E-citizen's charter, development gateway, 8 August. http://topics.developmentgateway.org/egovernment/rc/ItemDetail.do?itemId=1069158. Also check Dutch e-Citizen Programme. http://www.burger.overheid.nl/service_menu/english
36. DIT (Department of Information Technology), Government of India (2006) Guidelines for implementation of the common services centres (CSC) scheme in states, 9 October. http://www.mit.gov.in/cscguidelines.asp
37. Ministry of Personnel, Public Grievances and Pensions, Government of India, Right to Information Act (India) (2006) http://persmin.nic.in/RTI/WelcomeRTI.htm. Also check Central Information Commission (CIC)'s website at http://cic.gov.in/. Accessed 25 Dec 2006
38. Singapore Government (2006) eCitizen: Singapore. http://www.ecitizen.gov.sg/about_us.html. Accessed 24 Dec 2006
39. Singapore Government (2006) eCitizen: Singapore citizen's personal access: SingPass. http://www.ecitizen.gov.sg/singpass/index.htm. Accessed 24 Dec 2006
40. Pulliam D (2006) GSA launches upgraded FirstGov search engine, Govexec.com, Jan 24. http://www.govexec.com/story_page.cfm?articleid=33224&ref=rellink. Accessed 24 Dec 2006
41. Indian Railways Online Passenger Reservation (2006) http://www.indianrail.gov.in/. Accessed 26 Dec 2006
42. Bollettino JO (2002) The customer-centric digital department: e-Service in government. In: Rust RT, Kannan PK (eds) e-Service: new directions in theory and practice, Chap. 12, Prentice-Hall of India, New Delhi, p 276
43. DARPG (Department of Administrative Reforms and Public Grievances), Ministry of Personnel, Public Grievances and Pensions, Government of India, Citizens' Charters in Government of India. http://goicharters.nic.in/faq.htm. Accessed 25 Dec 2006

44. Cabinet Office, United Kingdom, Charter Mark, National standard of customer service excellence. http://www.cabinetoffice.gov.uk/chartermark/index.asp
45. Deakin M (2010) Review of city portals: the transformation of service provision under the democratization of the fourth phase. In: Reddick C (ed) Politics, democracy and e-government: participation and service delivery. IGI Publishing, Hershey
46. Davison RR, Wagner C, Ma LCK (2005) From government to e-government: a transition model. Inf Technol People 18(3):280–299
47. Lombardi P, Cooper I, Paskaleva K, Deakin M (2009) The challenge of designing user-centric e-services: European dimensions. In: Riddeck C (ed) Research strategies for e-Government service adoption. Idea Group Publishing, Hershey
48. Reddick CG (2009) Handbook of research on strategies for local e-government adoption and implementation: comparative studies, vol 2. IGI Global, Hershey. ISBN-13: 9781605662824
49. Deakin M, Allwinkle S, Campbell F (2006) The IntelCities e-Learning platform, knowledge management system and digital library for semantically rich e-governance services. Int J Technol Knowl Soc 2(8):31–38
50. Coleman S (2003) Speech at "crossing boundaries" conference. Ottawa 8 May 2003
51. Batty M, Axhausen KW, Giannotti F, Pozdnoukhov A, Bazzani A, Wachowicz M, Ouzounis G, Portugali Y (2012) Smart cities of the future. Eur Phys J 214:481–518
52. Gil-Garcia, R (2012). Enacting electronic government success. In: An integrative study of government-wide websites, organizational capabilities, and institutions. Springer, New York
53. UNESCAP (2007) Good governance. UNESCAP Available: http://www.unescap.org/pdd/prs/ProjectActivities/Ongoing/gg/governance.asp
54. Walravens N (2012) Mobile business and the smart city: developing a business model framework to include public design parameters for mobile city services. J Theor Appl Electron Commer Res 7(3):121–135
55. Schuurman D, Baccarne B, De Marez L, Mechant P (2012) Smart Ideas for smart cities: investigating crowdsourcing for generating and selecting ideas for ICT innovation in a city context. J Theor Appl Electron Commer Res 7(3):49–62
56. Caragliu A, Del Bo C (2012) Smartness and European urban performance: assessing the local impacts of smart urban attributes. Innov Eur J Soc Sci Res 25(2):97–113
57. Bǎtǎgan L (2011) Smart cities and sustainability models. Informatica Economicǎ 15(3):80–87
58. Kourtit K, Nijkamp P, Arribas D (2012) Smart cities in perspective—a comparative European study by means of self-organizing maps. Innov Eur J Soc Sci Res 25(2):229–246

Part I
E-Smart City Governance-State of the Art Studies

E-Governance State-of-the-Art Survey: Stuttgart, Germany

Satyendra Singh

Abstract E-Governance or Smart Government is buzzword across Europe to improve interaction between government and stakeholders. Governments both at national and local level are making efforts to transform themselves into a well-connected entity that responds efficiently to the needs of its citizens by developing an integrated back-office infrastructure. Cities are implementing smart and innovative means to improve quality of life and enhance competitiveness. Various measures are being taken to provide transparent, efficient, innovative and responsive government through adoption of various Information and Communication Technology (ICT) tools. Stuttgart, located in south-west of Germany is no exception which started its E-Government initiative a decade ago, has today crossed significant milestones. Various e-services platforms are created to inform and assist citizens and e-transaction platforms are created to reduce the burden on existing staff in various departments. These services have shown remarkable results in management of some areas such as waste, mobility, human resource, etc. These platforms are still evolving and being created and developed after a thorough research, planning and consultative process with various stakeholders. Stuttgart has been successful in extending the reach of services digitally to its citizens and promoting a sense of sustainability. This city takes care of wide spectrum of administrative, political and social services through diverse ICT and Geographic Information System (GIS) platforms. Stuttgart is hub of high-tech auto industry and also known for its advanced ICT industry and highly innovative green technologies. This sense of sustainability is also reflected in adoption of environmental friendly practices which are being supported by befitting e-democracy and E-Governance measures. The objective of this paper is to review process of formation of E-Government services in Germany and the key milestone related to development of E-governance in Stuttgart. An case of best practices in mobility is presented to ascertain if Stuttgart leads its way to be a smart city.

S. Singh (✉)
Visiting Lecturer, University of Applied Sciences, 101452, Schellingstr. 24,
Stuttgart 70174, BW, Germany
e-mail: ssingh27@hotmail.com

T.M. Vinod Kumar (ed.), *E-Governance for Smart Cities*, Advances in 21st Century Human Settlements, DOI 10.1007/978-981-287-287-6_2

Keywords E-Governance · E-participation · Information and Communication Technology · Smart Governance · Mobility · Sustainability Smart City

1 Introduction

The term E-Government is made up of two words, 'Electronic' and 'Government' which stands today as a synonym for a modern, transparent and efficient administration with Internet as a connecting medium between stakeholders and administration. Municipal, State and National Government use information technology and internet to support government operations, engage citizens and extend government services [1]. It is often referred in context to handling of diverse administrative processes with the help of modern Information and Communication Technology (ICT). The interaction may be in the form of obtaining information, filings or making payments and a host of other activities via the World Wide Web. The European Commission is actively supporting E-Government both at the national level and at its own supra national level. The European Innovation Partnership for Smart Cities and Communities combines ICT, energy management and transport management to come up with innovative solutions to the major environmental, societal and health challenges facing European cities today (http://ec.europa.eu).

Germany being a developed country has shown remarkable progress in some of the sub-key indicators though some of the services extended through E-government programmes are still undergoing optimization. Germany has started much early to formulate strategy for E-government and has implemented it at municipality level a decade ago. On September 2006, The German Federal Cabinet adopted the comprehensive strategy focused on the future: Innovations for the Administration (Zukunftorientierte Verwaltung durch Innovation), which aims to modernize the Federal State Administration, to downsize bureaucracy and improve the quality and efficiency of public sector services. The integral part of the strategy consists of the E-Government 2.0 Program. The programme has been developed in compliance with European Union Action Plan i2010 and utilizes already existing know-how on E-Government [2]. Current E-Government 2.0 programme framework, which is conceptualized keeping in mind the overall modernization of the Public Administration at federal level, has also laid down framework to be implemented at local (municipality) level. The idea is to create a fully integrated E-Government in Germany which connected federal government, federal-state governments and municipal administrations. Germany has made strides in the promotion of E-Governance in the federal government but still lacks the necessary legal infrastructure to further promote E-Government in the lower levels of government and in federal states [3]. This deficiency results in a lackluster showing in the E-Government promotion at the lower level of the government and the federal states.

Various reviews on quality of E-Government services suggest that portals created to serve the purpose of E-Participation extend only as a gateway to the government website but the portals often do not have interactive tools where public can contribute and share their opinion. These portals do provide information about the opinions of the citizens in decision-making process conducted by government. This paper is an effort to explore the state of the art of E-Government services in Stuttgart metropolitan region and also examine the role of e-Participation to promote inclusion and empowerment. An appraisal of the select best practice, i.e. Sustainable Mobility management in Stuttgart is examined as how it fits to the dimension of emerging smart cities and what challenges and issues it poses on planning a future oriented city. The inferences of this investigation are going to benefit development of emerging cities in developing world where efforts are being made to develop new smart cities. The basis of this paper is designed by providing answers to the following basic questions:

- How various administrative procedures are made available to citizens?
- How the information is disseminated to various departments and how the participation of citizens is ensured?
- How the functioning of administration is made more transparent, efficient, participatory and democratic?
- How management of all these processes has been made understandable to all?

Some salient points of Stuttgart Mobility Development 2030 (VEK 2030) [www.stuttgart.de] have been listed with focus on cleaner and environmental friendly practices and futuristic sustainable mobility concepts. Some of the key agendas which are being included are intelligent and social use of transport means along with smart control of traffic. There are efforts to promote policies for non motorized mobility, mobility for disabled and elderly as well. A small case of smart mobility management is presented, which is directed towards making Stuttgart a smart city.

2 European E-Governance Framework

European countries started E-Government initiatives, which were mainly focused towards improvement of governance at the national level. Some of the cross-border E-Government activities also took place. The European Commission has laid down a clear vision, policies and objectives to ensure modern and efficient administration. Several programmes have been initiated including i2010 Action Plan that defined the principles and directions of E-Government policy of the European Commission till year 2010. The i2010 project was designed to foster a fully inclusive information society, which is based on the widespread use of ICT in public services, small and medium size enterprises and household [4]. Later, Ministerial declaration on E-Government, which was presented in Malmö, Sweden, has laid down guidelines for development between 2010 and 2015. The next E-Government Action Plan that describes the directions for the period starting at 2011 has been announced by the European Commission under the name Europe 2020 (http://www.egovap-evaluation.eu/).

In the beginning, the focus of all these initiatives was to ensure Interoperability and promote the efficient use of ICT for cross-border services to citizens and enterprises in Europe. Delivery of European E-Government Services to public Administrations, Businesses and Citizens (IDABC) was a European Union Programme launched to meet this objective. It aimed at stimulating the development of online platforms delivering public e-Services across Europe (www.wikipedia.org). It used the opportunities offered by ICT to encourage and support the delivery of cross-border public sector services to improve efficiency and collaboration between European public administrations and to contribute to making Europe an attractive place to live, work and invest. IDABC supported the member state to develop clear objective with recommendations to developed solutions and services that enable national and European administrations to communicate electronically while offering modern public services to businesses and citizens in Europe.

3 German E-Governance Framework

Germany, called 'Bundesrepublik' Deutschland in German, is a federal State made up of 16 federal states or so-called 'Länder'. These states are comprised of 300 districts and around 13,000 municipalities. Local authorities are usually in-charge of most government services and considered key contact point for politics, administration and citizens. However, when it comes to E-Government policies, it is national federal government which formulates key policies and takes up a lead position. Therefore, any decision on E-Government policy follows a structure which is implemented at federal, state and local level.

E-Government started getting attention of Germany in the late 1990s when efforts were made by some experts from different segments of German society by formulating memorandum of E-Governance. The memorandum was prepared after inviting input from all stakeholders on common need of electronic administration, electronic democracy and the reorganization of structures and processes. In order to achieve this objective, a careful appraisal of ongoing administrative practices was done where administrative processes could be made more efficient with the help of ICT. A strong need of organizational re-engineering and restructuring of work processes in administration was felt. As a first step, this restructured face of administration was presented to citizen electronically by means of different ICT platforms. An IT supported interface between administration and stakeholders was tested with an idea to provide a chance for citizen to participate in political decision-making (Fig. 1).

As a first step, the task of distribution of information digitally through web was facilitated using web. After carefully analyzing the user's behaviour and how a citizen makes use of this information, integration of network-based systems was envisaged. Some authorities also initiated online transactions using electronic signature. In 1997, the functionality of legally accepted electronic signature for various online transactions was introduced in Germany after carefully assessing all the

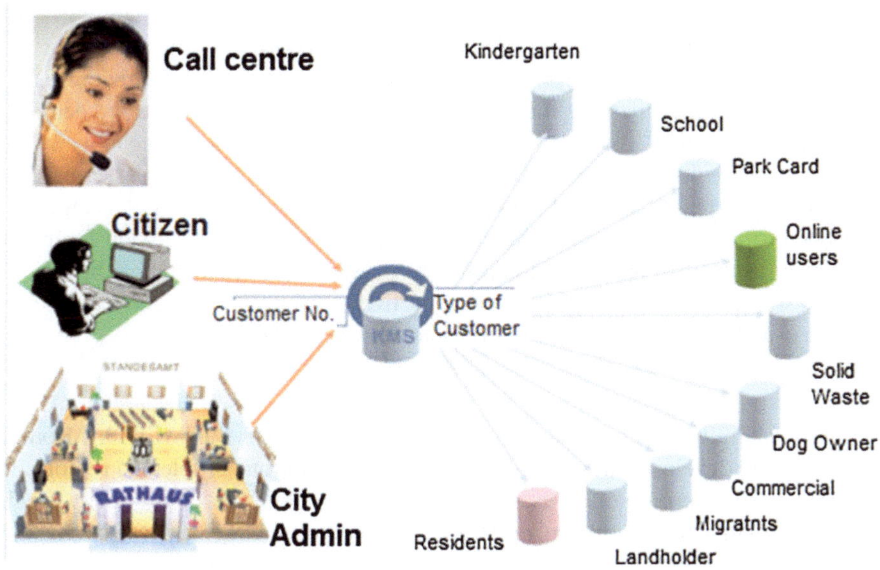

Fig. 1 E-government stakeholders. *Source* Report "Praxis des E-Government in Baden-Württemberg"

security issues. A first significant step on the federal level to facilitate the introduction of IT-solutions in the public sector was the 1997 ratification of the law on electronic signatures in attempt to provide a legal framework for an infrastructure that can be used for many different types of online transactions [5].

All the associated issues to ensure secure transaction were given a top priority. Gradually, after successful implementation of this process, a need was felt to add the functionality of 'feedback' in which a user can also communicate to administration and ensure participation in decision-making process. Efforts were also made to reach the communities which are not connected to net but have interest to participate in decision-making process.

3.1 The Media@Komm Project

Esslingen Municipality: Esslingen, a town located southeast of Stuttgart is home to high-tech companies and the automobile industry. Esslingen with a population of 100,000 people is also the administrative center for people living in neighbouring villages. The town came into limelight for its innovative approach to implement online administration and support E-Government and E-Democracy activities.

The purpose of the Esslingen Media@komm project was to improve the participation possibilities of citizen. The employed concept included providing a broad scope of information to citizens as the ground work for an increased democratic participation. The core of the project was a virtual market place in which a company register, a job market and a citizen/investor information service, among other things could be found [6]. The public budget dialogue in Esslingen is one of the first European projects to use an Internet based discussion forum as an instrument to inform and consult the public about the municipal budget [7]. It was noted that citizens participated actively online and provided valuable input to authorities to improve on service and make use of budget.

The first significant project done in this direction in 1998 was MEDIA@Komm project for the development of local E-Government solutions in selected regions. It was envisaged to promote the usage of electronic signature at municipality level. Esslingen town which is a part of Stuttgart metropolitan region was among the top three cities selected to implement the concept of electronic transaction with a budget of 8 million Euros. A pilot project was executed to apply variety of solution for electronic planning, electronic award of contracts, online reminder and access to electronic libraries among others. The foremost concern was on secure mean of data and information exchange (www.esslingen.de).

MEDIA@komm succeeded in its objectives and gave further impetus to development of electronic-based system development in other municipalities with diverse applications. Some of the municipalities came up with innovative and ambitious approach to implement the integrated system at larger scale by teaming up together. Mobile electronic administration which were developed in Berlin and Magdeburg cities is a perfect example of innovation and team effort which further found its ways in Poland and Russia for cross-border transfers [8] (Fig. 2).

3.2 Key Developments

In order to make Germany a leader of electronic governance, a number of public-private partnership (PPP) project were initiated with the help of representatives of enterprises, associations, political parties and political institutions. Global players such as Microsoft, Siemens, BMW, AOL, IBM also played a crucial role to take this efforts to next level. Several academic projects and real-time efforts have been initiated to develop e-Democracy system for decision-making and its integration to the election process. Election of the council of young citizens (Jugendgemeinderatswahl) in Fellbach, under Stuttgart metropolitan regions was one of such actions. Around the same time Electronic Health Card were issued under D21 initiative connecting patients, medical doctors, dentists, hospitals, pharmacies and health insurance companies for smooth exchange of information [9].

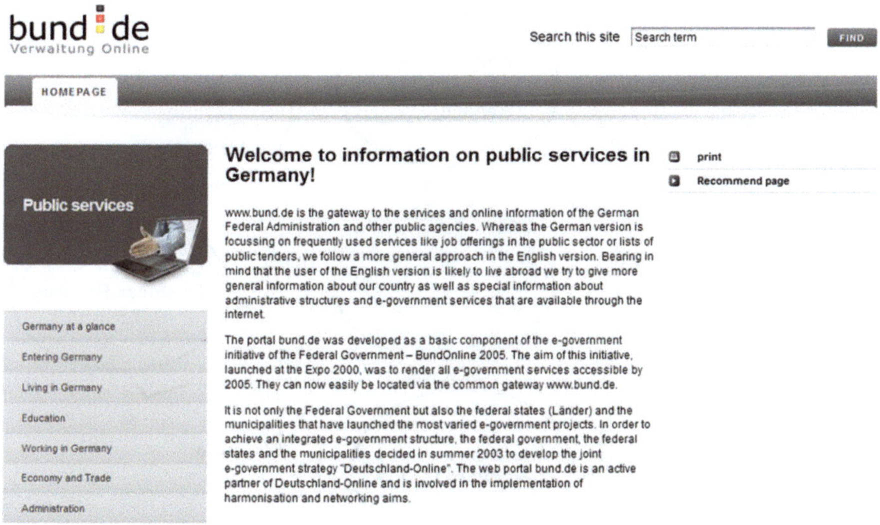

Fig. 2 Portal of federal government

Federal administration came up with another initiative BundOnline2005 for modern, service-orientated enterprise services. The portal bund.de was developed as a basic component of the E-Government initiative of the Federal Government—BundOnline 2005. The aim of this initiative, launched at the Expo 2000, was to render all E-Government services accessible by 2005. The vision was to follow a user centric approach by focussing on citizens and their needs. This initiative was launched with the main objective to modernize the federal public services with electronic delivery capability. After successful completion of this programme in 2005, vast range of services were made available online and two-way dialogue was established with stakeholders. They can now easily be located via the common gateway www.bund.de.

Development of common infrastructures with an objective of integrated E-Government development were met with Deutschland-Online initiative which started in year 2003. The emphasis was to develop Integrated Electronic Services for citizens and businesses and Interconnection of Internet portals. An emphasis was placed on provision of Common standards and Experience and knowledge transfer. Other notable programmes namely 'Federal IT strategy' and 'Broadband Strategy' provided a major boost to improve IT management within the government, as well as providing businesses and household with high end broadband services.

3.3 E-Government 2.0

The programme E-Government 2.0 is worth a mention which laid a special emphasis on E-Participation. E-Government 2.0 or Gov 2.0 refers to government policies that aim to harness collaborative technologies to create an open-source computing platform

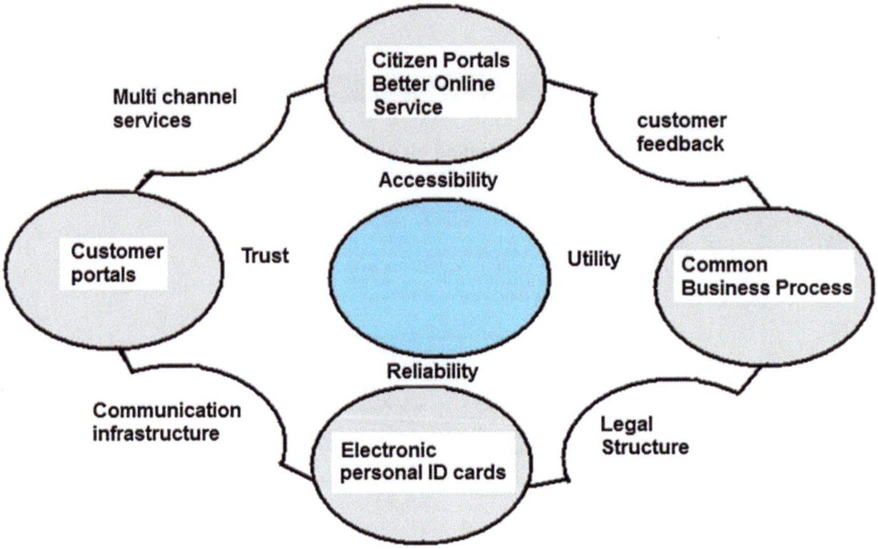

Fig. 3 Need for better transparency and participation. *Source* Interior federal ministry report

in which government, citizens and innovative companies can improve transparency and efficiency. Gov 2.0 combines Web 2.0 fundamentals with e-government and increases citizen participation by using open-source platforms, which allow development of innovative apps, websites and widgets. The government's role is to provide open data, web services and platforms as an infrastructure (www.wikipedia.org) (Fig. 3).

Trust and Reliability of system are key issues which were addressed by provision of strict security measures without compromising the accessibility and utility. In order to meet this objective, several projects have been initiated. Notable of them was provision of electronic identity Card in 2010 to facilitate identification of the owner. This Identity card with microchip contained holders' data in electronic format including biometrical data (digital facial image/fingerprints). The possibility of integrating digital signature was also explored.

Another significant development is 'De-Mail' aimed at facilitating the secure exchange of electronic documents among citizens, businesses and public authorities via Internet. The concept of providing unitary service number 115 was another notable development in which citizen gets a number (115) to contact public administration and gets all the related information.

4 E-Government Framework of State of Baden Württemberg

Stuttgart is the capital of state of Baden Württemberg in Germany. Baden Württemberg (BW), located in south west corner of Germany is surrounded by Switzerland, France and Austria. BW with a population of around 11 million is third biggest state in Germany in terms of area and also population. BW is considered among world's most successful regions in industry, science, education and culture. There are as many as 1.101 small and medium municipalities (called Gemeinden) and 93 of them are relatively bigger city and 312 of them meet the definition of city. Almost all of them have online presence with facility of atleast providing online vehicle registration facility, registration of business, population registration and marriage certificate online.

"www.Service-bw.de" is the main Service Portal for the State which also connects itself to almost all local authorities. The Portal offers detailed information about administrative services specifically designed for general public, companies and employees in administration. Apart from providing general information, it also extends service to easily access online forms and online services. The procedures are made simple and described explicitly with their service offerings adequately supported by range of electronic forms and online applications (Fig. 4).

'Service-bw' offers access to over 9,000 public authorities and institutions spread across state of Baden-Württemberg. Wide-ranging life situations in over 50 categories are described and connected to around 1,000 concrete administrative services. The use of these administrative process is made simple with the help of large number of useful tips and information. The administrative process worth mentioning are police, environment, justice.

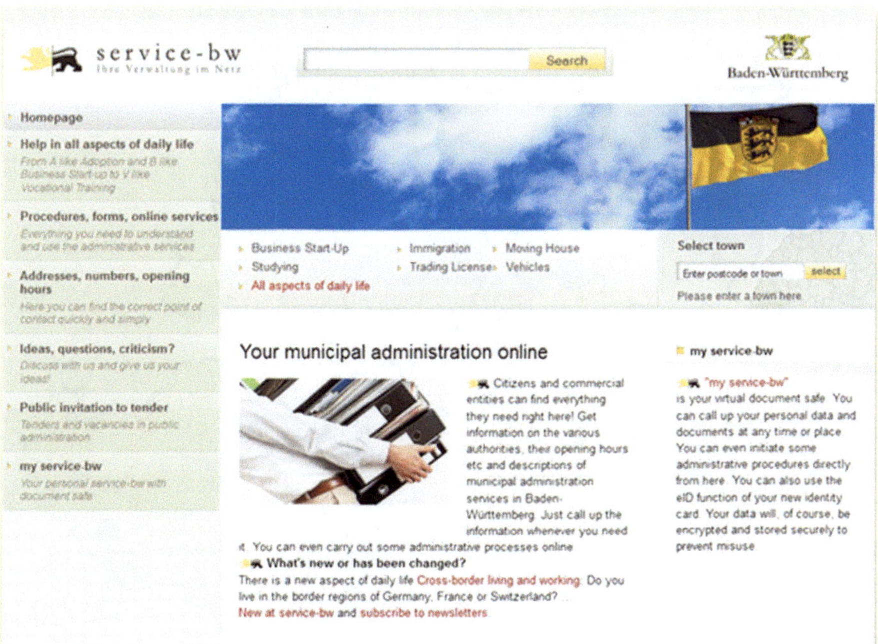

Fig. 4 Service portal of state of Baden Württemberg. *Source* www.service-bw.de

5 Stuttgart City: Introduction

Stuttgart, the sixth largest city of Germany is spread across hills, valleys and parks with an area of about 210 km^2. The population is Stuttgart is 5,90,000 while the greater Stuttgart metropolitan Region has a population of 5.3 million (2008). The main city, with a high-population density of 2,850 persons/km^2, is surrounded by small satellite towns which have relatively low population density. Various surveys conducted on quality of life in German cities indicate that Stuttgart offers best quality of life with highest per capita income among German cities. Being a place of origin of Mercedes and Porsche, it is considered by many to be the starting point of the worldwide automotive industry and is sometimes referred to as 'The cradle of the automobile' (Fig. 5).

Stuttgart is a hub of high-tech auto industry and is also known for its advanced ICT industry. The city offers a conducive environment to develop highly innovative green technologies. Stuttgart's healthy innovation climate is fostering the development of environmentally friendly and sustainable technologies for the future. To meet this agenda, by 2020, Stuttgart plans to produce 100 percent regenerative energy for entire population of Stuttgart without fossil fuel, CO_2 or nuclear means (www.stadtwerke-stuttgart.de). Major emphasis is given to ecological construction combined with high-tech methods and building types which do not emit harmful emissions, use no fossil fuels and are made of recyclable building materials. With this innovative, ecological focus, Stuttgart is a model for future sustainable urban development (Fig. 6).

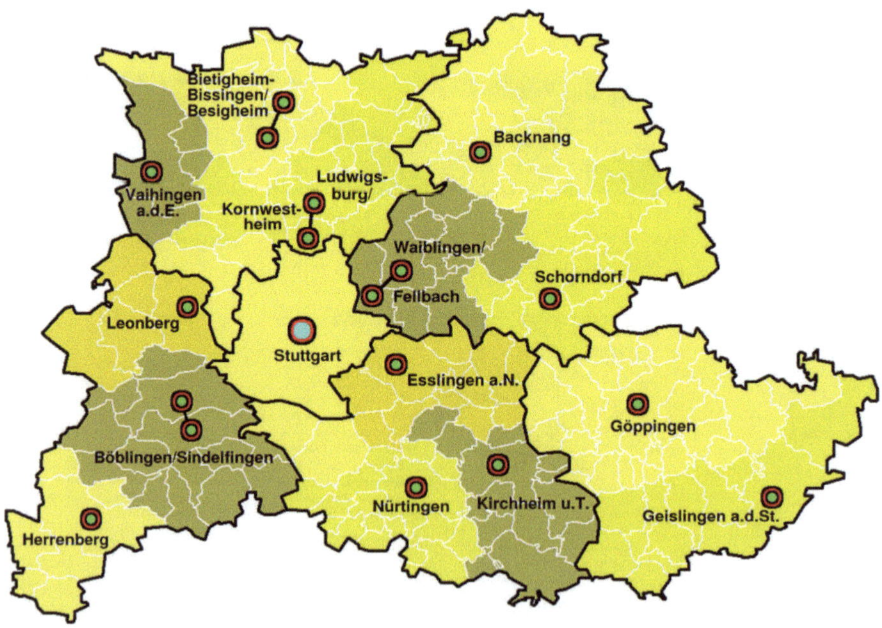

Fig. 5 Stuttgart metropolitan region. *Source* www.wikipedia.org

Fig. 6 Electric car2go initiative started by stuttgart-based daimler AG

This excellence is reflected in adoption of environment friendly sustainable practices which is also made an integral part of city planning and development. The process to manage these practices is adequately supported by befitting E-democracy and E-Governance measures. Various e-services platforms are created to inform and assist citizens and e-transaction platforms are created to enhance the efficiency of existing staff in various departments dealing with these issues. These services have shown remarkable results in management of some areas such as waste, mobility, human resource, etc. These platforms are still evolving and being created and developed after a thorough research, planning and consultative process with various stakeholders.

5.1 E-Government Framework in Stuttgart

The E-Government Stuttgart initiative which started in the framework of Media@komm project in 2000, has started with an idea of creating an information portal for all administrative services. It was envisaged to integrate also the private and socio-cultural aspects and develop specific applications to ensure online participation of stakeholders. This initiative has crossed significant milestones including establishment of Stuttgart competence centre and content management system 'InfopoolIBS'. The contents and applications are carefully designed keeping in mind the growing need of cooperation and collaboration among various departments. The applications are developed after thoroughly studying the life of its citizens, changing need of business community and fast changing culture of using mobile devices. The idea was to increase the reach of these services to all age groups digitally and subsequently promote a sense of sustainability.

Today, this centre offers wide spectrum of administrative, political and social services of city through diverse ICT and GIS platforms and extended applications. It offers an electronic channels to inform it's citizens about the service offerings as well as functionality to perform selected administrative tasks online. Different sections are being updated and developed after a thorough research on users' behaviours and usability. Adequate tools are made available for controlled workflows between the management (individual departments) and external partners (companies, non-professional organizations, associations, etc.)

Any online application which is developed caters to the need of specific community or location. They are developed keeping in mind the need of users and how it addresses the typical query of citizens. The applications have a strong link with geographic component but it's made simple and effective. These online services offered by Stuttgart can be categorized in three types for the ease of understanding.

Group 1: All the information about city administration, opening time, transportation, events, emergency services, etc. is made available on city portal www.stuttgart.de and also at the citizen kiosk (Bürgerkiosk). Another extension http://mobil.stuttgart.de is developed to cater to the need of growing mobile phone users where application (apps) are developed to provide real-time information on smartphones and PDAs especially useful for mobility part. Efforts are being made to provide free information for multi-lingual users (German, English, French and Spanish) to ensure greater electronic participation.

'Stuttgart-info' section deals with important information about planned and ongoing events, art and architecture, nature, etc. which is constantly being updated. A gamut of information is available at one place which deals with quality of life, city planning, economy and culture. The interactive element of E-Government can be seen in the collaboration platform, 'My Service Stuttgart' where City administration and private organizations such as day care centres or sports clubs work together. The departments which operate under 'my service' section with different level of elaboration are given in Table 1.

The organizations share key information with city administration which often relates to planned events or new offerings. The municipality keeps the information in one place and shares it with registered users of same category. The information is presented differently for city dwellers and for companies, a non-profit organization, a club or similar institution. After a quick registration, user gets access to 'my service' section and avail the listed services. Each citizen is entitled to have an account to Intelligent and dynamic online applications and data entry screens (Fig. 7).

Group 2: The other category is called Electronic citizen service Assistant (eBSA-Elektronisch Bürgerservice-Assistent) which extends the facility of 'online research on the proximity' to user. This is where geographic component also plays an important role. It is possible to find civic service/facilities in the given proximity such as nearest kindergarten, post-office, library, waste collection point, administrative services, etc., once information about users location is provided. The application often asks the user for its location and proximity orientation i.e. at what point user need this information and proximity requires geographic

Table 1 Departments under 'my service' section of Stuttgart city portal

Road and transport	*Personal information*
Transportation/Parking garage Resident parking permit Special parking permit for businesses, social services and all-electric vehicles Feinstaubplakette (related to environment) Ideas and complaints Vehicle license plates	Information on the birth time Marriage, birth and death certificate Local jurisdiction registry office Local jurisdiction city hall Next, citizens' advice bureau Next week market Online lost and found
Building and housing	*Facts and figures*
Information about the living environment	Polling place finder Statistics Stuttgart (shop) Survey Geodata (shop)
Trade	*Tendering and contracting*
Trade register information Commercial confirmation	Tenders eVergabe (electronic allocation)
Waste and disposal	*Sport and leisure*
Waste/recyclables container stock position Waste removal calendar Express bulky waste pick-up Bulky waste pick-up Green waste collection Ideas and complaints	Sports guide Play areas Club guide VHS course registration

source www.service.stuttgart.de (translated)

Fig. 7 My service section Graphical User Interface (GUI)

component in order to locate nearby Area of interest (AOI) whether it is about finding required activity, service or finding a location of school, hospital or other emergency service. Integration of Geo-referenced data is done in 'InfopoolIBS' which contains information about address, street ID, Gauss-Krüger coordinate system and information about ward, zone, town and city.

In order to facilitate the various administrative functions, various forms are made available online which can be printed or sometimes related CD or printed material can be ordered online. For example, it is possible to get emissions sticker or a resident parking permit from the given facility after filling up online forms which avoids unnecessary trip to the office. Special applications are developed for selected departments where online money transaction/payment facility is also integrated. The secure online money transactions facility is available with electronic signature. A co-operation platform with specially designed web application is provided to business and NGO for information exchange. The access is customized differently for each administrative or civic unit based on the needs of user.

6 Case Study: Mobility Management

As a leader in the automobile field, Stuttgart has provided a major thrust to meet the major challenges posed by the mobility situation. Industry has given new impetus to mobility and are currently developing many revolutionary technologies and processes which will change the way how mobility is managed in future. They are helping to design the mobility of tomorrow with intelligent technical solutions, and new ways to manage the traffic without loosing sight of their commitment to the environment.

The Integrated Traffic Management Centre Stuttgart (IVLZ) records the real time traffic situation for every minute without break to avoid congestion and slow moving traffic. This recorded information is processed and analysed for active intervention to regulate both road traffic and the public transport network. IVLZ has been constantly providing input for smooth management of traffic and maintaining punctuality of bus, tram or light railway. The IVLZ collects real time data from following sources:

- Permanently installed measurement points to record incidence of traffic.
- Mobile sensors.
- 700 taxis with GPS.
- City's Civil Engineering Department provide additional information about the traffic situation, in particular on the main arterial routes and at major junctions.
- The traffic surveillance system operated by the Police Department.

The real-time traffic situation can be calculated with the help of current speed of each individual taxi even in the locations where no measurement facility is installed. Any incidents or events likely to affect the traffic situation can be reported directly to the operators in the Control Centre either over the operations control computer or by police radio (Fig. 8).

Fig. 8 Real-time traffic situation in north Stuttgart. *Source* State capital stuttgart

The picture shows the realtime traffic situation in north Stuttgart where Red represents congestion, Yellow represents heavy traffic, Green is free-flowing traffic. The dynamic picture created of the traffic situation and the on-going evaluation of data permit evolving critical traffic situations or bottlenecks in the transport system to be detected at an early juncture. The operators working at the IVLZ evaluate all events of relevance for the traffic situation and work as a team to put in place a series of operational measures designed to guarantee a fast response.

Another important element of the overall traffic picture is provided by the Stuttgart transport Authority (SSB) with the help of CCTV cameras, which monitors the situation in the city's bus and tram network. This allows a fast and selective response to delays in the bus and tram timetable. Information on existing or planned construction sites, organized events and occupancy levels in the city's open-air and multi-storey car parks is collated centrally in the IVLZ using specially developed software (Central Traffic Information VIZ).

In order to meet the commitment of sustainability and environment, an integrated system of ticketing is implemented in which a single ticket is valid for all forms of transport. German national railway (Deutsche-Bahn) subsidiary DB offers 'Call a Bike' service to hire bicycle available at 65 locations spread across the Stuttgart city. The first half-hour of cycling is free and bike can be left at any of the stations. The whole process to get the bike can be completed with the help of mobile phone (Fig. 9).

Fig. 9 Call a bike service

Car2go is a fleet of electric vehicles being operated in Stuttgart with three hundred 2 seater smart Fortwos cars. This programme is started by Stuttgart-based Daimler and now being operated in whole Europe. Stuttgart is equiped with 500 charging points which enables drivers to recharge electric vehicles throughout the city. On their smartphones or the Internet, customers can see where the nearest unoccupied car2go vehicle is located (Fig. 10).

Bicycle and electric car are also integrated with yearly local travel ticket. The yearly ticket holder can have access to these services at anytime to reach the final destination. This move is supported by ministry of Economy and technology till the end of 2015. There are as many as 22 co-operation partner coming from research community, private sector assisting these initiatives.

All these mobility-related information are integrated with use of mobile devices. The car2go trips can be booked and managed by smartphones or on

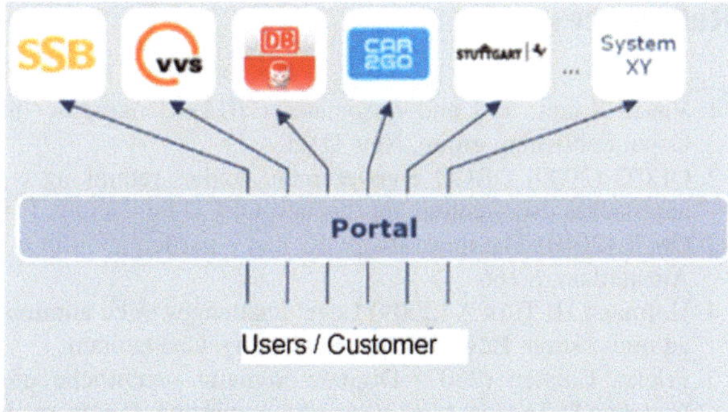

Fig. 10 Integrated portal for all mobility services. *Source* www.stuttgart.de

the Internet, customers can see where the nearest unoccupied car2go vehicle is located. Various other apps are developed to get real time information about delays and unexpected situation. The information about time plans and real-time status of each public transport means such as bus or trams is centrally collected, processed and offered to the users through mobile applications on mobile phones. Development in this direction shows the commitment to the sustainable environment. So far it is 'walk + public' or 'bicycle + public transport' but in future information about pedestrian, bicycle, car-sharing, rent a bike and taxi would also integrate with improved information about traffic jams.

7 Conclusion

This survey is an effort to decipher the facts which were largely unknown regarding Evolution of E-Governance in its functional aspect in Germany and Stuttgart. It is evident from the survey that Germany has taken a lead in implementation of E-governance at federal level but idea to create a fully integrated E-Government landscape which connects federal government, federal-state governments and municipal administrations still lacks the necessary legal infrastructure. This has affected the development mainly at municipality level which failed at enabling a stronger dialogue between citizens and government. "GIS for Smart Cities" defines Smart City as a 24x7 knowledge based city that communicates, extends real time knowledge to stakeholders with easy public delivery of services, comfortable mobility, conserve energy, environment and other natural resources. Stuttgart seems to be leading in some of the key building blocks of the smart city which deal with Smart mobility, Smart Environment and smart economy. There is also considerable improvement in mobile phone based intelligent Governance with digital infrastructure, maintaining low carbon footprints and sustainable resource management. There are still gaps visible in Integration and transformation aspects along with user-cantered experience and also integration of a wide range of services across a whole government administration.

References

1. Vinod Kumar TM and Associates (2014) Geographic information system for smart city. Copal Publishing group, New Delhi
2. OECD (2009) ORCD e-government studies rethinking e-government services user-centred approaches user-centred approaches. OECD Publishing, Paris, p 82
3. Obi T (2010) The innovative CIO and e-participation in e-government initiatives. IOS press, Amsterdam, p 166
4. Hofmann H, Türk A (2009) Legal challenges in eu administrative law: towards an integrated administration. Edward Elgar Publishing, Cheltanham
5. Fricke, Carsten (2002) Digitale Signatur—rechtliche und technische Anforderungen, In: Initiative D 21 (ed): Mit Internet Staat machen, Berlin, pp 50–55
6. Riedel EH, Wolfrum R (2006) Recent trends in German and European constitutional law: German reports presented to the XVIIth international congress on comparative law, Utrecht, pp 137–139
7. Böhlen M (2005) E-government: towards electronic democracy. In: International conference, TCGOV 2005, springer science & business media, Bolzano, Italy, 2–4 March 2005, proceedings
8. Winkel O (2007) Electronic government in Germany—a key future prospect, but expectations are exaggerated. Innovative Management. Linz, pp 163–186
9. Alcatel-Lucent (2010) Stiftung für Kommunikationsforschung Gemeindetag Baden-Württemberg, Innenministerium Baden-Württemberg Stiftung der Württembergische Gemeinde-Versicherung a.G. Praxis des E-Government in Baden-Württemberg Richard Boorberg Verlag, pp 232–235

Websites

10. http://www.egovap-evaluation.eu/
11. www.wikipedia.org
12. www.esslingen.de
13. http://www.service-bw.de/
14. http://service-bw.de/
15. http://ec.europa.eu
16. http://www.stuttgart.de/
17. http://www.dvvbw.de/servlet/PB/menu/1202078_l1/index.html
18. http://ec.europa.eu/eip/smartcities/index_en.htm

E-Governance and Smart Cities: Cases of Ahmedabad and Hyderabad

Vinita Yadav

Abstract Smart cities are smart not only in terms of higher level of services but also having an efficient and effective system. Its locationality brings balanced regional development. Better governed institutions are the one where procedures are transparent. E-Governance has increased the interlinkages between different departments. Through E-Governance, services provided are varying from birth/death certificate, booking of community facilities and town planning. The usage of mobile for M-Governance also provides last mile connectivity and its applicability streamlines not only the data collection but also analysis. The chapter attempts to understand the definitions of E-Governance and its parameters. Through two case studies of Ahmedabad and Hyderabad, it tries to find answers of: whether E-Governance or M-Governance's implementation led to transformative governance including increased participation or transparent administration or not? and where both the cities stand in the 5 stages of E-Governance evolution from basic web presence to well-connected smart governance and 4 model of non-transformation of government structure to creation of smart administration to rearranging the position of government within the urban system.

Keywords Smart City · E-Governance · Accountability · Transparency · M-Governance · Information and Communication Technology

1 E-Governance: An Effective Tool of Governance

Governance refers to the exercise of political, economic and administrative authority in the management of a country's affairs, including citizens' articulation of their interests and exercise of their legal rights and obligations as per the constitution. E-Governance or Electronic Governance is performance of this governance via the

V. Yadav (✉)
Faculty, Department of Regional Planning, School of Planning and Architecture, 4-Block-B, Indraprastha Estate, New Delhi 110002, India
e-mail: v.yadav@spa.ac.in

T.M. Vinod Kumar (ed.), *E-Governance for Smart Cities*, Advances in 21st Century Human Settlements, DOI 10.1007/978-981-287-287-6_3

electronic medium in order to facilitate an efficient, speedy and transparent process of disseminating information to the public, and other agencies, and for performing government administration activities. The terms E-Government and E-Governance are treated to be same by some authors, however, these are two different terms. E-Governance is a wider concept than E-Government, since it can bring about a change in how citizens relate to governments and to each other. On the other, E-Government is web-based services in which different level of governments uses information technology for operations and provision of services [1].

E-Governance implies a new definition and concept of public governance. It represents a paradigm shift to new information age. It involves promotion of Information and Communication Technology (ICT) and E-Commerce and adopting of these technologies in an open and transparent system over the Internet for government business, citizen interaction and engagement for development. On the other, E-Government is application of information technology to the process of government functioning in providing digital information and online transaction services to the citizen.

E-governance is the use of ICT by government to enhance the range, quality of information and service provided to the citizen in a cost-effective manner and different actors of the society to improve their access to information and to build their capacities. The interaction in E-Governance is between the government and citizen, employees or business by the various types of interaction like Government-to-Government (G2G), Government-to-Customer (G2C), Government-to-Business (G2B) and Government-to-employees (G2E).

E-Governance is defined differently by different organizations. E-Governance can bring forth new concepts of citizenship, both in terms of citizen needs and responsibilities. It's objective is to engage, enable and empower the citizen [2]. It represents the strategic and systematic use of modern ICT by a government to improve the efficiency, transparency and accountability in its functioning and interface with citizens [3]. E-Governance is a form of public administration which makes use of ICT to enhance the access and delivery of government services to benefit citizens, employees and management of Urban Local Bodies (ULBs)'. It aims to 'help strengthen government's drive toward effective governance and increase transparency to better manage social and economic resources for development' [4].

E-Governance is the public sector's use of ICT with the aim of improving information and service delivery, encouraging citizen participation in the decision-making process and making government more accountable, transparent and effective. E-Governance refers to the use of information technologies that have ability to transform relation with citizen, businesses and other arms of government [5]. These technologies can provide better delivery of government services to citizens. The E-Governance provides access to information to empower citizen and enable their participation in government and enhance citizen economic and social opportunities. It not only helps to make their life better but also of future generation.

The Central and State Government implemented E-Governance in Custom and Excise, Indian railway, Postal, Passport and Visa, Card-registration process (state government of Andhra Pradesh) and LOKMITRA (state government of Himachal Pradesh) and many more governance functions.

2 Municipal E-Governance: Efficiency and Participatory

Municipal E-Governance can be defined as 'the use of ICT by the government to improve efficiency, transparency and accountability in its functioning regarding information provision, service delivery and increased citizen participation'. It is the governments which carries out its duties and tasks in an efficient manner through the use of ICT [6, 7].

In 2012, Indian Government of India has taken up an initiative through National E-Governance Plan (NeGP) at the national level in accordance with the recommendations given by second administrative reforms commission in 2008. The local E-Governance initiatives were implemented at state level earlier. NeGP intends to implement E-Governance at the national level through 27 Mission mode projects at centre, state and integrated projects level. This includes income tax, passport, unique identification number (UID) at the central level; land records, municipalities, commercial taxes at the state level; and E-Courts, India Portal at the integrated level. The plan increases the access to Government services through common service delivery outlets and ensures efficiency, transparency and reliability of such services at affordable costs by using public–private partnership.

The implementation of E-Governance reforms is one of the mandatory reforms under the Jawaharlal Nehru National Urban Renewal Mission (JNNURM). It aims to provide easy access to services and information by single window clearance system and grievance redressal system. It helps the ULBs in enhancing interdepartmental coordination and integration of data and services of various departments, and also results in better decision-making and resource mobilization. Information, thus, collected can be used to set up a Management Information System for tracking the changes in revenue, expenditure, service delivery and implemented programmes to take faster and efficient decisions. A few of the successful examples of E-Governance are from Madhya Pradesh, Andhra Pradesh and Kerala state. In Madhya Pradesh, Indore Municipal Corporation (IMC) used ICT for providing data on major revenue sources including property tax and water supply charges and beneficiaries of urban poor schemes. Citizens can also submit suggestions and report grievances and receive response through email. The E-Governance system addresses information provision and citizen participation through invitation of suggestions and grievance redressal. In contrast, E-Seva Centre in Andhra Pradesh focused on service provision which includes payment of utility bills, reservation of train tickets, getting birth and death certificates, vehicle permits, driving licenses, etc. [8]. This has reduced the visit to different offices and reduced the time spent. Government of Kerala is pro active in implementing E-Governance at larger scale and launched several programme under E-Governance. The Kerala state government has started a programme named 'Akshaya' under E-Governance to provide E-Literacy [9]. Till now 2,300 akshaya centres have been started and the plan is to have two akshaya centres at each panchayat. The state government is allowing and helping people to get loan from banks for setting up akshaya centres and thus, setting of akshaya centres encourages people to become entrepreneurs. Further details related to Akshay centre are provided in chapter "E-Governance for Public Realm: Around Panniyankara Monorail Station, Kozhikode, Kerala".

According to a survey by the Center for Media Studies conducted with 4,500 citizens from five cities (Hyderabad, Delhi, Mumbai, Kolkata and Chennai), E-Governance has brought down corruption in India. The study covered basic services such as water supply, electricity and departments i.e. municipal corporations, urban development, transport, civil supplies and hospitals. Between 2000 and 2004, corruption went down from 63 to 27 percent in Hyderabad. The decline in corruption was attributed to the successful functioning of E-Governance projects (***Economic Times, 19 January 2004***).

Ahmedabad and Hyderabad are selected as cases to analyse the implementation of E-Governance in municipal institutions.

3 E-Governance in Ahmedabad

Ahmedabad Urban Agglomeration (UA) is a million plus UA with a population of 63,52,254. It is the seventh largest UA in terms of population. The city embarked on the path of ICT in 2002. ICT is used for public service delivery through 7 civic centres located in 5 zones in 2002. Each of the civiccentre is having 10 nodes. The number of civic centres have increased to forty-six in 2013 (refer Fig. 1).

E-Governance includes many services such as payment of dues for property tax, vehicle tax, professional tax and other dues. The registration of marriages and issuing birth and death certificate has become easier after E-Governance. The issuing of licenses for shop, establishments, health, hawkers and granting building permission online has increased transparency. The development of grievance

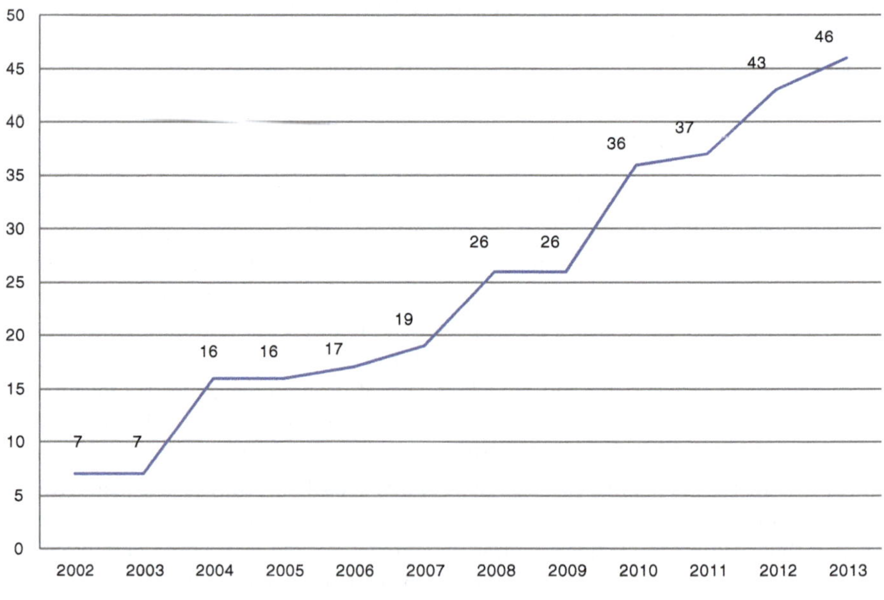

Fig. 1 No. of civic centres

handling mechanism has not only ensured recording of complaints but also their redressal in the shortest possible timeframe (refer Figs. 2 and 3).

The adoption of E-Governance in Ahmedabad Municipal Corporation (AMC) resulted into benefitting both the citizens as well as AMC. For citizens, it led to

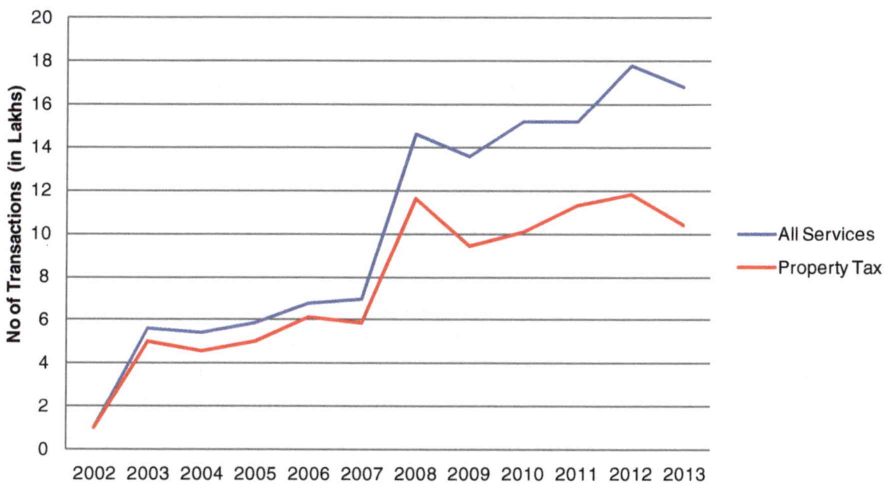

Fig. 2 Transactions versus property tax

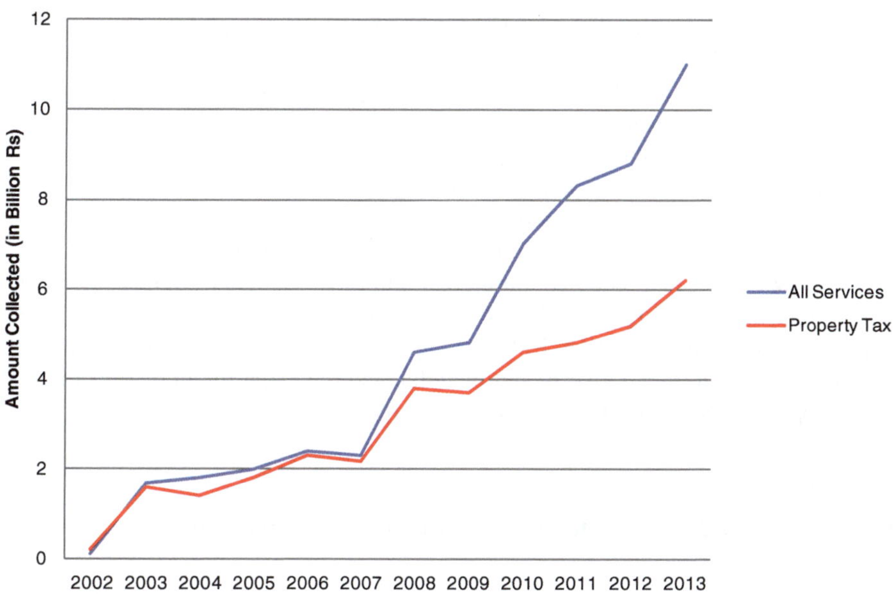

Fig. 3 Amount collected for all services vis-a-vis property tax by Ahmedabad Municipal Corporation

Table 1 Benefit to citizens from E-Governance at AMC

Application/service	Before	After
Birth and death certificates	Several days/multiple visits	Two days in most cases
Property tax payment	Hours, multiple visits	Under 10 min
Vehicle tax	Hours, multiple visits, evasion	Under 10 min
Building plan approval	Average of 6 months	Instantly if application is in order. Further refinements are in progress
Complaints	No way of knowing status, accountable person	Status known immediately, attended within 24 h in majority of the cases
Licenses for shops and establishment	Several days, multiple visits and middle men	Under 10 min

Source:: Center for Electronic Governance, IIMA, 2004

efficient and transparent processes leading to removal of middle men, easy access to information, saving the time and redressal in the shortest possible timeframe (refer Table 1). On the other, AMC's revenue has increased along with creation of database. With E-Governance, one can pay from the comfort of their home leading to more compliance and timely payment. The transparent system has increased citizens' trust on one hand and need to build up E-Governance infrastructure on the other. Recent initiatives of Indian government to develop five hundred cities under $\text{JNNURM}_{\text{under JNNURM}}$ and 100 smart cities in regional context will further propel the growth of usage of E-Governance [10]

AMC is the first local body to launch Comprehensive Complaint Redressal System (CCRS) through toll free number to register the complaints pertaining to civil services [11].

AMC has won series of awards in the area of governance especially pertaining to financial empowerment, civic centres and infrastructure design (refer Table 2). The organizational promotion of E-Governance is also through instilling different awards. Centre for Science, Development and Media Studies (CSDMS) launched E-India award in collaboration with the Department of IT, Ministry of Communications and Information and Technology, Government of India. AMC won the award in ICT enabled municipal initiative category. It has also won Gold Icon Award during the 12th National E-Governance conference in 2009. Department of Administrative Reforms and Public Grievance, Department of Information Technology, Government of India, and the Department of Information Technology, Government of Goa initiated the award.

3.1 Phases of E-Governance Evolution in Ahmedabad

Out of 5 phases of E-Governance as mentioned in chapter "E-Governance for Smart Cities" in Sect. 1.5, AMC was the earliest to begin with the first phase by making the information available about various municipal services on its website. In the

Table 2 Achievements of AMC

S. No.	Award	Agency	Reason for giving the award	Year
1.	CRISIL National Award 2003	Credit Rating Information Services of India Ltd.	Best Financial Management System	2003
2.	International Best Practices 2004	ICMA International City/Country Management Association, Washington D.C	Best Practices in City Civic Centers & E-Governance	2004
3.	Prime Ministers Award for excellence in 'Urban Design & Concept'	Prime Minister of India	Excellence in 'Urban Design and Concept to Sabarmati River Front Development Project'	2006
4.	UNHABITAT Dubai International Awards		Best Practices to Improve the Urban Environment (Slum Networking Project)	2006
5.	For excellence	Credit Rating Information Services of India Ltd.	Excellence in Municipal Initiatives 2007 (A Leader among Category A-Mega Cities)	2007
6.	India Tech Excellence Awards	India Tech Foundation	Main Streaming Urban Poor Slum Networking to improve habitat of Urban Poor and providing livelihood opportunities	2008
7.	National Urban Water Awards	Ministry of Urban Development	Effective Water Supply Management and successful development of market-based Financing System	2008
8.	Horizontal Transfer of ICT-based Best Practice GOLD Awards	Government of India	Best Practice in E-Governance	2008
9.	eGovernance, eIndia 2009	Centre for Science Development and Media Studies (CSDMS)	ICT Enabled Municipal Initiative of the Year	2009
10.	Urban Environmental Accord		Most Committed City in India for Urban Environmental Accord	2009
11.	Best Mass Transit System	Government of India	Janmarg-Ahmedabad BRTS Project	2009

(continued)

Table 2 (continued)

S. No.	Award	Agency	Reason for giving the award	Year
12.	International Awards on Sustainable Transport Award	TRB at Washington DC USA		2010
13.	National Award for Innovations in servicing the needs of the Urban Poor	Govt of India, India Urban Space Foundation, Swedish International Development Agency (SIDA) & the World Bank		2010
14.	International Award of Outstanding Innovation in Public Transport	International Transport Forum (ITF) and International Association for Public Transport (UITP)	Ahmedabad Janmarg Ltd-Ahmedabad BRTS	2010
15.	The Largest Metropolitan Region in Gujarat, perhaps the most market-oriented and business-friendly of Indian states and the third Fastest growing city in the world	Forbes Magazine		2010
16.	Knowledge and Research Award	59th UITP World Congress	Janmarg-BRTS Ahmedabad	2010
17.	Daring Ambition Award	59th UITP World Congress	Janmarg-BRTS Ahmedabad	2010
18.	Initiatives for Social Housing	41st Anniversary of HUDCO		2010
19.	Innovative Infrastructure Development	41st Anniversary of HUDCO		2010
20.	AIILSG Nagar Ratna Award	President of India	Best Performing City	2011

Source Accessed from http://www.egovamc.com/corporation/Achievements.aspx on 19 Aug 2014 second phase, interactive framework for comprehensive complaint redressal exists through 3 means, i.e. online, e-mail and phone/sms where problem, affected areas, ward and complainants details are entered. The option is also available to reopen the complaints, know the complaint status and feedback of customers. In the third phase, online services are available for issuing birth and death certificate in which registration, correction and issuing of duplicate certificate is also involved, marriage certificate, vehicle tax registration, registration of new plan or changes in existing building plan; property tax registration, registration and renewal of licenses for shops and establishments, registration of employees and business/establishments under professional tax, filing complaint, application for seeking information under RTI and book hall/party. However, transaction process for property tax and professional tax is online through credit card or online account by entering tenement number. In the forth phase, relationship between government, employees, community and citizens are established at AMC via registering the complaints, complaint redressal, issuing of various certificates and contractual obligations. In the fifth phase, out of 5 kinds of interfaces, as suggested in first chapter, horizontal connection between government agencies and vertical connections between central and local government agencies does not exist whereas connections between government and citizens through issuing different kind of certificates (birth, death, marriage, property tax, professional tax), grievance redressal mechanisms are established. The interface between stakeholders is not yet established through online mechanism. Thus, convergence of city to a smart city requires an effort to integrate horizontal, vertical and other connections so as to reduce digital divide and inclusion of E-Democracy.

3.2 Models of Governance in Ahmedabad

Out of all the 4 models mentioned in Sect. 3 in chapter “E-Governance for Smart Cities”, it is difficult to pinpoint on only one model which has been implemented in the case of Ahmedabad. Model 1 has been implemented in Ahmedabad in which through smart governance, transparency, efficiency and effectiveness has been enhanced. For this purpose, tender information has been given through website, efficiency level is increased by online receiving of application for issuing different kind of certificates and also providing an opportunity to check the status of application. In Model two, spatial decision support system (information consolidation and its review) for improved governance is being made possible due to different programming related possibilities of online data set. For Model 3, specialized staff has been recruited in AMC and technical person has been made in-charge of overseeing the operations of citizen service centres. The usage of technology must have reduced the number of staff employed for delivery of services. Within model 4, only one level of transformation, i.e. internal has been achieved and for external transformation, information has been made available to different kind of stakeholders for usage but external interactions with cities has been set up. The need is for AMC to move in the ladder and achieve the requirements of models 3 and 4.

3.3 M-Governance in Ahmedabad

M-Governance is also being made popular in Ahmedabad to support E-Governance. All the important official phone numbers are available on AMC website and a toll free number is also given to contact the officials of different departments. The booking as well as its status can be checked online and payments can also be made using mobile phone having Internet connection.

4 E-Governance in Hyderabad

Hyderabad with a population of 6,809,970 is fifth largest metropolitan and capital city of Telangana. It is highly urbanized city and has initiated a lot of steps to initiate good governance in Municipal Corporation. Greater Hyderabad Municipal Corporation (GHMC), erstwhile Municipal Corporation of Hyderabad is managing twin city of Hyderabad and Secunderabad. Its geographical area covers almost all the area managed by Hyderabad Metropolitan Development Authority (HMDA) and is second largest after Delhi. It was formed on 16 April 2007 through Government order 261 by merging 13 municipalities (L. B. Nagar, Gaddi annaram, Uppal Kalan, Malkajgiri, Kapra, Alwal, Qutubullapur, Kukatpally, Serilingampalle, Rajendranagar in Rangareddy district; Ramachandrapuram and Patancheru in Medak district and Bhongara in Nalgonda district) and 8 gram panchayats (Shamshabad, Satamarai, Jallapalli, Mamdipalli, Mankhal, Almasguda, Sardanagar and Ravirala).

In Hyderabad, services provided through E-Governance are provision of birth and death certificates, trade licenses, property tax, advertisement fee, rent payment of commercial properties and payment of contractors. The real-time records about street lighting, town planning, municipal park facilities and entomology are collected and analysed. There are multiple platforms, i.e. Citizen Service (E-Seva) Centres, Municipal Corporation offices, banks and web portal available 24×7 to provide such services. E-Seva is not only specific to GHMC but is implemented across Andhra Pradesh and Telangana.

E-Seva Centres provides the following facilities:

1. Payment of utility bills
   (a) Water & sewerage bills
   (b) Property tax
   (c) Prepaid parking tickets
2. Permits/licenses
   (a) Renewal of trade licenses
   (b) Registration of new vehicles—Transfer of ownership, change of address of owner
   (c) Registration—Birth, Death certificates

(d) Reservation of water tanker
(e) Issue of caste certificate.

In Hyderabad, Property Tax dues can be paid online through Property Tax Identification Number (PTIN). In order to E-Register for property tax, one has to obtain a unique PTIN and then add personal details (circle, person's name, mobile number, E-Mail id), building location (area, road, street) and building details (occupation, plinth area and building type). Other services provided are application for trade licenses (personal details, nature of trade, property tax no.), birth and death registration, advertisement space of both agency and hoarding details, access to approved building plans, seeking building permission and building plans, cell (telecommunication infrastructure) towers, grievance handling and E-Town planning.

The facility for registration under town planning involved approval of all building plans and layouts by GHMC within its jurisdiction according the powers vested under GHMC Act 1955. Online system of seeking permission helps us to ensure easy implementation and transparency and reduces personal interaction between citizen and the GHMC.

Online GHMC grievance redressal has the options for lodging the grievance both through using phone and online web-based system, checking its status and giving opinion. The options are also available for entering the details of grievance. After registering the grievance, one gets a registration number. It is worthwhile to note that the kind of details asked on the GHMC website helps us to analyse the grievances lodged by circle, ward, locality, by type (category, sub-category and actual complaint) and automatically reaches to the concerned staff member to handle. It helps officials to assess the nature of complaints as well as time taken in dealing with those complaints by circle, ward and locality.

The benefits of E-Governance in Hyderabad are:

- Improvements in the efficiency and effectiveness of service delivery
- Creating a new municipal culture of citizen-friendly service provision
- Improvisation in workforce attendance (inputs) improved from 85 to 98 % and bin lifting (outputs) from 76 to 98 % using Offsite Real-time Monitoring System (OSRT)
- Building permissions issued within four days without inspection
- Making mangers and citizens more responsible and managers more responsive
- Rs. 73 lakhs recovered till May 2011 from contractors due to Service Level Agreement violations such as irregular attendance, non lifting of bins, improper sweeping of roads, etc., and cell phone cost and monthly rentals
- 50 unauthorized buildings demolished
- 90 criminal cases filed against property owners for constructing buildings without permission
- Substantial reduction in citizen complaints on sanitation and street lighting in newspapers, e-mail and paper grievances receipts
- System-generated real-time reports
- Information access to citizens and non-officials
- Simplifies corruption-prone municipal processes and improves their functioning.

4.1 Phases of E-Governance Evolution in Hyderabad

In continuation of AMC positioning in 5 phases, let us see the progress of GHMC in five phases. It has moved up from phase one of introducing the website for providing the citizens' information to second phase of an interactive framework where citizen's can seek the information, register their grievances, pay property tax online and download various forms. In the third phase, GHMC website does not reflect options of receiving the payment but it seeks all the information for sanctioning the building plans. This has led to larger transparency in the system. E-Seva centres help to access different services under one roof. Stakeholders also applied online for different kind of certificates with only limitation that it mainly benefits those who know how to access Internet. The forth phase requires an effort from the local municipal governments to interlink municipal corporation with other government departments. Though different departments have their individual websites to provide the information, efforts have not yet been made to open an interactive forum for interaction. As far as fifth stage is concerned, out of all 5 types of interactions, GHMC established only an interaction between government and citizen whereas efforts are required for horizontal and vertical, interoperability and inter-relationship of different kind of stakeholders through web.

4.2 Models of Governance in Hyderabad

Model 1 is the base of implementation of smart governance in GHMC. Without much of structural change, efficiency and effectiveness is initiated through providing the information on the website, options for registering to get certificates and building sanction plan approval. In correspondence with model 2, decision support system has been developed for collection as well analysing the information related to solid waste management, explained in detail in next section. It is difficult to gauge whether Smart administration has been fully achieved or rather aimed to achieve in Hyderabad as per Model 3. Downsizing the staff has reducedthe operation cost since liberalization but lower level staff is still requiredto exchange the files andsending the information from one department to another. Higher usage of computer has decreased the dependence on human resources but still it is required for doing ground level checks, verifying documents and providing the important information to those who are not well versed with computers.

Model 4 of interconnections between cities in city region is only established in limited manner by interconnectivity of 13 municipalities and 3 gram panchayats. The question that remains unanswered is how we can use the data collected from city region to increase the interlinkage between urban and rural areas and sectoral as well as spatial integration.

4.3 M-Governance in Hyderabad

M-Governance using mobile have been used for reporting disaster, coordinate political protest and finding the rates of a product to increase profitability but camera was not used for public domain [12]. OSRT provides information about the distant events in real time. OSRT is a tool to ensure accountability with internal, upward and downward dimensions in public domain. The camera of mobile devices is used efficiently to register the workers' attendance as well as status of cleaning the bins within the GHMC spread over 625 km^2 area with 7.40 million population since 2011. The supervisors check the attendance of 14,000 workers on the basis of pre-scheduled jobs. The images of bin, their coordinates, bin and its surrounding area are captured and analysed to study the effectiveness of its implementation.

5 Conclusions

Both the cities have implemented E-Governance and M-Governance but its level of implementation varies. Ahmedabad scored high on implementation of various infrastructure projects whether it is Bus Rapid Transit System (BRTS) or river front development. Hyderabad used M-Governance effectively for increasing efficiency of Solid Waste Management. In both the cities, E-Civic Centres are helping to provide access to different kind of services under one roof to those who are computer illiterate and information on website helps in reducing the time taken in delivery of services and bringing governance to the door step.

References

1. Palvia SCJ, Sharma SS (2007) E-Government and E-Governance: definitions/domain framework and status around the world. Conference on E-governance. csi-sigegov.org. https://www.csi-sigegov.org/1/1_369.pdf. Accessed 7 Aug 2014
2. UNESCO (2005) E-Government Toolkit for Developing Countries, UNESCO and National Informatics Centre (NIC). www.unesco.org
3. National Institute of Urban Affairs (NIUA) (2012) Pearl: offsite real-time monitoring system, State-City: Andhra Pradesh—Greater Hyderabad municipal corporation, BP Code: URF-# # -09-0101-0511
4. Jawaharlal Nehru National Urban Renewal Mission (JNNURM) (2011) E-Governance, *ULB Level Reform,* JNNURM Primers. http://jnnurm.nic.in/wp-content/uploads/2011/01/Mandatory_Primer_1-e-governance.pdf. Accessed 27 May 2014
5. World Bank (2004) E-Government. http://www1.worldbank.org/publicsector/egov/. Accessed 30 July 2014
6. National Institute of Urban Affairs (NIUA) (2004) Urban Finance, Newsletter, vol 7, issue 3, New Delhi. http://www.niua.org/Publications/newsletter/urb_fin_sep_04.pdf. Accessed 27 May 2014

7. United Nations Educational, Scientific and Cultural Organization (2005) http://portal.unesco.org/ci/en/ev.php-URL_ID=3038&URL_DO=DO_TOPIC&URL_SECTION=201.html. Accessed 27 May 2014
8. Suthrum P, Phillips J (2003) Citizen Centricity: E Governance in Andhra Pradesh, the Department of Corporate Strategy and International Business, The University of Michigan Business School. Accessed 12 Dec 2013
9. Krishnan C (2013) E-Governance in local governance: the case of Kerala, Vol. 23, No. 2, http://www.iimahd.ernet.in/egov/ifip/jun2013/krishnan.htm. Accessed 30 July 2014
10. TNN (2014) States with ready land may get first lot of smart cities, 30 July 2014, 04.34 AM IST. http://timesofindia.indiatimes.com/india/States-with-ready-land-may-get-first-lot-of-smart-cities/articleshow/39265108.cms. Accessed 20 Aug 2014
11. Mehta H (2013) Ahmedabad municipal corporation's complaint call centre launched, 10 July 2013. Posted in GENERAL, Technology. http://theahmedabadblog.com/ahmedabad-municipal-corporations-complaint-call-centre-launched.html. Accessed 19 Aug 2014
12. National Institute of Urban Affairs (NIUA) (2012) Pearl: offsite real-time monitoring system, State-City: Andhra Pradesh—Greater Hyderabad municipal corporation, BP Code: URF-# # -09-0101-0511

Part II
E-Smart City Governance-Domain Studies

E-Governance and Its Role in Infrastructure Services of UAE, Case Study—Dubai

Ashmita Karmakar

Abstract As per World Bank (www.worldbank.org) definition (AOEMA report): 'E-Government refers to the use by government agencies of information technologies such as Wide Area Networks, the Internet and mobile computing that have the ability to transform relations with citizens, businesses and other arms of government'. These technologies are meant to serve a variety of different ends that includes: (1) Better delivery of government services to citizens. (2) Improved interactions with business and industry. (3) Citizen empowerment through access to information. (4) More efficient government management etc. Many countries are adapting to the E-Governance and its applications in the field of infrastructure to improve the governmental services and providing empowerment to the citizens. Dubai is an example of one such city in the UAE which has been at the forefront of adopting the advanced technologies to improve the efficiency of governance and including itself in the list of smart cities. This chapter is effort to represent Dubai as the city which has transformed itself to form smart city and adopted the technologies for E-Governance in the field of infrastructure. With the effort of visionary leadership of UAE has initiated numerous E-Government programmes that aimed at effective policy making, and service delivery. The city is constantly growing with substantial influx of expat population every year to contribute to the city's trade and commerce. To achieve the target, the city is making efforts to reach to every individual through information and network especially in the field of infrastructure facility provision creating a strong bond with the government. The smart technology in E-Governance becomes active from the time

A. Karmakar (✉)
Dubai Internet City, C/O Ashish Gupta, 3d Floor, Injaz Building, PO Box 11549, Dubai, United Arab Emirates
e-mail: ashmita.karmakar@gmail.com

T.M. Vinod Kumar (ed.), *E-Governance for Smart Cities*, Advances in 21st Century Human Settlements, DOI 10.1007/978-981-287-287-6_4

when a person enters the city as well as in day to day services. Today the UAE is considered to have one of the most advanced and world-class information and communication technology infrastructures. This chapter has listed few smart services that are used in Dubai as a part of E-Governance. With the Information technology picking up, the e-services are becoming mere necessity in the human life. Use of the smart facilities to interact with the Dubai residents is offering convenience in better service delivery, people empowerment, easy access to information, improved productivity and cost savings in business and participation in public policy decision-making in the governmental activities. Although there are some grey areas and gaps for which the city is making best efforts to come up with an efficient solution and set up an example of E-Governance in front of the developed world. With the success of its E-Government initiatives and its overall popularity for its strong infrastructure and standards of living, it will achieve its goal in public sector integration, and successfully deliver its services to all residents.

Keywords E-Government · Infrastructure · Information technology · Smart city

1 Introduction

Information and Communication Technologies (ICT) have affected the ways in which people, governments and businesses interact with each other. The rapid diffusion of the Internet, mobile telephone and broadband networks demonstrate how pervasive this technology has become. ICT has become the fundamental building blocks of modern societies and digital economies. Yet, the revolutionary pace in countries worldwide is dependent on the preparedness of several factors of both social and political environments. From a government standpoint, E-Government adoption is becoming an unquestionable task. E-Government deals with facilitating the operation of government and the distribution of governmental information and services. The ultimate goal of E-Government is to be able to offer an increased portfolio of public services to citizens in an efficient and cost effective manner. Anticipated benefit of E-Government include efficiency, improved services, better accessibility of public services, and more transparency and accountability [1].

The objective of the chapter is to study the development of E-Governance in Dubai City in the last few decades and its future goals/targets. The chapter provides with the details of E-Governance and how the city if Dubai has transformed itself to adopt various means in order to establish itself as a smart city. The E-Government of Dubai has implemented various ways to interact with its residents/visitors and provide them with easy access to government-related information, power to participate in government-related activities and also in day to day use of services. This chapter has highlighted especially on the smart infrastructure services which forms a major link between the residents and the E-Government of Dubai.

1.1 The Term E-Governance

E-Governance can be defined as the performance of a government via electronic medium in order to facilitate an efficient, speedy and transparent process of disseminating information to the public, and other agencies, and also for performing government administration activities. E-Governance is generally considered as a wider concept than E-Government, since it can bring about a change in the way how citizens relate to governments and to each other. It can bring forth new concepts of citizenship, both in terms of citizen needs and responsibilities. Its objective is to engage, enable and empower the citizen.

World Bank (www.worldbank.org) definition (AOEMA report): 'E-Government refers to the use by government agencies of information technologies (such as Wide Area Networks, the Internet, and mobile computing) that have the ability to transform relations with citizens, businesses, and other arms of government. These technologies can serve a variety of different ends: better delivery of government services to citizens, improved interactions with business and industry, citizen empowerment through access to information, or more efficient government management. The resulting benefits can be less corruption, increased transparency, greater convenience, revenue growth, and/or cost reductions'.

United Nations (www.unpan.org) definition (AOEMA report): 'E-Government is defined as utilizing the Internet and the world wide web for delivering government information and services to citizens'.

1.2 Objectives of E-Governance

The strategic objective of E-Governance is to support and simplify governance for all parties—government, citizens and businesses. E-Governance uses electronic means to support and stimulate good governance. In other words the objectives of E-Governance are the ways to achieve the objectives of good governance. Good governance can be seen as an exercise of economic, political and administrative authority to better manage affairs of a country from national to local levels.

Regarding E-Government, the distinction is made between the objectives for internally focused processes (operations) and objectives for externally focused services.

External strategic objectives. The external objective of E-Government is to satisfactorily fulfil the public's needs and expectations on the front-office side, by simplifying their interaction with various online services. The use of ICTs in government operations facilitates speedy, transparent, accountable, efficient and effective interaction with the public, citizens, business and other agencies.

Internal strategic objectives. In the back-office, the objective of E-Government in government operations is to facilitate a speedy, transparent, accountable, efficient and effective process for performing government administration activities. Significant cost savings (per transaction) in government operations can be the result.

Thus, E-Governance is more than just a Government website on the Internet with Political, social, economic and technological aspects forming its integral part.

2 Development of Dubai

Dubai is located on the southeast coast of the Persian Gulf and is one of the seven emirates that make up the country. Abu Dhabi and Dubai are the only two emirates to have veto power over critical matters of national importance in the country's legislature.

Dubai was formally established on June 9, 1833, by Sheikh Maktoum bin Butti Al-Maktoum when he persuaded around 800 members of his tribe of the Bani Yas, living in what was then the Second Saudi State, to follow him to the Dubai Creek by the Abu Falasa clan of the Bani Yas. It remained under the tribe's control when the United Kingdom agreed to protect the Sheikhdom in 1892 and joined the nascent United Arab Emirates upon independence in 1971 as the country's second emirate. Its strategic geographic location made the town an important trading hub, and by the beginning of the twentieth century, Dubai was already an important regional port. The city has made its global identity for its skyscrapers and high-rise buildings, world's tallest building, ambitious development projects including man-made islands, hotels and some of the largest shopping malls in the region and the world.

3 Need for E-Governance in Dubai

3.1 Dubai and Its Global Identity

Today, Dubai has emerged as a cosmopolitan metropolis that has grown steadily to become a global city and a business and cultural hub of the Middle East and the Persian Gulf region. It is also a major transport hub for passengers and cargo. Although Dubai's economy was historically built on the oil industry, the emirate's Western-style model of business drives its economy with the main revenues now coming from tourism, aviation, real estate and financial services. Dubai has recently attracted world attention through many innovative large construction projects, sports events and the sole host for World Expo 2020.

3.2 Practice of Traditional Governance

Similar to other Emirates Dubai had a Traditional Government. Traditional governments were always small, both in size and scope. This was natural, given the size of the communities and the difficult economic environment in which they existed. However, this environment valued consensus as well as participation in a traditional form that would exist within the context of a majlis or council. In this framework, issues relevant to the community were discussed and debated. Opinions were expressed and the sheikh would take these opinions into consideration prior to taking a decision. Traditionally, the ruler of an Emirate—the Sheikh—was the leader of the most powerful tribe, while each tribe, and often its subsections, also had a chief or Sheikh. The Sheikhs maintained their authority only as long as they were able to retain the support of their people. This in essence, was a form of direct democracy. Part of that process was the unwritten, but strong, principle that the people should have open access to their ruler, and that he should hold a frequent and open majlis, in which his fellow citizens could voice their opinions. It is now evident that it is these elements of governance that has served as a solid foundation in maintaining the unique identity of the country against a backdrop of rapid economic and social changes.

3.3 Initiation of a Change in Governance

Such a direct democracy, which may be ideally suited to small societies, becomes more difficult to maintain as the population grows. Simultaneously, the increasing sophistication of government administration means that many people now find it more appropriate to deal directly with these institutions on most matters, rather than seek personal meetings with their rulers.

3.3.1 A Balanced Approach

The changes envisioned and undertaken by the UAE leadership represent an indigenous initiative reflecting the need to transform the country's traditional political heritage—based on consensus, the primacy of the consultative process and gradual social change—into a more modern system that takes into account the rapid socio-economic advances made since the establishment of the federation.

3.3.2 Adaptation of E-Governance by Dubai

ICT have affected the ways in which people, governments and businesses interact with each other. The rapid diffusion of the Internet, mobile telephony and broadband networks demonstrate how pervasive this technology has become. Today, ICT is considered as one of the fundamental building blocks of modern societies and digital economies.

From a government standpoint, E-Government adoption is becoming an unquestionable task. E-Government deals with facilitating the operation of government and the distribution of governmental information and services. The ultimate goal of E-Government is to be able to offer an increased portfolio of public services to citizens in an efficient and cost effective manner. Anticipated benefits of E-Government include efficiency, improved services, better accessibility of public services and more transparency and accountability. The primary drivers of E-Governance is explained in Fig. 1 [1].

In a smart city initiative, the citizen and Government merge together and cannot be distinguished from one another other as the end user is just a service consumer and is heavily dependent on both the E-Government and private sectors.

Dubai has been investing heavily in adopting and implementing ICT in its government and private sectors. The Global Information Technology Report 2010–2011 indicates that the UAE leads the MENA region in leveraging ICT for increased economic diversification and competitiveness. The E-Government programme in the UAE is a key initiative of the UAE-Government Strategy 2011–2013 that lays the foundation to achieve UAE Vision 2021 [2] (Fig. 2).

The strategy is divided into seven general principles as follows:

- Enhance the role of federal entities in devising effective regulations and integrated policies by successful planning and enforcement.
- Enhance effective coordination and cooperation among federal entities and with local governments.
- Focus on delivering high quality, customer-centric and integrated government services.
- Invest in human resource capabilities and develop leaders.
- Promote efficient resource management within federal entities and leverage dynamic partnerships.

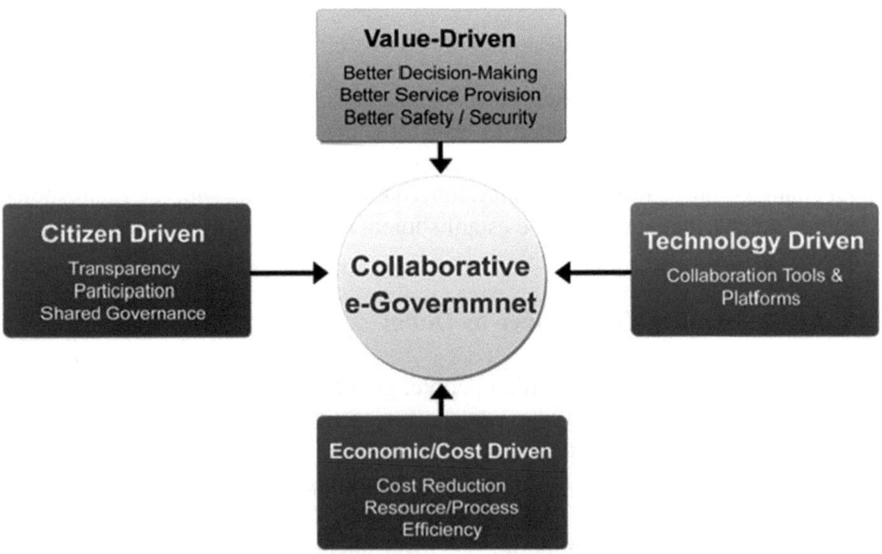

Fig. 1 Showing the importance of E-Government. *Source* [1]

Fig. 2 Showing the important strategic domain of action in E-Governance. *Source* http://www.emiratesegov.ae

- Pursue a culture of excellence through strategic thinking, continuous performance improvement, and superior results.
- Enhance transparency and accountable E-Governance mechanisms throughout the federal entities.

4 Achievements of Dubai as a Smart City

According to 'Internet World Statistics', the UAE has 5,859,118 Internet users as of June 12, 2013, 70.9 % of the population of the country has access to internet either through web or mobile devices. Usage of E-Services has become a daily norm for Dubai citizens and this high percentage of users makes Dubai implementing smart city-related initiatives and projects.

Although Dubai is in the initial phase of setting up the building blocks for a smart city transformation, the city has made a remarkable progress in the use of smart devices/technologies in the field of infrastructure. The E-Government has tried to introduce maximum number of services with user friendly IT applications thus establishing a strong network between Government and the citizens. In case of Dubai use of smart card and smart phone has proved to be successful in connecting to the masses (Fig. 3).

This chapter has tried to highlight few prominent smart technologies adopted by Dubai city so far on the way to smart city development. These technologies are as listed below.

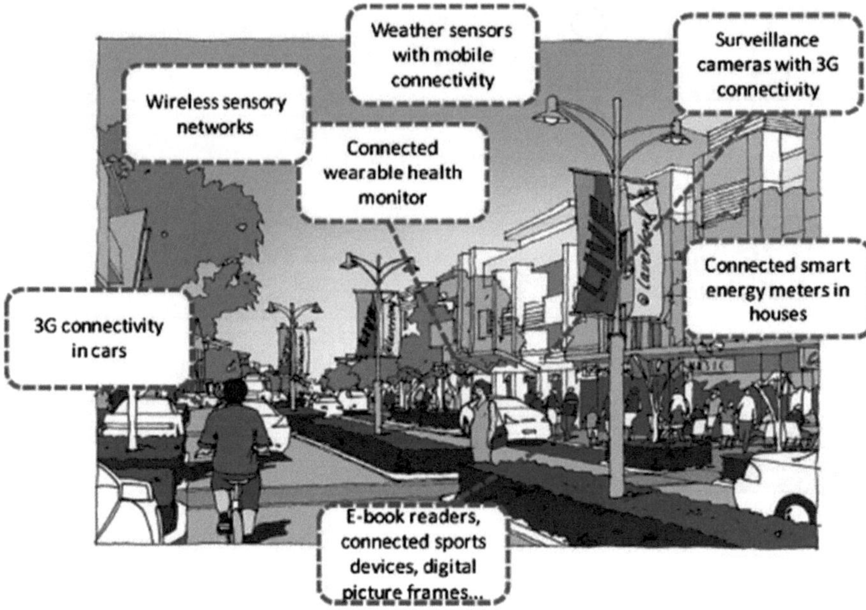

Fig. 3 Showing the concept of a smart city where citizens, objects, utilities services connect in a seamless manner using ubiquitous technologies. *Source* [9]

4.1 Use of Smart Card at the Entry Points to the City

The smart city has made a commendable progress in the field of security. The city records the identity of each person crossing the border through eye scanning and starts maintaining the database thereafter. And once the person re-enters the database is checked at the security point against the person's criminal records, outstanding fines and immigration details registered within in any emirates becomes accessible at the touch of a button.

For the people who are frequently entering and exiting the UAE can use E-Gate card which makes the passport control procedure as easy as swiping card. Emirates ID cards are soon going to replace the E-Gate cards which will be capable multiple uses and also continue to bypass the long passport control procedure at the airports. Around three passengers will be able to be processed each minute using the new system and Smart Gates—a vast improvement on the current average wait of about an hour.

At a later stage smart Gates are the second-generation 'electronic gates' in which the travellers only need to scan through the Smart Gate with their passport, Emirates ID or E-Gate card.

4.2 Emirates ID Card

The UAE national identity management infrastructure is a strategic initiative to enhance homeland security and develop a federated identity management system enabling secure E-Government transactions. A federated identity is the means of linking a person's electronic identity and attributes, stored across multiple distinct identity management systems. Such systems would allow individuals to use the same user name, password or other personal identification as a part of the programme, the UAE issues smart identity cards for all of its population. The UAE national identity card is one of the world's most advanced and secure smart cards. The card is provided with identification parameters stored securely in the smart chip. It thus enables establishing a person's identity on-site (physically) and remotely (virtually), enabling secure and trusted transactions. The multi factor authentication which provides both match-on-card4 and match-off-card5 features, facilitates validation, verification and authentication of any given identity. The cardholder can then access all identity-based services sign into the networks of more than one enterprise in order to conduct transactions (Fig. 4).

The UAE ID card capabilities of on-site identification, remote identification and authentication are available and are used across the different applications enabling various forms of electronic transactions e.g. G2C and B2C these are facilitated by PIN code selected by the user. The UAE national identity management system eliminates the need to maintain distinct user credentials in separate systems and instead a single user credential can be used for multiple facilities. In an E-Government context, this has resulted in greatly simplified administration and streamlined access to resources, data verification, biometric authentication (match on card and match off card features) and digital signatures.

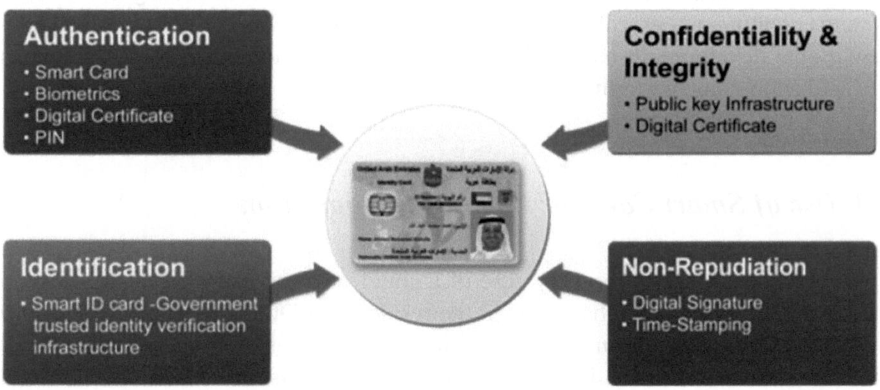

Fig. 4 Shows a sample of UAE national ID card. *Source* [1]

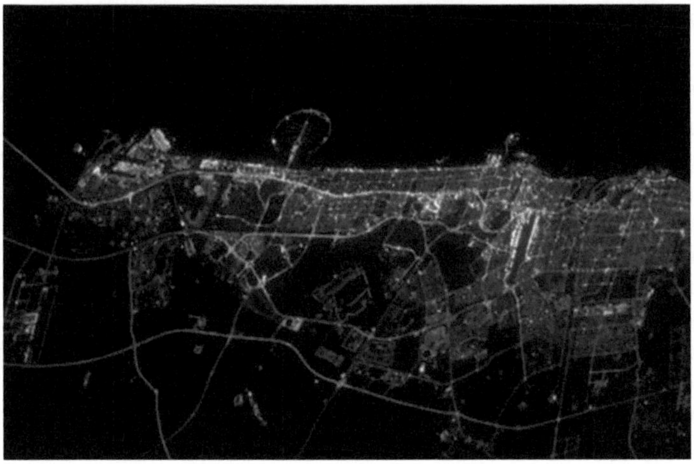

Fig. 5 Showing the physical urban limits of Dubai city. *Source* Dubai Strategic Plan, year 2014

4.3 Internet Access for Government Activities

As per the recent governmental strategy of Dubai, wireless internet access will become available across the city under the latest plans to turn Dubai into a 'smart city' (Fig. 5).

Sheikh Mohammed bin Rashid, Vice President and Ruler of Dubai, has announced a plan to increase Wi-Fi penetration across the city. The project aims to provide all Dubai residents with high-speed internet in public places, and live services and information. This Smart City project involves remote sensor devices all over Dubai. Education, health care and general security will be managed via smart systems creating a new reality for our people, a different life for our children and a new global development model. The strategic planners are studying various plans and will soon come up to the mega launch.

4.4 Use of Smart Card for Public Transportation

Since technology drives the very core of a smart city, the usage of smart devices, the smart card technology has enabled the residents of Dubai to carry a single card to reach to any destination through public transportation service. These services include the metro railway, Dubai taxi, Dubai water transport and buses. The card is available at any transportation node, supermarkets, retail shops and railway stations.

4.5 Use of Power and Water Services

The escalating scale of urban development outlined above is impacting significantly on the demand for electricity. Annual demand growth peaked in the UAE at 14.1 % in 2004, and continued to be strong in 2005 with an increase of more than 12 %. In fact, the UAE has the highest projected increase in demand within the Gulf region, which is expected to continue to grow at a minimum rate of 10 % per annum until 2010, far outstripping the world average of 3 % per annum. Plans are, therefore, being formulated to increase the UAE's electricity generation capacity by as much as 60 % by 2010.

Dubai Electricity and Water Authority (DEWA) is taking major steps in saving the resources by involving the residents in various awareness programmes. The launch of smart grid technology, smart meters for the consumption of the resources, various types of bill generations showing the consumption rate for each property and various award ceremonies for residents saving the resources are the steps taken by the E-Governments. The application which can be downloaded on mobile and internet for the payment of the water and electricity and the services charges for a property, checking the history has been introduced by DEWA. This application can also be used for various enquiries and suggestions from the public side to the E-Government [3].

4.6 Use Traffic and Transportation Services

4.6.1 Map Guidance for Commuters

Use of Geographical information system (GIS) has helped in mapping all the parcels of the urban area with all associated information such as the geographical and physical features, transportation corridors, land uses, land marks as well. All the land parcels with any land use are to be registered in the Dubai land department and with various governmental authorities and eventually the information obtained by various authorities are compiled to be reflected in a centralized mapping system available with Dubai municipality. Use of GPS in Dubai is linked with the GIS to guide a person to any corner of the city. The application has the capability to track the path of the commuter while they are on the way to their destination.

There are various measures taken by the Road and Transport Authority (RTA) to control the flow of traffic within the city. The measures include the installation of toll gates, speed checking cameras along with the design measures for the roads such as access points, road diversions on and off street parking provisions as well. The devices used to maintain the traffic are controlled by the authority and are made connected to all the commuters through internet and mobile phone apps.

The apps help to check the vehicle registration details, traffic fines, payment of toll taxes, fines and invitation of management improvement suggestions from the users.

4.6.2 Guidance in Vehicular Parking

Dubai has introduced a website to locate the parking areas within the city. The parking areas are categorized as short term and long term and the parking fee is fixed accordingly. The commuter can select the parking area as per their choice even before they start for their destination.

The payment for parking has been made easy and the commuter has been given the choice of payment

- at the parking spot by paying cash at parking metres at hourly basis,
- buy a parking card from the municipality centers and use the provision at monthly basis,
- use their mobile phones to pay the parking fees via m Parking app as described below.

The RTA has launched mParking service (mParking) which is a new value-added service that will allow motorists to pay for their virtual parking permit using their mobile phones by simply sending an SMS in a pre-defined format to a number thus eliminating the need to walk to the Payment Display (P/D) Machines and search for coins. The mParking service also alerts the motorist via an SMS prior to virtual permit expiry and if needed the motorist can extend their parking period from their office or home without having to walk to his/her vehicle.

In order to make the SMS format easy, registration will be not be required for Dubai registered private plate numbers starting with the alphabet character A-K & M whilst all other motorists who own a non-Dubai private registered plate number registration is made mandatory in order to use the mParking service. Some of the benefits of m parking are as given below [4].

- No hassle to find a parking metre and coins.
- No worries about remaining parking time, with reminder alerts via SMS.
- No need to run to buy another parking permit when your paid parking period is about to expire. Simply extend your parking period from the luxury of your office or home.

4.7 Police Service

Dubai Police has launched its official mobile application as part of the mgovernment initiative announced by His Highness Sheikh Mohammed bin Rashid Al Maktoum, Vice President and Prime Minister of UAE and ruler of Dubai earlier this year. The Dubai Police app will provide access to various services including the followings:

- Traffic Services—Fine Payment, Traffic Clearance Certificate, Traffic Wallet and Road Status Notification.
- Security Services—Good Conduct Certificate, Police Report Inquiry, Prevent Crime, Tourist Security.
- Community Services—SOS Button, Police Leaders contact, Social Media, Pharmacies on duty, Critically ill patient service, Police Station location, Job Vacancies.
- Other Services—Appointment scheduling, News, Events Media [5].

5 Communication

Mobile phones are playing a vital role in the part of life of every residents of any country. The device has been taken into consideration for providing information and communication between the E-Governmental departments and the residents of Dubai city. Mobiles have proved to be a complete success in order to reach out to every resident and create overall transformation equation in the smart world. According to "Our Mobile Planet" Website, some interesting statistics are available to further understand the overall behaviour of Dubai's citizens. The following statistics from "Our Mobile Planet" portray the potential success for Dubai as a smart city (Fig. 6, Tables 1 and 2).

Many government entities had already launched smartphone applications to offer their services over mobile phones and portable devices. However, the number of apps increased considerably since the launch of the mGovernment Initiative. It became necessary to guide the users about the genuine and official UAE mGovernment apps in order to facilitate them secure access to the services.

A number of government entities are currently using this service, which covers Salik (Toll Tax) recharge from RTA, payment of traffic fines from Dubai Police, payment of electricity bills from Dubai Electricity and Water Authority, payment of telephone bills and donations to various foundations as well.

To use this service, which guarantees the highest degrees of security, users must register first with their credit card details in mPay, enabling payments to be withdrawn. More details about the service and how to register make payments and SMS can be found in on the mPay website. Alternatively, iPhone users can download the mPay app from the Apple Appstore, while Android users can download from the Google Play store.

The UAE mGovernment launched the UAE's official mGovernment App Store (www.apps.gov.ae), which is a repository of all the official mobile applications that provide UAE-Government services. The UAE is the first government to launch an official app store.

Fig. 6 Usage of smart applications by mGovernment. *Source* mGovernment, UAE, year 2014 [2]

Table 1 Smartphone usage frequency in Dubai

Device typology	Usage in (%)
Smart phone penetration rate	56
Frequency mobile internet usage via Apps	75 daily

Source mGovernment, UAE, year 2014 [2]

Table 2 Smartphone usage levels at various activity centers—Dubai

Location	Usage (%)
At home	94
At work	76
On the go	73
Café of coffee shop	72
Public transport	56

Source mGovernment, UAE, year 2014 [2]

6 Importance of Smart Applications in Dubai's Urban Governance

As the city gains momentum in the field of development through business, trade, commercial activities, the magnetic pull for population towards the city becomes stronger. As the population starts growing, the city starts expanding physically both in horizontal and vertical directions. Decentralization of the E-Government is a practice which is common for the developing urban areas. The main city no longer remains as the sole focal point in terms of business and hence is divided into various smaller urban centers to control a certain limit of urban activities and population around. When the city expands further the central and the subcentres need a support to maintain the network and that support is offered by smart technologies. The technologies

Table 3 List of few smart applications and practices in different sectors of Dubai

Category	Areas where smart devices introduced	Examples
Environmental improvement	Smart metres	Environment monitoring Energy billing
	Smart grid	Power distribution
	Solar panels	Power generation
Economic growth	Smart educations	Using of e books and tablets in schools as a replacement for physical books
	Green growth initiatives	Recycling of water
Cost efficiency	Replacing paperwork in government departments	Using a common and shared inter-departmental information database
	E paper work	Online applications and submission in various industrial sectors
Safety	Redirecting transport around a collision	Use of walkie talkies to receive instructions from road and transport department to the taxi drivers giving them road status information and necessary information.
	Traffic speed monitoring	Use of speed radars for vehicles and connecting to e information system for the same
	Toll tax collection for vehicles	Automatic toll tax sensors and payment
Quality of life	Feedback loops in urban panning from data across the city	Use of internet and smart devices in sending suggestions to the E-Governmental departments
Connected citizens	Transport apps for a connected commute	Use of M payment by citizens in paying the toll taxes and transport fines
Smartphones business models	Using data from smartphones across a city to create new advertising and revenue stream for local businesses	Use of various apps by smart phones in business-related activities
Expats settling	Connected immigration clearance to bank account openings, residence search, best schools recommendations	Use of E-ID card for various government-related operations

Source Author

with the help of smart devices help in eradicating the physical presence in various government transactions. The technology binds the citizens with the E-Government for any physical size of the city through smart applications. *Source* [10].

Dubai is a growing city and is continuously showing signs of further growth in future. This is the reason that adopting smart technologies at the initial phase of growth will boost the growth of Dubai in the field of sustainability. Implementation of smart technologies in different aspects resulted in a strong network between the E-Government and the residents.

On the operational front, having smart infrastructure in place has increased the life of each property, translating into fewer upgrades, less maintenance, faster maintenance, a smaller number of maintenance staff, and fewer inspection visits. With planned maintenance programmes in place, city management has enjoyed lower costs in the long run. Centralized teams are there to work to manage the city from one command and control center. Cost savings will eventually come from needing fewer maintenance and management team members, purchasing spare parts on schedule, monitoring the city remotely, utilizing energy saving programmes and leveraging other smart innovations [6] (Table 3).

7 E-Governance in Gulf Cooperation Council (GCC) Countries

Overall, GCC countries have maintained leadership in E-Government readiness among Arab peers by taking serious steps to support the diffusion of E-Government in their societies. Several UNDP reports confirmed that the growing efforts of GCC governments to promote digital transformation and literacy have helped further enhance. Governments of the GCC countries are considered to be in intense competition with each other to develop a new knowledge-based economy, away from the current dependence on oil, and to make their products and services competitive on a global scale. GCC countries are proceeding at a rapid pace to use more service-oriented and citizen-centric operating models. This rapid reform is bringing a paradigm shift in the way citizens in the GCC are interacting with their governments. There are serious efforts in these countries to develop electronic operating environments, with advanced capabilities to build the right conditions for the E-Citizens concept to evolve [1].

8 Conclusion

The above-mentioned aspects give a picture of Dubai as a smart city through various smart applications implemented and practiced in the field of infrastructure. The booming UAE economy is further fuelling the need for infrastructure development on an unprecedented scale. This has been depicted as a ‘new era of economic transition', characterized by a public-private partnership that is gradually taking over the role traditionally held by government in infrastructure development. Housing, tourist, industrial and commercial facilities, education and healthcare amenities, transportation, utilities, communications, ports and airports are all undergoing massive redevelopment, radically altering the urban environment in the UAE [7, 8].

Dubai stands a fair chance to achieve its ambitious goals of transforming the city into the Middle East's first smart city. With the success of its E-Government initiatives and its overall popularity for its strong infrastructure and standards of living, it will achieve its goal in public sector integration, and successfully deliver its services to all residents. However, parallel the governing body should consider is how to bridge gaps between its government departments and private companies. There should be initiatives defining innovative ideas, service alignment, end-user experience, synergy between public and private sector roles and adaptation of new technologies. At the end of the day, the residents of Dubai are ready to adopt their city as a smart city.

References

1. Al-Khouri AM (2012) E-Government Strategies the case of the United Arab Emirates (UAE). http://www.epracticejournal.eu
2. Hamad Obaid Al Mansoori, Director General, UAE, mGovernment and Deputy Director General, Information and E-Government Sector Telecommunication, Regulatory Authority, UAE (2014) An overview of E-Government in the United Arab Emirates-2014
3. Dubai Water and Electricity Authority Data, Dubai, UAE (2014). Available at http://www.dewa.gov.ae
4. Road and Transport Authority Data, UAE (2014). Available at http://www.rta.ae
5. Dubai Police department Data 2014
6. Anthopoulos LG, Vakali A (2011) Urban planning and smart cities: interrelations and reciprocities. Springer, Berlin
7. Government of UAE (2007) United Arab Emirates year book-2007, Infrastructure. Available at www.uaeinteract.com
8. Rauf A, Enterprise Architecture, Innovation & Strategy (2013) Dubai 'smart city': are we there yet?
9. Dohler M, Vilajosana I, Vilajosana X, LLosa J, Smart cities-an action plan. Barcelona, Spain
10. A Smart Card Alliance Physical Access Council White Paper (2011) Available at http://www.smartcardalliance.org

E-Governance for Public Realm: Around Panniyankara Monorail Station, Kozhikodc, Kerala

T.M. Vinod Kumar and P. Bimal

Abstract Public Realm of city of Calicut in Kerala has a great role in enriching the quality of life of people. A well designed and maintained public realm with the full participation of citizen, business and various governmental agencies can have a great role in making Calicut more liveable. Most public realms of Calicut are totally neglected, characterized by encroachment, not designed properly and are faced with the danger of total disappearance by unauthorized uses. Having a Master plan or Zonal plan alone cannot create dynamic public realm. End of Plan period review of past Master Plans of Calicut shows disappearance of open spaces in the previous Master plan provision. Yet, no action is taken. The answer is E-Government for public realm which is a website design where all aspects of Governance are executed 24 h and 7 days a week. This is illustrated by an exploratory study of one public realm identified near a monorail station in Calicut, Kerala. Kerala state level policy of Spatial Governance is presented in Kerala State Urbanisation Report and Vision Kerala 2030 document Chap. 13 and is used to guide the Governance of public realm in Calicut. This website allows all stakeholders to participate in all aspect of planning, design and development, and management of public realm and administration connected with that. This web design is analogous to e-commerce, which allows businesses to transact with each other more efficiently (B2B) and brings customers closer to businesses (B2C). E-government aims to make the interaction between government and citizens (G2C), government and business enterprises (G2B), and inter-agency relationships (G2G) more friendly, convenient, transparent, and inexpensive in designing, managing and administering public realm in Calicut. One of the goals

T.M. Vinod Kumar
Besant Nivas, Opp. Calicut Orphanage, Jayanthi Road, P.O. Kolathara, Kozhikode 673655, Kerala, India

P. Bimal (✉)
Department of Architecture, NIT Calicut, Kozhikode 673601, Kerala, India
e-mail: bimalp@nitc.ac.in

T.M. Vinod Kumar (ed.), *E-Governance for Smart Cities*, Advances in 21st Century Human Settlements, DOI 10.1007/978-981-287-287-6_5

of 74th Constitutional Amendment of India is participatory governance and above models effectively provide for it. First, public realms are identified and potential and constraints are analysed and comments are asked from general public through website. Then, zonal plans of nodes near monorail station in Calicut are prepared, comments elicited from the web, followed by designing of hybrid Form-Based Codes. New Urbanism principles and transect analysis were conducted, followed by formulation of regulation plan, built form standards, public space standards, architectural standards, landscape standards, environmental resource standards and administration for the study area and are presented in the web for responses from public and to seek alternative design from citizen or friends of the city living world over. Interactive website allows active discussion on potentials and problems of public realm, SWOT analysis, participation of interested citizen, and also issue of building permits around public realm.

Keywords E-Governance · Public realm · Issues · SWOT analysis · Vision plan · Zonal plan · Regulation plan · New urbanism · Form-based coding regulation · E-Government · Evaluation

1 Introduction

This chapter for the first time, through an empirical study of a locality Panniyankara, explores how E-Governance can be utilized to shape and sustain public realm of an emerging 'Smart City' Kozhikode, in Kerala by 'Smart People' for the sake of 'Smart living'. The term smart used in the last sentence has been fully explained in chapter E-Governance for Smart Cities. In Smart living, public realm assumes prime importance since people live most of their life in waking state in public realm, in outdoors, when they are not at home or at workplace. Therefore, it is their right to shape these areas suited to the specific local culture of people of Kozhikode and the distinct ecosystem prevailing in these specific areas. Both these aspects differ from place to place and E-Governance challenge is to find suitable solution by the people and for the people. Unfortunately, public realms are highly endangered spaces in Kozhikode and vanish as reported in the last two Master Plans and the latest third Master Plan under finalization. Sometime even if public realm is available, they are not used since it has not been developed for various types of uses. Encroachment by poor homeless slum dwellers, corruption of the rich and super rich encroachers who are politically influential, greed of the people and lack of social and community sense and inefficient governance are mostly responsible for this vanishing game of public realm being enacted in Kozhikode in particular and all towns in Kerala in general. E-Governance for public realm is in response to these major issues are largely ignored in Kerala. The third Master Plan of Kozhikode being finalized now admits helplessly that they have no land to allocate for public realm as per the Town and Country Planning Guidelines and standards accepted in the State of Kerala, for the provision of public realm as a percentage of total land use. There were no remedial measures suggested in this draft plan under preparation. All these reasons prompted two authors of this chapter and their 14 students in the Department of Architecture, Urban Design Studio, National Institute of Technology and Kozhikode to undertake this without seeking any financial grants to undertake this research and development work based on detailed field surveys. Before presenting the case study based on field research, few questions need to be examined for Kozhikode. They are:

1. Is Kozhikode a 'Smart city' in making with 'smart people' which can open up the possibility of 'smart living'?
2. What is the status of urban E-Governance of Kozhikode? Is there a possibility of converting the existing city E-Governance system into 'Smart E-Governance' required to shape the cultural and ecosystem of public realm?
3. How do we conceptualize the E-Governance System for public realm and operationalize it in Panniyankara based on detailed field study?

After presentation of one case study based on a newer conceptualization of E-Governance for Public realm, the research experiences are evaluated in the concluding part of the chapter to find out how far the study achieved success in E-Governance based on many well-known criteria.

2 Making of a Smart City, Kozhikode

Draft Perspective Plan Kerala 2030 [1] considers, smart cities as knowledge-based cities, 'a city well performing in a forward-looking way in economy, people, governance, mobility, environment and living, built on the smart combination of endowments and activities of self-decisive, independent and aware citizens'. Cities, companies and universities are exploring Smart City opportunities in Kerala. In chapter E-Governance for Smart Cities 'GIS for Smart Cities' [2] based on empirical analysis has found potential of Smart Cities in all seven (including Kozhikode), one million plus urban agglomerations of Kerala as per 2011 census and other statistics. This is further explored in the last chapter of this book. Hence, Kozhikode is a candidate Smart city. Kozhikode urban agglomeration (UA) with 2,030,519 populations in 2011 census has got all attributes of a potential Smart City. Indeed, "Perspective Plan of Kerala 2030" had mentioned Smart City, but with no further thinking on an integrated and comprehensive policy instrumentality of achieving this in Kerala. Such Smart City Policy shall be aimed to realize Smart Economy, Smart Environment, appropriate application of Smart City Technologies and setting up of suitable institution mechanisms to realize Smart City in Kozhikode. These are not explored in the Draft Perspective Plan 2030 Kerala.

Smart City in Kozhikode may trigger smart economy that results in rapid economic development of Kerala. Per capita Kerala State income in 2013 is about $1,000 as per economic survey of state of Kerala in 2014 and it may grow rapidly in the event, Comprehensive Smart City Policies are formulated and then implemented in 7 million plus agglomerations in Kerala utilising economic activities with internal and external backward and forward linkages. Kozhikode historically had external foreign contacts from ancient time from Middle East, South East Asia, North Africa and Europe serving as main source of spice market, thereby establishing in Kozhikode, a vital linkage in spice route globally. Today, Kozhikode UA operates on an international connection as before but in a different manner mainly as exporter of human resources of all levels of skills and expertise for employment outside India. In return, Kerala is getting large inflow of money remittance from these countries but unfortunately these investments have not gone for economic development or industrialization of Kerala. This will change again if the potential for Smart Economy in Kozhikode is realized through deliberate promotion of Smart cities, which can attract high Capital inflow by more remittance from abroad of NRI citizens for productive purposes and also FDI for specific ventures of economic development. In recent time, there has been evidence of flexibility of labour market in Kerala that allows large immigration of labourers from Eastern regions of India with no social tension. They are here to benefit the higher labour wages prevailing in Kerala and also Kerala benefits from shortage of labour force. They live and work harmoniously with the locals without disturbing the existing communal harmony of the highest order. Side by side, Kerala shall also explore immigration of entrepreneurs from Europe, United States and East Asia irrespective of their race and Nationalities, giving them suitable work environment to prosper and live in the Gods own country as being practiced in China, Singapore, North America and many European countries like, for example Switzerland with well laid out policies. People of Kerala in the past had exhibited innovative spirit and had shown ability to transform the city by their own initiative. With a policy of encouraging more immigration from economically productive persons from abroad, Kerala can transform itself with the seven emerging Smart Cities transforming itself to a powerhouse for economic development. Labour unrest as demonstrated by periodic strikes is getting less popular in economically most active city like Kochi. Smart cities and periodic strikes resulting in loss to industry cannot exist side by side. Kerala already has got Smart people with high Human Development Index (HDI) (Kerala rank second with 0.921 if we compare the HDI in the world and first in India) and high rate of urbanization (67.2 % urban population in Kozhikode district in 2011) [2]. It has shown elements of Smart Governance as exhibited by participatory peoples planning being practiced since mid-1990s *(Janakiya Asuthranam)*, which makes Governance participatory and transparent for public investment. Smart electric grid is being established in Kozhikode. It is aspiring for Smart mobility with plan to implement high-speed rail system from north Mangalore to south Thiruvananthapuram of Kerala State and now monorail system is being implemented in the city of Kozhikode and Thiruvananthapuram and Metro Rail in Kochi to spread high mobility of the State at city level. The challenge is to make this emerging high mobility at state to city level to be accessible to all in the hinterland of Kozhikode and the subject matter of this chapter aims to do that. Smart city is characterized by the presence of Smart Infrastructure. For electric supply,

Smart Grid is being implemented and most of the houses are fitted with Smart electric meters. The major drinking water supply under implementation is built on the basis of Smart Water Supply. Both in water supply and Electric supply SCADA is utilized for smart infrastructure management. People accept a Smart Environment in Kerala fully aware of the value of environmental protection and sustainable resources such as water and energy, which will be further explored locally in this study. Kozhikode aspires for Smart living giving highest value to culture as exhibited by the fact that it won the highest marks/score and the golden trophy for the eighth consecutive years in 2014 for State Youth Art festival. This annual event has been a recruiting ground for future cine artists, singers of different styles and, dancers of different art forms in Kerala for the highest positions in art. People of Kozhikode give high value for healthy living, individual safety, housing quality and educational facilities as indicated by indices of quality of living. Kozhikode had shown high social cohesion with no history of communal violence ever since the State of Kerala was born. These are all evidence of Kozhikode as emerging Smart City. All these development confers that Kozhikode is a smart city in making.

3 Urban Governance in Kozhikode

India had pre-historic tradition of good urban governance, if relics of pre-historic, Mohenjo-Daro and Harappa can speak for their population and recorded statement on Lichhavi democracy when Buddha was alive is studied in depth. Today, Governance at local urban level is embodied in the 74th Constitutional Amendment which has been adapted by all States and Union Territories of Federal States of India including Kerala. It aimed at the creation of local governments, periodic election and financial commission to direct inflow of Government Revenue to these institutions and devolution of financial and governing powers to them in the respective states, metropolitan regions, districts, municipality and municipal ward level. It outlined a framework of institutions that would act as the local governments which would ultimately bridge the gap between the government and the governed. The people would take part in the issues that affected them directly, thereby having a say in decision-making. The word governance derives from the Greek verb κυβερνάω [kubernáo] means to steer and was used for the first time in a metaphorical sense by Plato [3]. In its most abstract sense, governance is a theoretical concept referring to the actions and processes by which stable practices and organizations arise and persist. These actions and processes may operate in formal and informal organizations of any size; and they may function for any purpose, good or evil, for profit or not. Conceiving of governance in this way, one can apply the concept to states, to corporations, to non-profit organization, to NGOs, to partnerships and other associations, to project teams and to any number of humans engaged in some purposeful activity.

Kerala had adopted participatory Governance enshrined in Indian constitution. Participatory governance focuses on deepening democratic engagement through the participation of citizens in the processes of governance within the state at multi-level. The idea is that citizens should play more direct roles in public decision-making or at least engage more deeply with political issues. Government officials should also be responsive to this kind of engagement which we see here with some reluctance. In practice, participatory governance can supplement the roles of citizens as voters or as watchdogs through more direct forms of involvement [4]. Ordinance No. 51 in the year 2013 of Kerala Government called 'Kerala Town and Country Planning Ordinance 2013' [5] lay down the legislative provision of urban governance in multi-level planning from State, Metropolitan, District, Municipal, ward and Spatial Planning Scheme level. Kerala had also adopted regulatory Governance. Regulatory governance reflects the emergence of decentred and mutually adaptive policy regimes which rests on regulation along with service provision, taxing and spending [6, 7] .This term captures the tendency of policy regimes to deal with complexity with delegated system of rules. It is inevitable in arenas and nation and sub-nation which are more heterogeneous, complex, more global, more contested and more liberally democratic [8]. Indian constitution builds upon and extends the terms of the regulatory state on the one hand and governance on the other. While the term regulatory state marginalizes non-state actors (NGOs and Business) in the domestic and global level, the term governance marginalizes regulation as a constitutive instrument. The term regulatory governance therefore allows us to understand governance via regulation from the Town Planning Scheme areas to higher spatial hierarchies.

3.1 E-Governance in Kerala

World Bank [9] defines 'E-Government' as the use by government agencies of information technologies (such as Wide Area Networks, the Internet and mobile computing) that have the ability to transform relations with citizens, businesses and other arms of government. These technologies can serve a variety of different ends: better delivery of government services to citizens, improved interactions with business and industry, citizen empowerment through access to information or more efficient government management. The resulting benefits can be less corruption, increased transparency, greater convenience, revenue growth and/or cost reductions. It is analogous to e-commerce, which allows businesses to transact with each other more efficiently (B2B) and brings customers closer to businesses (B2C). E-Government aims to make the interaction between government and citizens (G2C), government and business enterprises (G2B), and inter-agency government relationships (G2G) more friendly, convenient, transparent and inexpensive. More important, it aims to help strengthen government's drive toward effective governance and increased transparency to better manage a country's social and economic resources for development.

E-Governance is the unending direction and matrix of mind of Smart City. The existing state of the art of E-Governance shall be extrapolated to reach the most suited E-Governance System for Smart City to give direction and intelligence to

Smart City. This chapter attempts it in one domain of public realm. Each domain within Smart City needs to have e-domain smart city governance driven by contours of substantive issues and appropriate solution space. This book addresses it. There are several E-Governance issues such as inclusion and exclusion and privacy needs to be sorted out to give final shape to the intelligence of Smart City.

Public systems created by Colonial British in India and sustained by so-called iron frame of Indian Administrative Service are characterized by ineffectiveness, lots of red tape, mindless procedural formalities, corruption and delays. IT-enabled systems are sleek, smart and effective befitting a Smart City. They can be accessible from Personal Computers from home, from mobile phones or dedicated public facilities like *Akshaya* Centres of Kerala. E-Governance would drastically change public systems. It is the lethargy and resistance on the part of employees that hold back E-Governance. It improves citizen friendliness. Today, Citizen interacts with government on three major aspects which in no way is sufficient and we have to go beyond that. They are payments to be made to government such as taxes, duties; entitlements to be received from the government certificates, permits and licence; and information required from government jobs, schemes and programs. Citizen focused minimal governance needs to address at least these three aspects primarily with the objective to induct a philosophy of efficient service delivery in Government and to treat the citizen as a valuable customer and not colonial subject, the so-called 'natives'. They in fact pay for services and are to be treated with respect. These can be considered only as first step in E-Governance and many more steps to go to reach smart E-Governance and some of them will be explored in this chapter.

Shri. Omman Chandy, Chief Minister of Kerala in his inaugural address at the 17th National E-Governance Conference in Kochi on 30 January 2014 outlined the digital vision of Kerala. The State had chartered a course of action in many years to become the first digital state in India. Kerala is the first state to implement online right to information (RTI) facilities. Kerala had implemented, E-district largely of rural orientation, e-office for administrative efficiency and transparency, e-tendering for procurement in all departments, and e-payment system for many government institutions. There are many more steps to be ventured to attain a status of fully digital state of Kerala. All of these innovations will form the foundation of E-Governance for Smart Cities in Kerala. There are but no urban-oriented E-Governance system similar to E-District in Kerala, while Kerala is marching towards nearly 100 % urban state around 2031 or 2041 with its emerging urban corridors as detailed out in State Urbanisation Report [11]. There are miles to go in Kerala to make E-Governance for Smart Cities, which means to get Government closer to people for active participation. Some initiatives of Kerala in this direction are encouraging.

FRIENDS of Akshaya in Kerala [10] is an integrated electronic interface, where citizens can access Government services, pay bills, taxes and obtain information from Government. Multiple computer counters deliver services in a pleasant setting, with highly trained and motivated young staff. FRIENDS' centres are set up in all 14 districts of Kerala. It works 7 days a week, 12 h a day. In any average FRIENDS Centre of Akshaya, there are about 1,458 Visitors per day.

Their average expenditure per month is Rs. 34,688 and Collection per month—Rs. 96 lakhs as per recent statistics published in Akshaya website. The Benefits of FRIENDS are improved perception of Government. Ninety nine per cent of Customers are fully satisfied with FRIENDS and there exist an enhanced revenue growth rate of 150 % per year as per Akshaya statistics.

Kozhikode and Kerala stands high in the status of male and female literacy in India. However, for better E-Governance, e-literacy is a must. E-literacy project undertaken by Akshaya is the foundation of E-Governance in Kerala. This e-literacy (Akshaya) project, first started in the adjoining Malappuram district near Kozhikode of Kerala with 4.11 million population in 2011, 44.2 % urban population and about 650,000 families. Out of which 1.7 million lives within Malappuram urban agglomeration. It is a socially and educationally backward district, with poor ICT infrastructure, and low telecommunication penetration and only 200 internet connections when this project started. E-literacy project changed all of these. This was the first district-wide e-literacy project in India and one of the largest known Internet Protocol (IP) based wireless networks in the world. In November 2002, the state government of Kerala put into place a project, piloted in Malappuram, with the goal for one person in every family to be computer literate in that district. Malappuram is now what is said to be India's first E-literate District in India. The mission continues to make Kerala the First E-literate state in India including Kozhikode.

Akshaya Model is for 100 % e-literacy, with a network of ICT centres and broadband connectivity. Akshaya accessibility for Malappuram district was based on 630 access points, one centre for 1,000–1,500 families and one centre for every 5 km. It is an entrepreneur-driven business model. In Malappuram district alone, Akshaya has conducted one of the world's largest computer literacy drives, claiming to reach over 600,000 households then, representing more than 3.6 million people in less than 6 months. The project has created a unique brand of state-funded computer access centres, and simultaneously led to a massive wireless infrastructure, providing a wide range of services and making way to many future E-Governance opportunities. Now all districts of Kerala are covered by 4 G WiMAX service by Bharat Sanchar Nigam Limited (BSNL), a public sector telecom enterprise. The Akshaya project offers a lot of services since 2008 like, E-Pay (electronic payment of utility bills like electricity, land, phone, drinking water, university fees, etc.); E-Krishi (for farmers to provide online agriculture trading and information portal, A to Z Solution); E-Vidya (advanced IT learning for e-literates and others); E-Ticketing (online train, flight, bus ticket reservations); online passport registration; a village kiosk for transparent 'collectorate' programme, online communication providers for expatriate Indians; and online medical transcription course, with extension programmes for all the above-mentioned services. Project Akshaya started with the noble intention of making at least one member in each of the 64 lakh families e-Literate in Malappuram District. The pilot project started in Malappuram, and was inaugurated by His Excellency Dr. A.P.J. Abdul Kalam, former President of India on 18 November 2002. Hundred per cent e-Literacy was achieved in eight districts of Kerala in Phase 1 and the project has been a roaring success. Nowhere has such a feat been attempted or accomplished. The vision and scope of the project is beyond comparison. The operations of Akshaya as unfolded so far consist of three distinct phases. The first phase is that of enabling the masses by imparting basic skill in computer operations and making them familiar with Internet. This phase is the period of e-literary training. In the initial phase, the Akshaya e-centre imparts basic training that not only familiarizes people with the basics and the scope of IT, but also ensures hands-on skill in operating a computer, using the Internet to at least one person in the 64 lakh families in the state that makes it the largest rural e-literacy training project worldwide. Malappuram is declared as the first e-literate district in India. Malappuram and Kannur districts were already declared 100 % e-literate. Kollam, Kozhikode, Thrissur and Kasaragod districts achieved e-literacy above 90 % and are ready for declaration. So far around 33 lakhs beneficiaries are being trained.

3.2 E-Governance for Public Realm

The public realm constitutes all exterior places and built form elements that are physically and/or visually accessible regardless who owns the property. These elements can include but are not limited to streets, right of ways, pedestrian ways, bikeways, bridges, plazas, nodes, roundabout, squares, transportation hub, taxi and auto rickshaw stands, gateways, parks, formal or informal hawker places, weekly markets, water fronts, natural features, view corridors, landmarks and building interfaces. The visual elements of streets include roads, sidewalks, street furniture, trees and open spaces that form the street character. All visible public spaces used for public gathering, pubic interaction or assembly come under public realm. Parks and organized playgrounds or civic space for various types of recreation for different age groups can be designated as public realm. All adjoining areas of water fronts, places enveloping temples, mosques, churches, heritage sites *and gurudwara* can also be considered as public realm. Urban Design is a part of Governance of public realm and falls within Indian Constitution Amendment number 74 which in turn strengthens the hand of governance by Town Planning Acts and Ordinance and gives legal instruments such as Master Plan, Zonal Plan, Urban Land Management Schemes, Site Planning schemes and Urban Design and constitutional bodies like State and Central Election Commission, State and Central Finance Commission, Municipality, Municipal Corporation, District Planning Committee and Metropolitan Planning committees to implement it. Seventy fourth Constitutional Amendments require higher participation of people in all aspects of Governance of public realm which can be easily achieved by E-Governance and E-Government of public realm.

Governance for public realm is all about the processes by which public policy decisions are made and implemented for provision, planning, design, implementation and management of public realm. The vast reservoir of creative ability of community (for example recent experience of establishing sculptures of renowned Indian Artists in different locations in Kozhikode and Mural paintings in Kottayam) and interested citizens and organizations are used in all governance aspect of public realm. These creators of public realm can be initially locally identified and finally sourced or made available all over the world in 24 h and 7 days a week through E-Governance. An interactive web page can act as E-Government for public realm which will result in interactions, relationships and networks between the different sectors (government, public sector, private sector and civil society) and involves decisions, negotiation and different power relations between stakeholders to determine who gets what, when and how. The relationships between government and different sectors of society determine how things are done, and how services are provided. Governance is therefore much more than government or 'good government' and shapes the way a service or set of services are planned, managed and regulated within a set of constitutional framework, political, social and economic systems. Seventy Fourth Constitutional Amendment of India is a strong advocate of democracy in Governance. Democracy means participation in Governance just like ancient Greek cities where existed a face-to-face participatory democracy and governance of about 6,000 plus people. E-democracy refers to the processes and structures that encompass all forms of electronic interaction between the Government (elected) and the citizen (electorate).

Attempt of this chapter on E-Governance for Public Realm in Smart Cities is to expand the existing practice of E-Governance in Kerala given above to encompass E-Governance for Public realm. Kerala Town and Country Planning Ordinance 2013 [5], Perspective Plan 2030 of Kerala (Chap. 13) [1] and State Urbanisation Report Kerala 2012 [11] are the basis of formulating E-Governance for Public realm. Town and Country Planning Ordinance 2013 specifically dealing with Town Planning Scheme and postulates functions of Urban Arts Commission are of importance to Public realm. Draft Kerala Perspective Plan 2030 envisages development of smart, green and compact cities and hence public realm with various Regulation Plans and Form-Based Codes Standards have great role in realising the goals of the Kerala 2030 Perspective Plan. Perspective Plan 2030 Kerala, frames several elements of spatial strategy such as Element 1: Land management, Element 2: Compact cities, Element 3: Smart cities, Element 4: Mobility and transport (strong infrastructure and sustainable transport systems), Element 5: Quality of roads and transport development, Element 6: Cultural facilities, Element 7: Community facilities to enhance the working, living and business environment of the Cities, and Element 9: Provision of utilities are of immense value for public realm. State Urbanization Report 2012 [11] which extrapolate urbanization in Kerala 2031 had made several recommendations of value to urban realm and advocates' urban compaction. As per Draft Perspective Plan Kerala 2030, the compact city does not have an exact definition, but three characteristics define a compact city. (1) Dense and contiguous development patterns. (2) Urban areas linked by public transport systems and (3) Accessibility to local service and jobs. The urban design of public realm shall achieve urban compaction.

As per Draft Perspective Plan Kerala 2030, physical planning will be based on the principles of high density, better options for walking or public transport, less need for transportation by cars, increased accessibility, preservation of green areas, promotion of existing infrastructure, better quality of life, better relations with neighbours and greater safety. In other words, it is clear advocacy of Smart Urbanism Principles to achieve perspective plan goals and indirectly recommends use of zonal plan and regulation plan to achieve regulation of public realm. However, zonal plan is not yet practiced in Kerala which we feel is a must for Kerala to incorporate spatially the main thrusts of National and State five-year plans and to implement incrementally budget allocation of annual plans. This is a great omission in the planning practiced in Kerala. Also, urban designs guidelines mentioned shall be replaced with form-based codes based on citizens' preferences and mainstreaming ideas of sustainability, gender and even arts. Form-Based Codes goes miles ahead of urban design guideline (which is mandatory but normally ignored during implementation). While newer areas need to develop, older neighbourhoods need guidelines, often ineffective to re-develop, while preserving their cultural heritage. All the elements together should improve the quality of life for the citizens. A common element that is suggested is the encouragement and preservation of public spaces especially community areas. Here, we find a contradiction since public realm disappears sooner or later through encroachment in cities of Kerala. Developing parks within cities are important elements of healthy living. Therefore, we advocate Hybrid Form-Based Code which takes into account zonal plan to achieve these Perspective Plan goals. Cultural facilities are a determining factor in how individuals choose where to live and work. Promotion of festival, marketplaces, entertainment districts, fine arts' galleries, performing art centres, sports arenas, convention centres and office complexes are important components in urban revitalising strategies. Urban design of public realm has a great role in it.

There are two integrated approaches to help in the E-Government of Public realm namely substantive and procedural. The substantive part we are presenting here is based on a manual for participatory Spatial Planning and design which has been developed out of experience of conducting many studies since 2005 in the National Institute of Technology, Kozhikode [12]. The objective of the manual was to train on whole spectrum of participatory urban design and spatial planning, application of new urbanism principles and form-based coding regulation in an easy to understand manner in as graphic manner as possible. The Manual tried to assemble various plant and tree types applicable to Kerala and spatial standards which can be adapted to Kerala as per regulations. The primary assumption of this manual is spatial planning and urban design can be created with the help of this manual by mobilising interested person coming from any related area of interest for this work. They need not be chartered engineers, city planners, architects, landscape architects or urban designers. There are two levels of training envisaged in this manual. First, officers of Town Planning Department of Kerala and teachers of Architectural and Planning Schools are trained. Then, each of the trained person mobilizes local team and conduct training for target spatial planning areas by identifying local teams locally and execute procedures of the manual dovetailed with the ongoing State Planning Board driven, annual planning process at ward level as well as preparation of Town Planning Scheme or land management envisaged in 2013 Town and

Country Planning Ordinance. Extensive use of new urbanism principles, zonal planning and form-based coding standards were used here. Every year Faculty of Architecture, National Institute of Technology (NIT-C), Kozhikode modified the approach with minor variation as dictated by the area and subject matter thrust of study. We found the external jury who examined the student project based on this Manual had great appreciation for the result achieved using this manual. Also, urban design students of 2013 had submitted last year's studio project of one of three groups in NIT-C for University of Westminster London (UoW) Trophy at 56th National Association of Student Architects (NASA) annual meet in Hyderabad. They scored the highest marks for their entry and are selected for an exhibition in London in summer of 2014 along with nine other architecture schools getting lesser marks than them. This demonstrates the quality of output the manual can bring out.

The procedural aspect conceives E-Government of public Realm as an interactive website which meticulously follows the step-by-step process of the manual. This website is assumed to be part of the Municipal Corporation Website of Kozhikode Corporation. The result of work of the locally mobilized team discussed above for spatial planning and design of study area is published in the web as first iteration and only for eliciting opinions, suggestion and alternate design at every step of the process. Participants from any part of the world, any time can send their inputs which may be their opinions or design alternatives through website. Finally, administrative team selects the best design ideas for implementation. Also, there is a maintenance system; there is provision to submit building approvals which will be transacted by this website and payment portal for all types of payment. Although the web designed as E-Government for Public Realm of Panniyankara was of considerable details, a brief overview is presented here to explain the core model.

4 Study Area: Panniyankara and Zonal Plan

The study area is located in Kozhikode (Kozhikode), which is one of the three major cities in Kerala, India, the other two being Kochi and Thiruvananthapuram, the state capital. As per the census 2011, the population Kozhikode urban agglomeration (UA) is 20,30,515. If one rank Kochi UA and Kozhikode UA, it is of the same high rank as per population. In 2011, urban content of Kochi in Ernakulum district was 68.07 % and Urban Content of Kozhikode district was 67.15 %. Unlike many other states, the district populations in Kerala have decreasing growth rate, but the urban area in these districts shows a positive numerical growth but not high rate. Interestingly, most of these increases are not due to people moving to cities in a massive scale, but it is due to the conversion of more and more rural area into urban. This results in loss of agricultural land and open spaces, and makes the distribution of infrastructure expensive. Densification of the urban area is the most suitable solution to such a scenario. Densification in an existing city will definitely choke its roads, unless an effective public mass transit system is not in place.

In Kozhikode, there is a proposal for monorail system, connecting its important nodes of the city in two phases; the first phase being a connection from Kozhikode Medical College to Meenchantha with 15 stations covering about 14 kms. The second phase may connect Meenchantha to Ramanattukara, Kunnamangalam or even Airport in Karippur. Kunnamangalam is the Gateway to Wayanad hilly district and Ramanattukara is a growing industrial centre on way to Karippur Airport. This gives an opportunity to increase the population density of these innumerable nodes in an orderly planned fashion that are connected by the system of monorail stations, and thereby releasing more area for public realm and also more number of residents gets easy access to monorails. This requires interventions at various levels of planning. However, the issues faced by public realm cannot be solved only by Master planning exercise since public realm do not even appear in the Master Plan scale and issues are not discussed or analysed in Master Plan level and solution suggested. Town Planning Scheme, which is one approach to Urban Land Management, practiced in Kozhikode does not address many requirements of provision of public realm such as zonal codes and form-based codes detailed in this chapter. Therefore, next lower level of planning, zonal planning is required below Master Plan at these nodes. Zonal plans do not give greater spatial details of planned interventions of the urban spaces and hence form-based codes are required in addition to land development codes supplied by zonal plans. Therefore, zonal plan practiced in many states of India are required in Kerala. Public realm will become extremely important as it has to be optimized to facilitate higher density. Public realm is generally an outcome of history and sociocultural background, and it evolves according to the aspirations of the users. It involves every aspect of the physical environment, and the user experience while using public spaces.

The study area identified for this chapter is a place called Panniyankara, in Kozhikode Municipal Corporation. Panniyankara is located towards the south side of the city and is on National Highway (NH) 17. The study area is divided into two parts by the NH and railway line running north-south. The place has traditional Hindu community and Muslim community habitat, existing close by on the east side. Roughly, the population on the eastern side are economically well off, while the west side population is relatively weaker in household income. Truly, the railway line acts as physical barrier between east and west side. Panniyankara is going to have a monorail station and a road overbridge to replace the railway crossing is already under construction. Hence, the first exercise is to develop a zonal plan with a vision to accommodate a much higher density through proposed detailed urban design strategies based on Form-Based Codes for various existing and proposed public realms to enable them to efficiently work in the higher density proposed by the zonal plan (Fig. 1). This is accomplished first by an in-house planner but later allowed to be modified on the website by anyone. Finally, one zonal plan is selected.

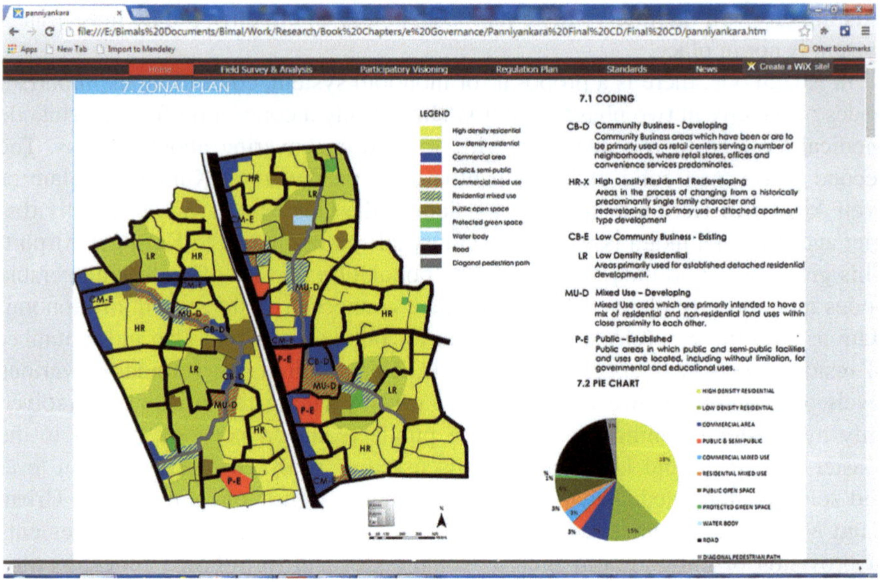

Fig. 1 Zonal plan

5 Public Realm Website for Panniyanakara

The website of Public Realm Panniyankara shall be part of Municipal Corporation of Kozhikode website. In reality, this website is designed to act as the 'Public Realm Government' and functions as informative, interactive and transaction-oriented website. It is designed for ongoing, participatory planning in Kerala incorporating for the first time, a missing link namely spatial aspects in people's planning (*Janakiya Asuthranam*). The website is designed for multiple uses such as for maintaining public spaces, monitoring of urban development, for annual planning and e-budgeting of Kozhikode Corporation, providing administrative sanctions around this area for building plans, and also help in electronically remit taxes and fees for all services of Municipal Corporation Kozhikode and other agencies. About 10–15 min walking distance from the proposed monorail station in Panniayankara will be the areal limit of the website in this chapter. It comprises of part area of municipal wards having definite boundaries and elected members, which are the lowest constitutional spatial entity in India. It also has got few Residential Welfare Associations in this area with designated areal boundaries. Updating the detailed maps of the area is the responsibility of these two entities with the help of local volunteers from schools and colleges nearby or others. They can use hand-held GPS or Smart Phones and input the data to a GIS maintained by them. There are also cameras which can geotag latitude, longitude and time reading. For example any public investment as part of Annual Planning may be located in the GIS with date stamps, latitude and longitude and other attribute data. Freeware GIS in Linux platform is also available to Residential Association who does not get any Government Grant. Also, pot holes of roads may be located with GPS coordinates and date stamped in the GIS, including photo from mobile/camera, for repair services by concerned road maintenance department. These inputs will be highly useful for e-budgeting in Chap. 10 of this book. This website shall be the permanent record keeper of all existing and proposed GIS planned entities with relevant entries from proposed Master Plans once in 20 years, zonal plans once in five years and annual plans once in a year for all years to come by. Existing and proposed geographic entities are recorded as soon as it is available here with the help of Ward Committees and residential Welfare Association, thereby achieving the self-awareness which is a primary requirement for an emerging smart city Kozhikode.

Any interested party can register and log in with limited identification such as name, email ID, mobile phone number and qualification. This website facilitates E-Government interaction between government and citizens (G2C), government and business enterprises (G2B) and inter-agency relationships (G2G) in friendlier, convenient, transparent and inexpensive way 24 h and 7 days a week. Citizen by definition need not be citizen of Kozhikode or Panniyanakara but global citizen transcending nationality, caste, religion, sex and race who are interested in the locality to extend personal and professional capabilities to develop this area 24 h a day and 7 days in a week from anywhere in the world. For example someone abroad of a different nationality, who wants to paint the walls of the public realm with his/her own mural, or give away his/her own sculpture at public square, he or she is a global citizen interested in the study area.

The public realm interactive website clearly indicates all steps in spatial planning/designing and administrative process involved in the creation of public realm and maintaining and financing it. It shows results of one design of at every stage in a spatial planning process and then open up for suggestions and discussion for alternate design ideas from users of the website with proper prompts. Before implementation, the best idea is selected by a designated committee.

An interactive and transactional website for active participation of all stakeholders in every stage of development of public realm is envisaged in the design of website with the sole responsibility to conduct E-Governance for Public Realm which includes issue identification, spatial planning and design, implementation, administration and maintenance. It is hoped that public realm hereafter will not disappear unnoticed through prevailing corrupt encroachment and mindless acquisition. Website is the 'Government of Public Realm' and no need for stakeholders to visit any Government office for any transaction since this website is meant to facilitate the planning, design, maintenance and administration of public realm. The website records the spatial planning/urban design issues identified in the extensive field study and propose the solution at four stages, the first being SWOT analysis, which brings out strategic interventions based on detailed field survey and study, second visualization of urban form appropriate to the strategic interventions, and thirdly zonal Plan based on Master Plan which details out the spatial requirement as per town planning standards adopted in Kerala and spatial policies, and finally transect analysis, and regulation plans are applied and each public realm is detailed out for implementation, based on application of New Urbanism principles and Form-Based Coding Regulation to enable the public realm to efficiently work in a high-density scenario. The website envisages developing the design largely through public participation and hence the interactive and transactional web-based system, which will allow both-way communication, not only as feedback or complaint but as a way to bring in the genuine local knowledge base to the planning, designing and managing of public realm. Integration of all these aspects together in a website makes it E-Governance.

Traditionally, public spaces were planned and designed by experts, and later managed by the municipal agencies. Seventy Fourth Amendments to the Indian constitution give all these powers to the Municipal Corporation. The practical implementation of this constitutional power, often results in difficulty in mobilizing required expertise. But the fact is that, very much authentic and context-sensitive expertise is available within the community and the difficulty is to realize its presence and mobilize it. Interestingly, such expertise is often available voluntarily with no cost to Municipality. What is required is a tool to mobilize and consolidate such collective wisdom in a cost-effective manner. When the common men need to come forward for series of meetings, and then spend time to work on the problem, at the expense of their working days, they will be hesitant to participate. On the contrary, if the system allows them to participate in the process at their convenient time, and location, people will be willing to participate in the process. E-Governance becomes a necessity in participatory planning, design and effective maintenance of public realm, especially in a state like Kerala, where a large fraction of the population work outside the state all over the world [13].

Unfortunately, finer aspects of public realm namely urban design are often neglected in the modern urban development scenario in Kerala as well as in other states. Urban design interventions are limited only to selected places of importance (either historical sites or places of political importance) due to various reasons including lack of expertise or lack of funding for consultants fee. Urban design interventions are considered as extravagance for public spaces used by the common man. This sounds undemocratic and not befitting India. There is no effective legislation in enforcing the essential aspects of urban design for urban spaces except the guidelines of urban arts commission for selected places. The urban Arts Commission in the Planning Ordinance 2013 of Kerala should mandate this issue for common people as their terms of reference and adopt implementation of zonal plans and Form-Based Codes as part of annual planning in Kerala bringing it into the annual budget cycle. We feel that Urban Design is required in every important urban space, which is only possible if it is participatory exercise and not solely depended on few hundred urban designers with academic qualifications available in India for about 8,000 urban centres with many millions of urban realms.

5.1 Life Cycle of Public Realm

Public realm is an outcome of the culture (i.e. regularity of human behaviour in geographic space), history and socio-economic characters of the population. Though it is firmly rooted in the culture and history, it is dynamic in nature. It responds to the aspirations of the population, and responds to the socio-economic changes from time to time. For example there has been a drastic change occurring in the housing form and structure in the last few decades in Kerala, in which the large multi-family houses gave away to a large number of small nuclear family dwellings. Architectural styles of houses seem to have imprints of the period the house that came into existence. Also, Kerala houses in general and houses in each districts of Kerala are significantly different from the residential structures found in other states and even here separated by particular styles of decades. Hence, design for a public realm will be continuous process ever changing and the design should refine itself and adapt to the changes in the scenario. Hence, the cyclic process of Governance for Public Realm can be roughly classified into four stages, Design evolution; Execution; Maintenance and Review. This is repeated in the subsequent cycles and website is the witness to change.

5.2 Types of Public Participation in Life Cycle of Public Realm

The nature of public participation required in each of the stages enumerated above is different. Design of a public realm involves a number of sub-systems and each of these requires different kind of expertise; but all such systems need to be designed in such a way that they work in harmony with each other. This overall planning requires urban design expertise which need not come from university education alone. Urban design expertise is available to those who are deep rooted in their culture since the public realm shapes these cultural requirements in three dimensions in a particular geographic space. This has been happening several thousands of years in India even before the oldest Universities like Nalanda in India was functioning, for example in Mohenjo-Daro and Harappa cities. There are many to foot this bill with their insight locally, nationally and internationally with no academic qualifications. Three or four urban design departments in India are not the sole repository of this skill. The local level expertise will be easy to find in the locality. Local nursery or planters will have clear idea about the kind of vegetation that can be grown in that region in that particular geography, under a particular microclimate and adjoining trees; very easily. They also know how land shall be shaped for ecological vulnerabilities based on practical knowledge their ancestors had taught them. Shop owners will be able to contribute largely on planning facilities for the market area to improve business. Taxi drivers will know the best locations for the taxi stands and other amenities required. To integrate all such sub-systems, work out alternate scenarios and to analyse 'what if' a specialists in urban design, planning and/or architecture may be required. The best course available is taking these resources from Government Town and Country Planning Department and College teachers nearby. Such expertise may not be always available locally, but can be sourced remotely through the E-Governance infrastructure. Hence, the participation at this stage will be between three groups of people, pooled in experts from public, local governing body and the local academia. The design can be gradually evolved by the collective action of such experts through E-Governance.

Execution of the evolved design will invariably include many of the government agencies and authorities as it involves constitutional mandate, public space, public money and public welfare. Private agencies will also take a major role in this process, may be as contractors or companies who can construct facilities like bridges, foot paths, roads, parks, individuals developing their own property as commercial structures, residences, or as parties supplying materials or labour for construction activities. E-Governance can become the medium to integrate the activities of such stakeholders. E-Governance for execution of public realm should facilitate interactions between [14].

(a) Government agencies (Coordination of works between various government departments like water authority, Public Works Department and Electricity distribution agencies).
(b) Government and Public (Public need to get approval for their proposed developments, applications for electricity and water. Government may have to communicate policy changes, procedure changes and also the decision on individuals request for approval).
(c) Government and Business (invitation for various e-tenders, expression of interest, coordination between government agencies and private contractors, request for approval, investment in government projects).
(d) Business and Business (Subcontracting of labour and materials, coordination between private agencies).
(e) Public and Business. (Invitation for tenders for individual's development works, offers for development from private companies to individuals).

Maintenance stage of the process will include routine tasks related to the upkeep of the developed public realm. This involves payment and collection of various bills and fee, receiving and resolution of complaints regarding utilities and services, and inviting tenders regarding maintenance activities. Individual users of public realm will be logging on to the E-Governance site with their applicable credentials, and they should be able to access required facilities at ease.

The existing public realm needs to be studied thoroughly before getting into designing it. Existing sociocultural requirements and its manifestations in the physical development has to be understood. The spatial possibilities along with ecological implications are to be considered [15]. Review stage will constantly aggregate the complaints and suggestions regarding the existing public realm and will act as the starting stage of the next design cycle. This can be initiated as soon as the design is implemented, and the users/stakeholders can record suggestions or complaints about the facilities. The system should aggregate it and present it to a design monitoring team, and this team will classify such inputs according to their importance and applicability. Crucial and easy to implement suggestions may be implemented quickly, but the remaining can be preserved categorically until the next design cycle.

5.3 Role of Participants

Different type of participants can be summarized as follows; these participants have different roles to play in the design cycle. Important aspects of the expected roles are discussed below.

5.3.1 Local Government

Local government initiates the process of public realm design by appointing a Web Design Agency and implementing training program and action programs as stated above. During the process of design the people's representative from the locality will monitor the functioning of the agency without interfering in the technical process and facilitates the required resources by representing them at the council meetings of the corporation. Once website is designed Local Government have role in all aspect of E-Governance of Public Realm as detailed in the web design.

5.3.2 Web Designer

The website is expected to collect opinions over the examples presented in the website and initiate discussions and result in a better design. Hence, the website acts as the medium of communication and will initiate discussions while the web designer facilitates the design. Unlike many other examples of wiki like Wikipedia and Wiktionary, this public realm design process will have very small set of contributors who are limited by the geography concerned. People who are connected with the concerned area only will be interested in contributing; whereas wikis draws their contributors from all over the world and is limited by only the expertise. Due to this reason, in a public realm wiki depends on informal online social networks such as Facebook and LinkedIn to bring relevant persons to the website for their contribution. Local beneficiaries organize and develop their contributors by conducting local awareness campaigns and interactive sessions. Also, it will require to actively inviting certain expertise which are locally unavailable in order to direct the development by means of suggestions and reviews. Such interactions are required to keep the development process fast and minimize errors. Website initiates discussions on certain aspects by actively developing example scenarios or sketchy proposals,

Table 1 Types of participants for public realm web

Types of participants					
Sl. no.	Type	Pool	Qualification	Scope of participation	Participation
1.	Local government and government machinery		Elected by public, through formal election as per Indian constitution	Selection and appointment of DFA overall performance assessment and control of activities departments	
2.	Web designer	Private firms or public institution	Appointed by local government	1. Organize website 2. Facilitating e-governance website 3. Basic data collection/ preparation Initial proposals from participatory designers are presented	By appointment
3.	UD experts	Urban designers, urban planners, architects, urban administrators etc., from all over the world	1. Urban designers 2. Architects 3. Urban planners 4. Urban administrators	1. Expert's comment/suggestions on integration of subsystems	1. By invitation 2. Voluntary
4.	Specialists	Experts of the relevant field from the study area and the surrounding region or from all over the world	1. Educational qualifications of the relevant field 2. Experience in the concerned fields Experience in the concerned fields 3. Local artists, artisans, etc.	1. Supply required information 2. Perform SWOT 3. Propose solutions and alternate solutions 4. Comment/suggest collected information 5. Comment/suggest on SWOT of relevant field 6. Scrutinize proposals suggest alternatives	1. By invitation 2. Voluntary 3. Peer invitation

(continued)

Table 1 (continued)

Types of participants					
Sl. no.	Type	Pool	Qualification	Scope of participation	Participation
5.	Resident public	1. Residents of the study area 2. Residence associations 3. Commercial/Industrial/Government establishments of study area	Proof of stake holding	1. Comment/Suggest on proposals 2. Propose solutions and alternate solutions 3. Complaints 4. Applications for permits, utilities and services 5. Fee/Bill payments review of public realm	Voluntary
6.	Business	Suppliers of material/equipment/services contractors	As decided by the design facilitating Agency/Local government	1. Comment/Suggest on proposals 2. Attend concerned complaints 3. Attend concerned requests for service/utilities etc.	Voluntary

which will trigger comments from the users. Hence, the role of the Public realm Web designer is as summarized in Table 1.

1. Developing and maintaining of web platform.
2. Organize participants by various means and ensure their participation.
3. Initiate discussions by example, solutions, case studies, alternate solutions, etc.
4. Ensure that the development is on the right direction and pace, and finish the process in a time bound manner.

5.3.3 Spatial Planner/Urban Designers

Designing Public realm is a complex process under participatory regime. It involves a large variety of systems and designs. The interesting fact is that many of these decisions appear to be trivial and simple to arrive at, but they have a complex interrelationship with other components. Hence, it is very easy to arrive at simple solutions to smaller issues but it is probable that such decisions may lead to failure of some other systems. In case of Panniyankara, hard landscaping of public spaces being implemented in residential front yards are the obvious and simplest solution for beautification and avoiding puddles during long rainy days. But this worsens the drainage condition of the public realm, as it need to manage an additional volume of run-off water, and it is practically difficult to take water to longer distance in a flatter terrain under gravity due to topography of Panniyankara. Comprehensive planning and design is required in such conditions, and often requires experience and training to do so.

5.3.4 Specialists

Specialists are category of participators, who have expertise in one of the field. They are required to suggest probable alternatives to the issue to be tackled and its possibilities and limitations. Such specialists may include architects, artists, sculptors, carpenters, masons, engineers and people working with different utilities. They also help in preventing expensive mistakes with their prior knowledge and experience. Engineers will be able to comment about the feasibility of proposed water supply or drain system through an uneven terrain than the common public. A carpenter from the locality will be able to suggest a better detailing to make the wood work easier with the help of new tools.

5.3.5 Resident Public

Their role extends from providing required information to start the design process to reviewing the functioning of the evolved public realm. Resident public represents the general users including residents, commercial or industrial facility owners, and also includes the local experts. They comment on the proposals, raise demands for maintenance and pay for using various facilities.

5.3.6 Business

Business category represents the group who has business interests in the given area, which includes local shop owners, suppliers of materials, services and builders. They contribute by commenting about different alternatives for materials or technology available in the market. They interact with the government and individuals through e-tendering to apply and secure orders for construction, services and maintenance.

5.3.7 Roles of Participants

There is no one model to show how this website will work and the roles of many type of participants. There are examples of effort in creating Wiki or working on Linux distribution system from where inspiration can be drawn, but one cannot copy them in totality. The website is expected to work in a wiki style or style of Linux software distribution development. The website acts as the medium of communication and will initiate discussions prompted by network of people. Unlike many other examples of wiki like Wikipedia and Wiktionary, this public realm wiki will have very small set of contributors with diversity of expertise, which is limited by the interest in geographic space concerned. People who are connected with the concerned area only will be interested in contributing; whereas other wikis draws their contributors from all over the world based on interest in particular topic, and limited by only the expertise. Informal network of interested parties can identify potential contributors and can invite them to participate. Also, it will require to actively inviting certain expertise which are locally unavailable in order to direct the development by means of suggestions and reviews. Such interventions are required to keep the development process fast and minimize errors. The website will also initiate discussions on certain aspects by actively developing example scenarios or sketchy proposals, which will trigger comments from the users. Their main tasks would be

Developing and maintaining of web platform.

1. Organize participants by various means and ensure their participation
2. Initiate discussions by example, solutions, case studies and alternate solutions.
3. Ensure that the development is on the right direction and pace, and finish the process in a time-bound manner.

6 E-Governance Website Implementation for Panniyankara

Implementation of public realm website was done as a part of the academic exercise mentioned earlier. The web was implemented by a group of 14 students [16], following the procedure prescribed by the manual [12]. Though the web was publically available, it was not expected to receive a large public participation as it is an academic exercise. Hence, the students were simulating the interactions between different kind of participants.

6.1 Website Structure and Organization

The Public realm web is implemented with a simple structure. The Home page gives the authentication and login facility, and it links to municipal corporation web for the payment gateways, e-tendering work, and recent orders and so on. The menu structure on the top links the user to the design processes, discussion area, development and maintenance. Regulation plan has active links to each polygon and these polygon pages have links to their corresponding standards. The structure is given in Fig. 2.

The E-Governance web on its home page gives a brief about the place under consideration. Figure 3 gives the home page of E-Governance website for Panniyankara. Regular users can quickly navigate to the required section through the menu system as well as buttons on the main form, while the home page introduces the whole process to the first time visitor. The location is marked on a dynamic map, which helps the first time users to identify and locate the study area and its boundaries.

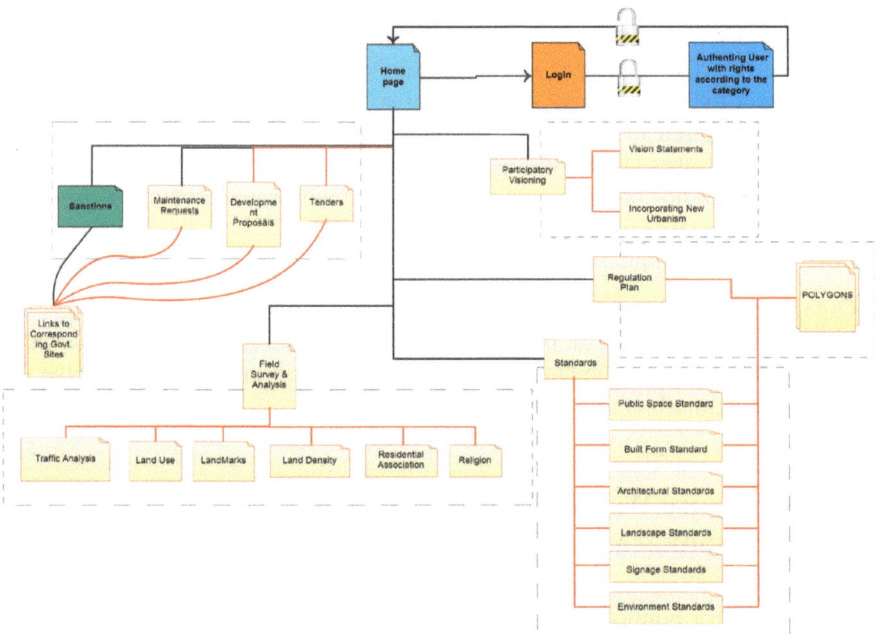

Fig. 2 Public realm web site map

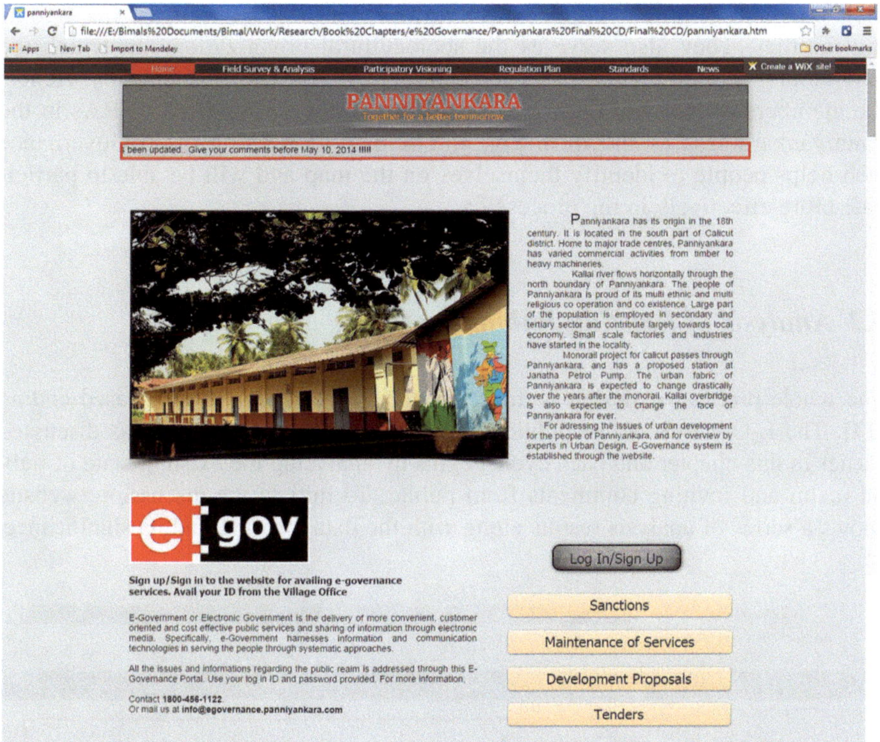

Fig. 3 Home page

The website allows anyone to interact with the site, but it gives special attention to the stakeholders of the study area and invited expert members. To avail these privileges, one need to login with their credentials. On this, with the details given at the sign up, the website grants appropriate privileges to the visitor.

The lowest constitutional/administrative division is a municipal ward in India. One ward consists of close to 1,500 or more houses, and each ward has a democratically elected member to the municipal council. As per panchayati raj, people were supposed to put forward the proposals in ward conventions at designated intervals or through the ward member to the municipality any time. However, for the purpose of public realm development, this division of ward is not practical due to two reasons. Primarily, the wards were divided on the basis of population, hence many cases, single public realm lies in two wards. Secondly, the size of a ward is so big that the citizens may not know each other or all ward geography in totality. Due to such practical considerations, the existing residential associations (RA) were considered as the best grouping unit. RAs are voluntarily formed by a group of contiguous houses in order to address their common requirements and resolve minor issues between the neighbouring families. They also serve as the sociocultural organizations, in which the neighbours meet on regular basis and conduct cultural festivals annually. Hence, the members of RAs are known to each other closely. There are 18 RAs in the Panniyankara area as shown in Fig. 4. The map of RAs on the E-Governance web helps people to identify themselves on the map and will be able to participate more effectively in the process.

6.2 Analysis of Existing Public Realm

The whole process of design is done through zonal plan and form-based coding [17]. The E-Governance for public realm design is a cyclic process as discussed earlier in this chapter and each cycle begins by analysing the existing state of public realm and inviting comments from public. To initiate the discussion, website shows a series of analysis results along with the data, process and its significance,

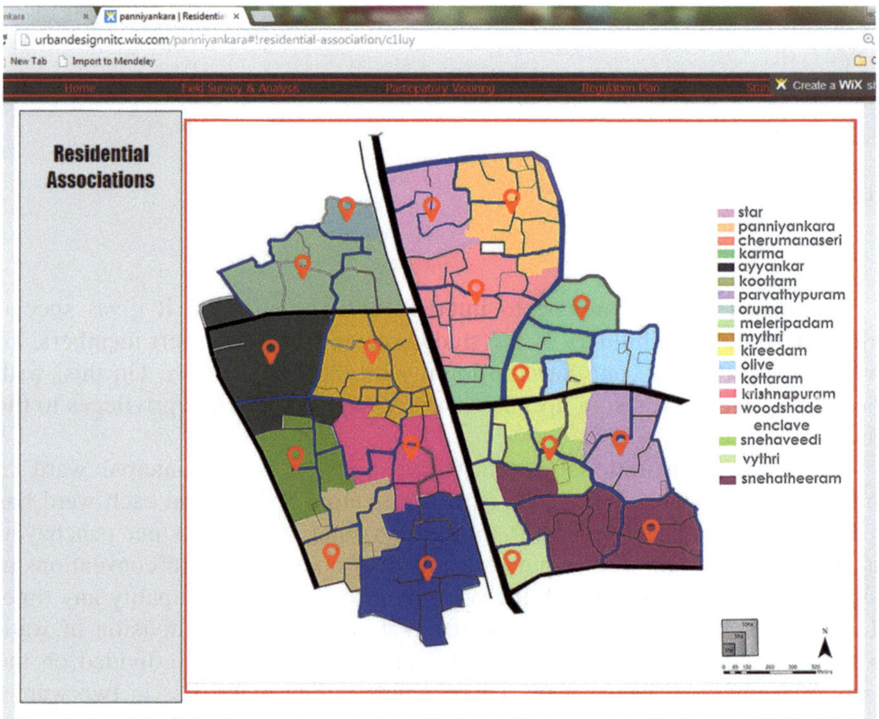

Fig. 4 Residential association map

over the public realm [18]. An example case is illustrated here in the form of traffic analysis for the study area in Fig. 5. The analysis represents different roads classified on the basis of the traffic it handles at various times. The bar graph gives a comparison among important roads of the area about the traffic load during different periods during the day. This is used to identify important junctions or nodes,

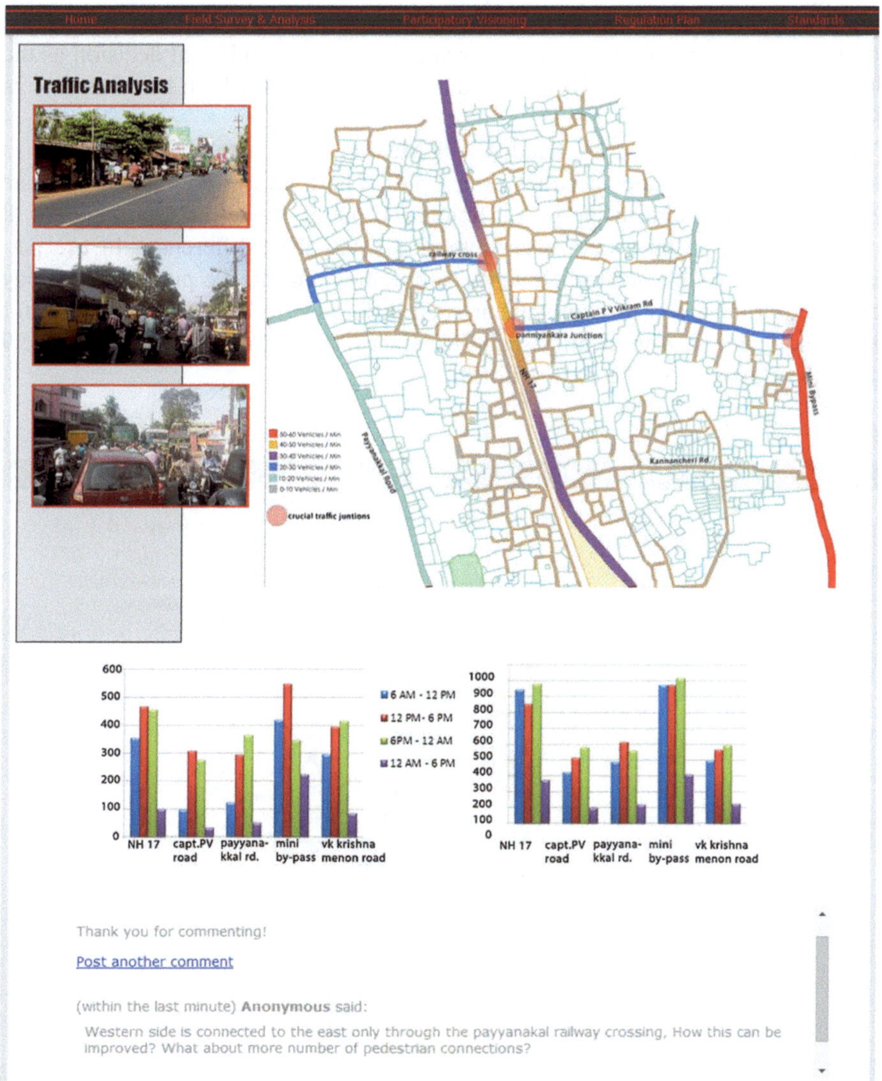

Fig. 5 Traffic analysis

which will later become central points for public realms within the study area. The public comments on the important areas were invited through the comments box, and any type of participants were invited to comment, and the comments were used to reassign the significance of roads and junctions.

The analysis continued to other aspects of public realm such as land use, building use map and land value, which are shown in Figs. 6, 7 and 8. Such information was used further in other analysis to assess the possibilities of modification of existing structures to make space for public realm. A poll was chosen as a method to collect opinion about the conversion of Capt. PV road to one way, as a solution for road congestion due to narrow road. This was a significant decision as the alternate option will be to widen the road by taking over the high land value plots along the road.

The place is occupied by Hindus and Muslims, and naturally they live in a clustered manner around mosques or temples. There are no communal or caste issues in Panniyankara in the past. Though the public space may look similar at both places on the first look, each community has their differences in use pattern of land for public realm. However subtle they appear, the public realm they developed around was to complement such difference in customs. Proposals for the public realm need to identify such differences and address them [19]. Figure 9 shows the clustering of the communities.

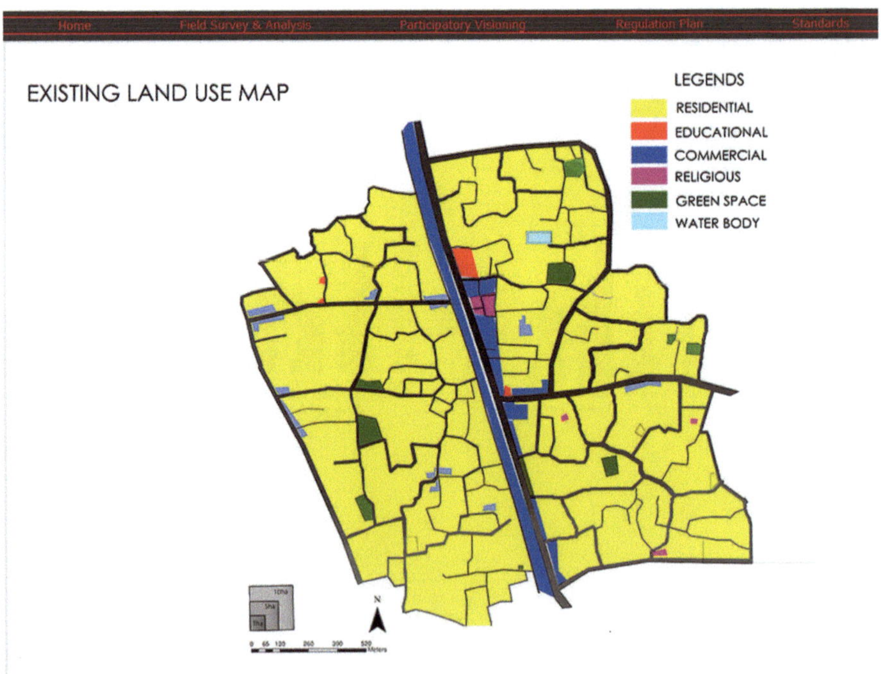

Fig. 6 Land use map

Fig. 7 Building use map

6.3 Identification of Public Realm

Public realms were first, identified with the help of the above information and a brief of the spaces were prepared. Public realms were also ranked from major to minor and potential for new public realms were also identified. The proposed monorail station will become the most important node of activity and it has all the possibility to be a major public realm in future. Figure 10 shows the identified public realms and its ranking. Brief of all these public realms were prepared in the website which represent the important issues, potentials and probable solutions derived out of a detailed SWOT analysis in a crisp pictorial form. Some examples of the brief are given in Fig. 11.

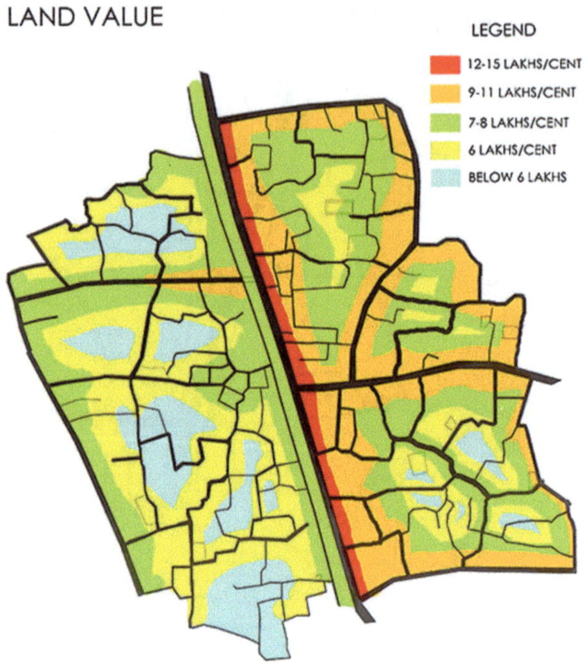

Fig. 8 Land value map

6.4 *Transects*

Transect identification conceptualizes the total Urban Form of a city and classify it into differing immersive urban environment. For example core of a city is characterized by high proportion of man-made high rise built up areas and very few areas of natural environment, while in periphery of the city, natural environment predominates in contrast to build up environment which are generally highly scattered and sparse single floor dwellings. In between core and periphery, there is distinct urban form of differing characteristics with proportionate mix of built-up and natural environment. Transect identifies these urban forms using google maps and site visits and designates it to T1, T2, …, Tn. There can be immersive environment within say T4 showing T3 characteristics or T2. We call them as T3/T4 or T2/T4. These areas are also studied for their built-up and natural environment. Any further interventions in public realm design is made harmonious to that urban environment to preserve the diversity and natural form. This is achieved by using differing built forms and natural landscapes appropriate to Transect. In summary, Transects were identified on the basis of built-up and natural environment characteristics of the land. Major transects are on the basis of taking city at macro level and minor transects were identified at a micro level in study area. This approach is adopted to capture the reality more precisely. This division helps all the development proposals to take care of the existing ecological context. Figure 12 gives transects in the study area.

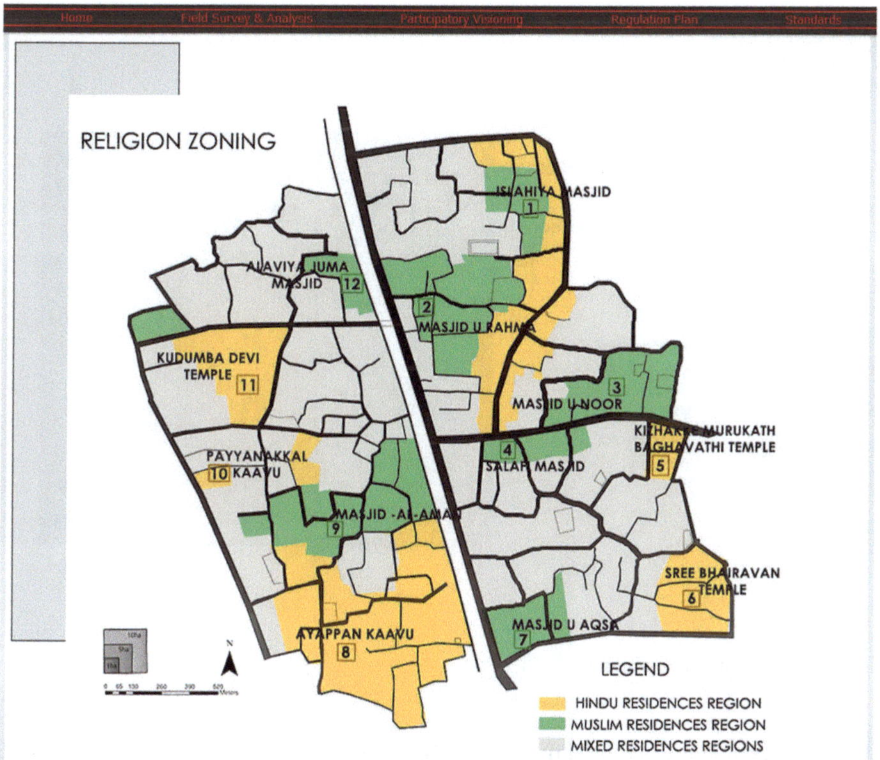

Fig. 9 Existing religious clustering

6.5 Participatory Vision and New Urbanism

To ensure development process of the whole public realm and its components to be continuous and unidirectional, a vision is required. This vision unifies the whole development and serves as direction for every detailing of its components. It will be a difficult task to form such a comprehensive vision directly from the community. Hence, the relevant experts develop a sample vision, which gives a base for individuals to comment on, and later website evaluates the comments and modifies the vision accordingly. Three-dimensional vision diagrams were prepared by SketchUp software, which is again a freeware. Figures 13 and 14 shows the initial concept and the vision. This preliminary vision was later elaborated with ten principles of new urbanism incorporated in this vision. These principles are given below.

1. Walk-ability
2. Connectivity
3. Mixed Use and Diversity
4. Mixed Housing

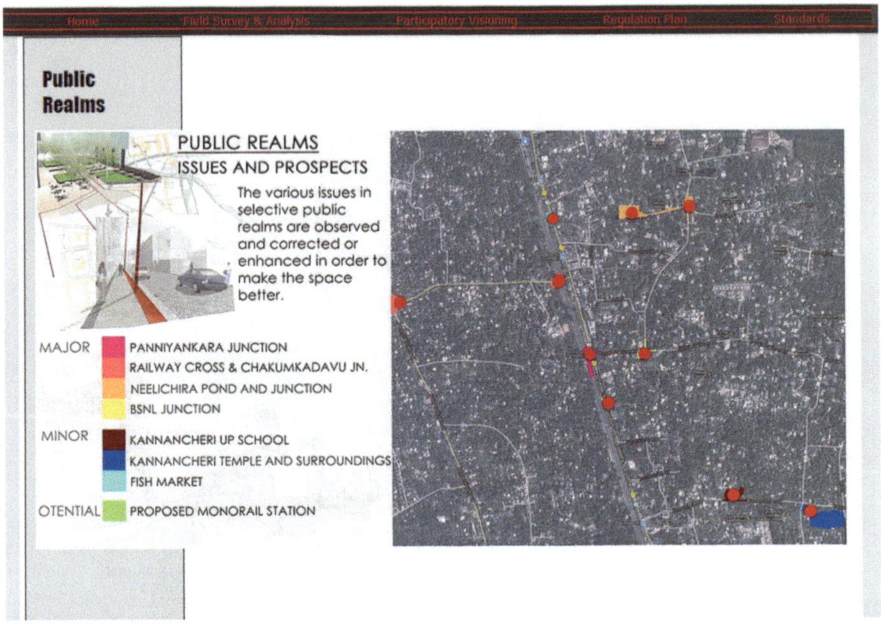

Fig. 10 Identified public realms

5. Quality Vernacular Architecture and Urban Design
6. Traditional Neighbourhood Structure
7. Increased Density
8. Smart Transportation
9. Sustainability and
10. Quality of Life

This check was essential to ensure the integration of the sub-components and also to test the scalability of the vision to various sub-components. Figures 15, 16 and 17 shows the detailed vision as per new urbanism principles. The final vision was evolved after lot of deliberations and public comments were frozen on a previously agreed day. It was generally accepted that the vision will not be modified beyond this date. The vision was developed after giving due consideration to characteristics of transects and existing physical, ecological and socio-economic features. The vision is further translated into various standards pertaining to urban design components specified by form-based coding.

6.6 Regulation Plan

Regulation plan defines the area that is coming under the proposed development regulations. Not all areas are under regulation but some around public realm. This is because; the public realm deals with entities coming within the public

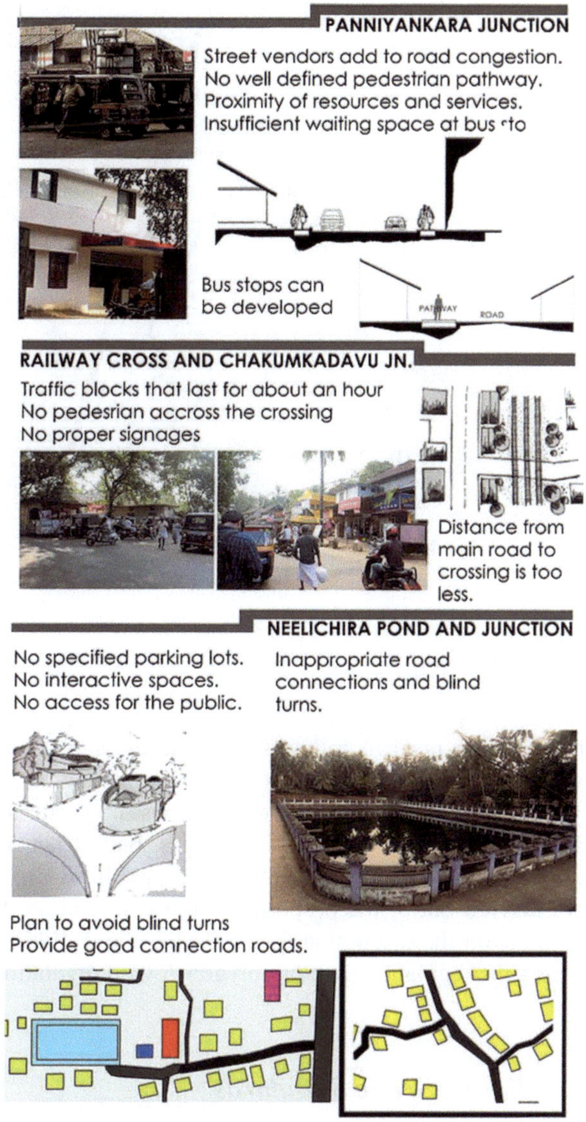

Fig. 11 SWOT analysis of public realm

experience only. This may include facades of buildings that abut public spaces, parks or plazas, roads and play grounds, but may not include remaining portion of the building or site. Entities that are not affecting public realm need not be regulated. Hence, the regulation plan defines a polygon around each of the identified public realm, which demarcates the region under regulations. These polygons are further divided by the micro and macro transect polygons (say Tn/Tm), which demarcates the balance between built and natural environment. So the stakeholder needs to identify the regulation plan polygon and transect that his plot belongs to as shown in Fig. 18. Any construction or development process happening within

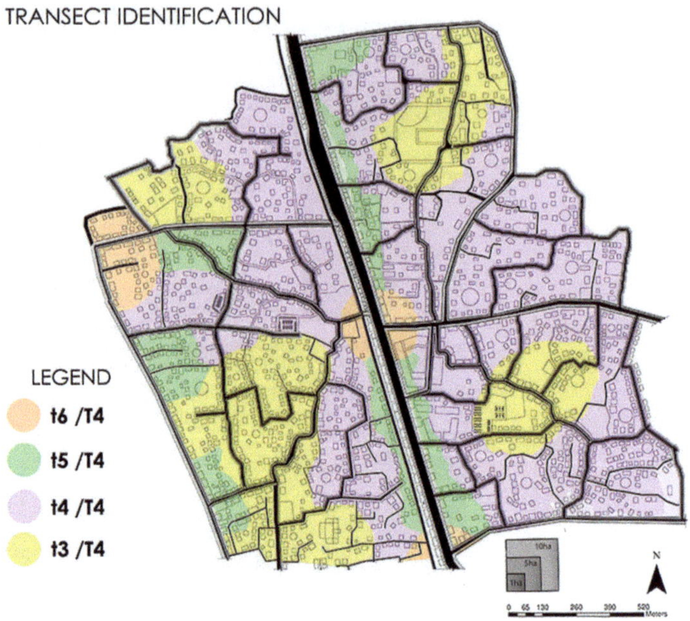

Fig. 12 Transects

the polygon need to follow the standards, and the stakeholders can refer to the standards by clicking on the respective polygons.

Regulation plan also defines positions of certain important lines unequivocally. These include Built to lines, tree lines, pedestrian or cycle lanes. Such lines are introduced to control the details of the public realm at a micro scale. These lines are derived out of the proposed vision plan and are finalized after detailed deliberations on the discussion forum during the evolution of the standards. Figure 19 shows the regulation plan of polygon A with regulation lines.

6.7 Public Space Standards

Public space standards regulate the developments on the public spaces coming under the regulation plan polygon. Public space here may include roads, lanes, parks, tot lots space and in general spaces largely used by public. Public space also includes civic spaces such as football grounds, tennis court, basket ball ground and others. Expected users of these standards will be government agencies, and construction contractors and strictly governed by its standards. During the development stage of the standards, all kind of members are expected to comment on the standard.

Public space standard of Polygon A in regulation plan is given in Fig. 20.

Fig. 13 Concept and vision objectives

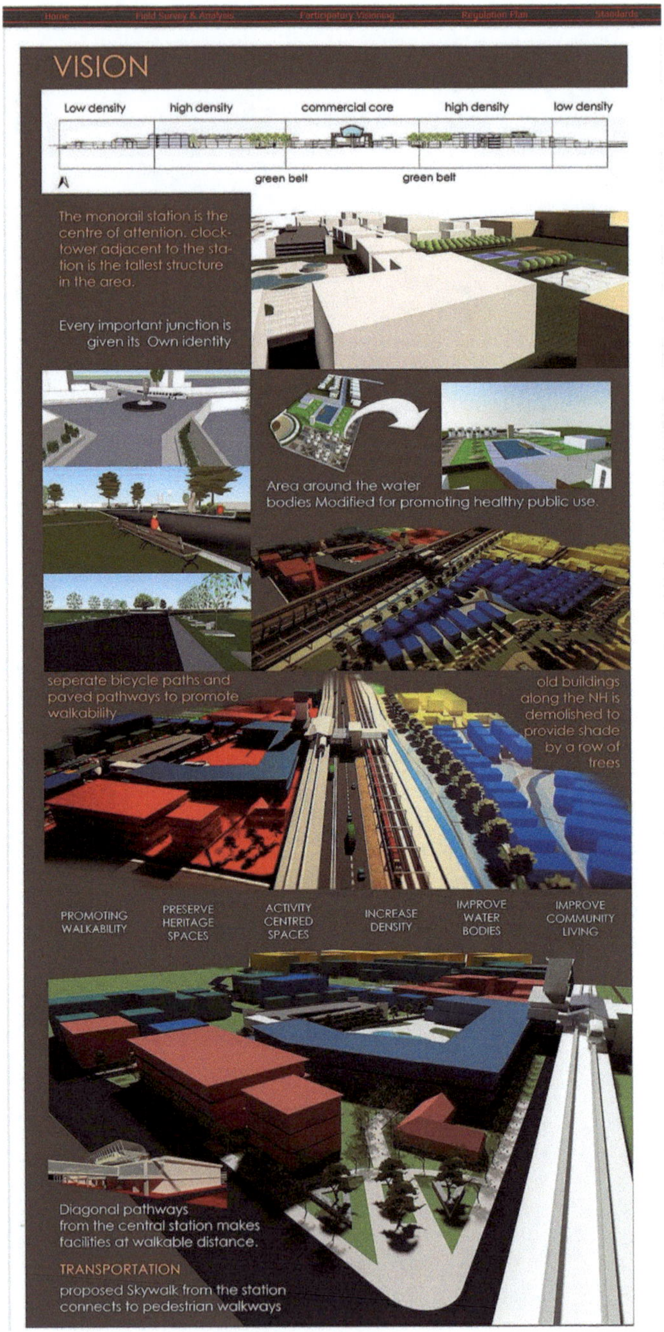

Fig. 14 Vision for public realms

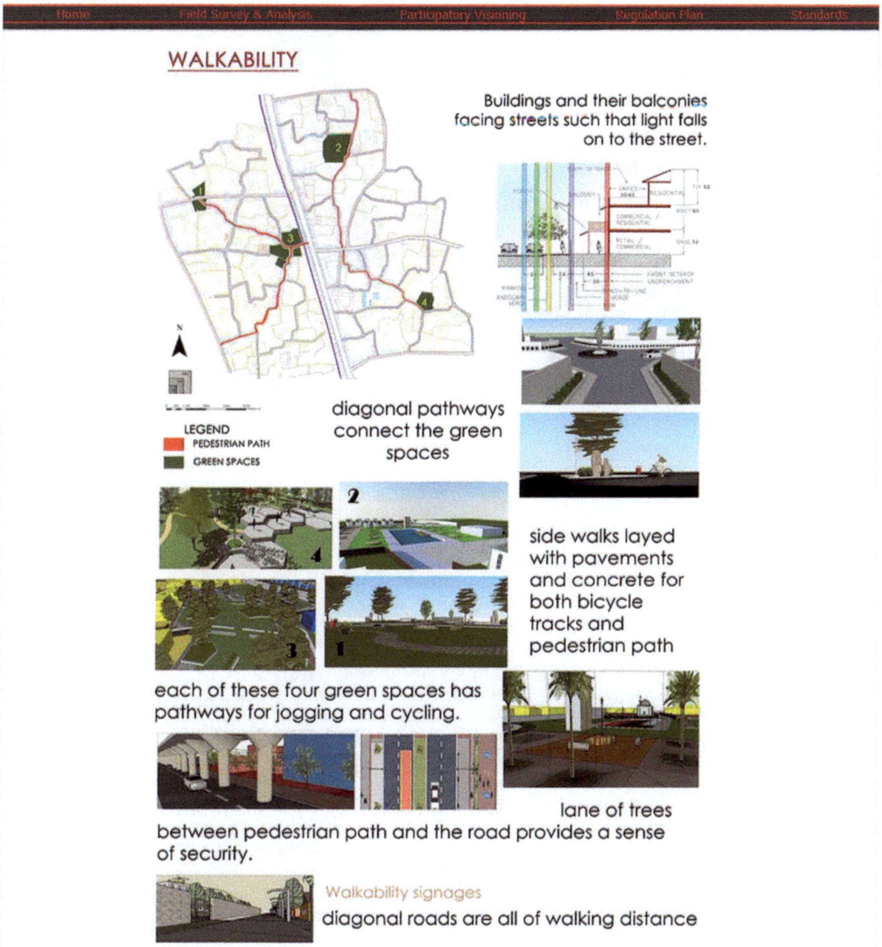

Fig. 15 Detailed vision as per new urbanism principles 1

6.8 Built Form Standards

The built form is a major entity in experiencing public space. Built form denotes mainly the buildings surrounding the public space and this gives the character, and identity of the public space. It also defines the limits of the public realm. Hence the surrounding massing needs to be regulated. It is important that not only the exterior surface but also the 3D mass need to be regulated. Some of the roof forms require the massing to be in a given proportion. Figure 21 shows the built form standard of Polygon A.

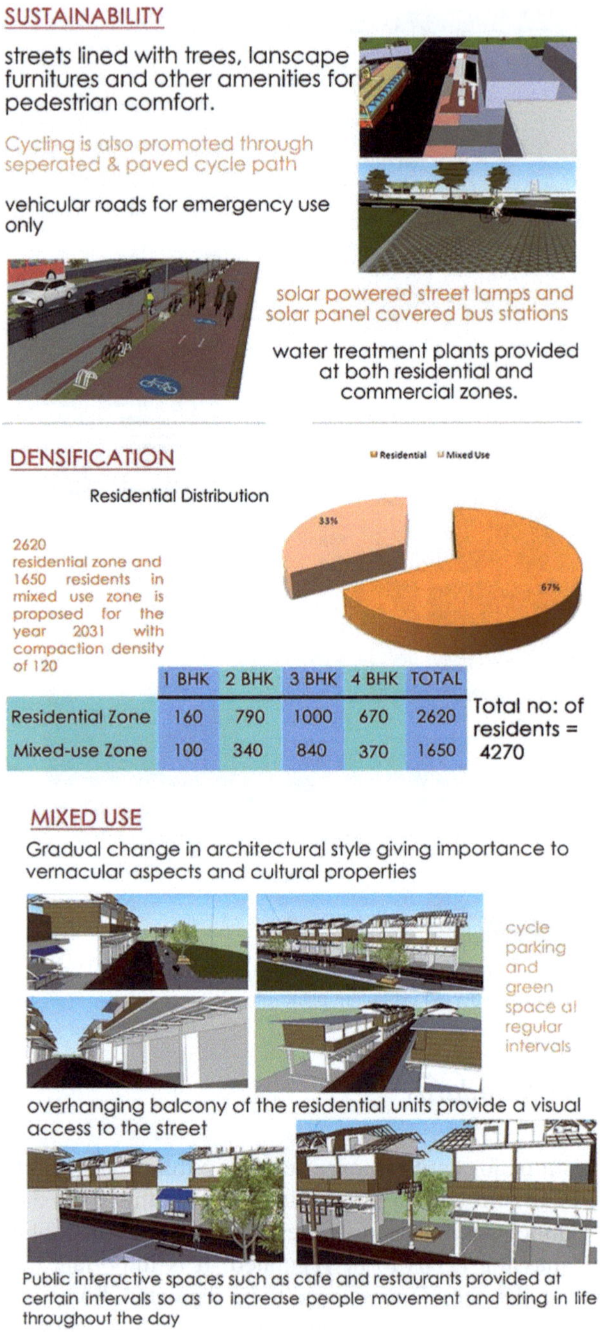

Fig. 16 Detailed vision as per new urbanism principles 2

Fig. 17 Detailed vision as per new urbanism principles 3

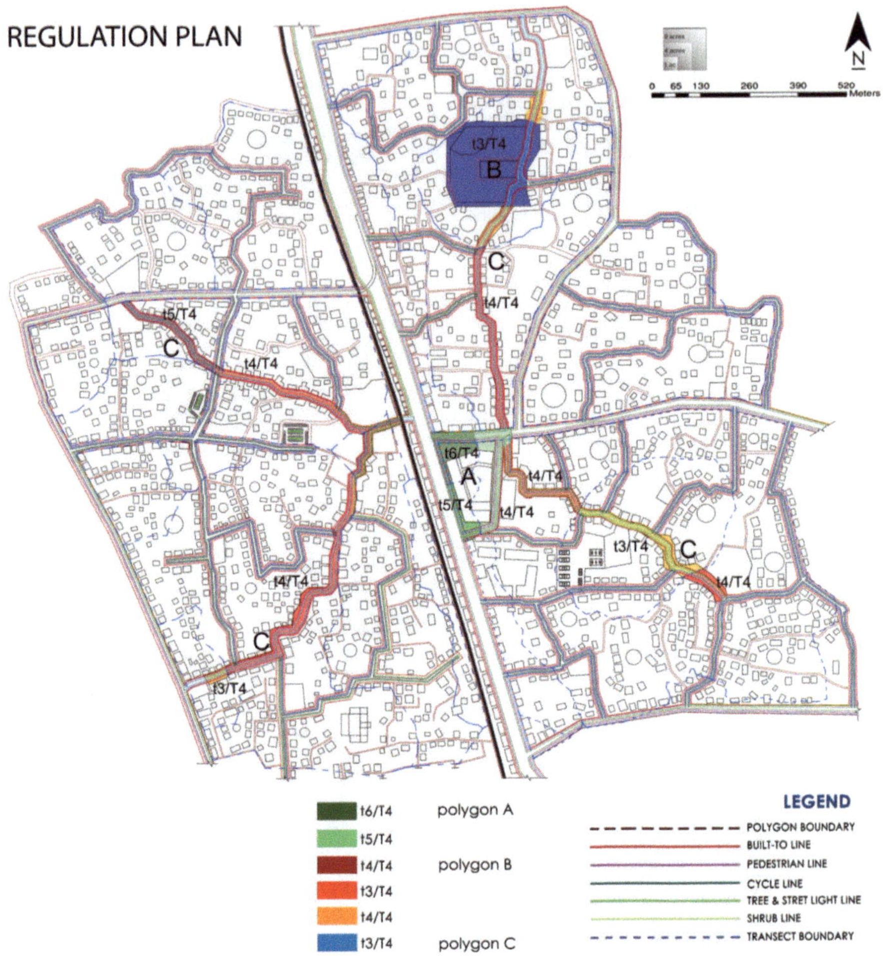

Fig. 18 Regulation plan

6.9 Architectural Standards

Architectural detailing of the built-up mass is regulated by architectural standard. The inspiration comes from locally available architectural features in Panniyankara.This standard is derived out of the existing architectural features around the place and that is typical in the region. Identifying and choosing these detailing can be through public opinions and discussions. A technical input from an architect or artist will be highly beneficial here to see that the selected features are in harmony with each other. Business entities can contribute here by commenting on the availability and practicality of making or installing such features in the given situation. Some of the custom designed features can be manufactured by the local industries or craftsman. This will benefit them by ensuring a decent

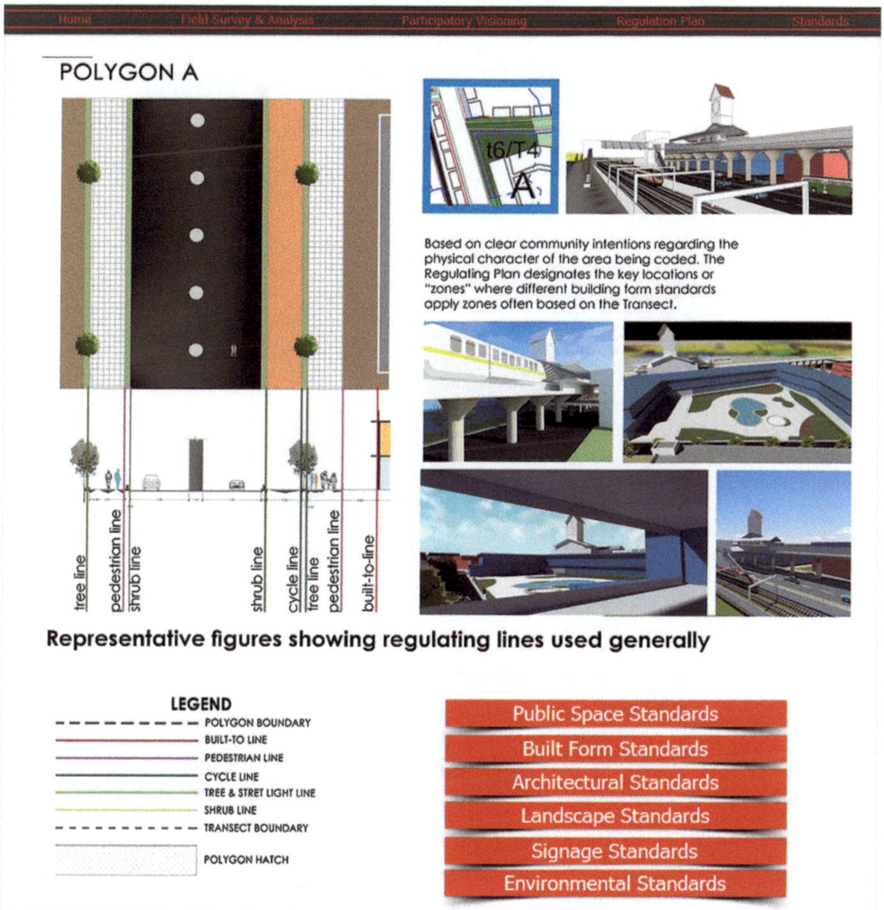

Fig. 19 Polygon a regulation plan

business volume, for their investment, whereas the building owner will benefit by getting an element at a cheaper rate compared to the case where they need to make it cast-in-situ. The community will benefit by having a common design element throughout the locality. Figure 22 shows a set of architectural standards developed for Panniyankara.

6.10 Landscape Standards

The vegetation and hard landscape that can be used in the public realm is suggested through the landscape standards. Pavement design for foot path and cycle track is also specified here. The citizens and local nursery persons will be able to

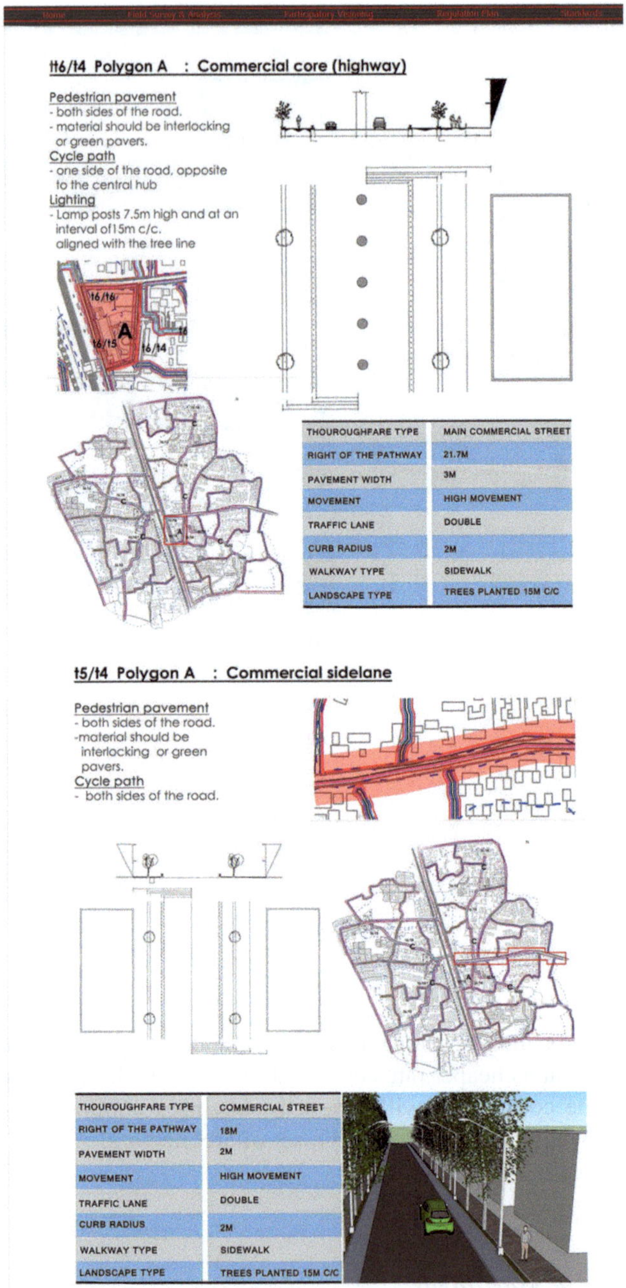

Fig. 20 Public space standard

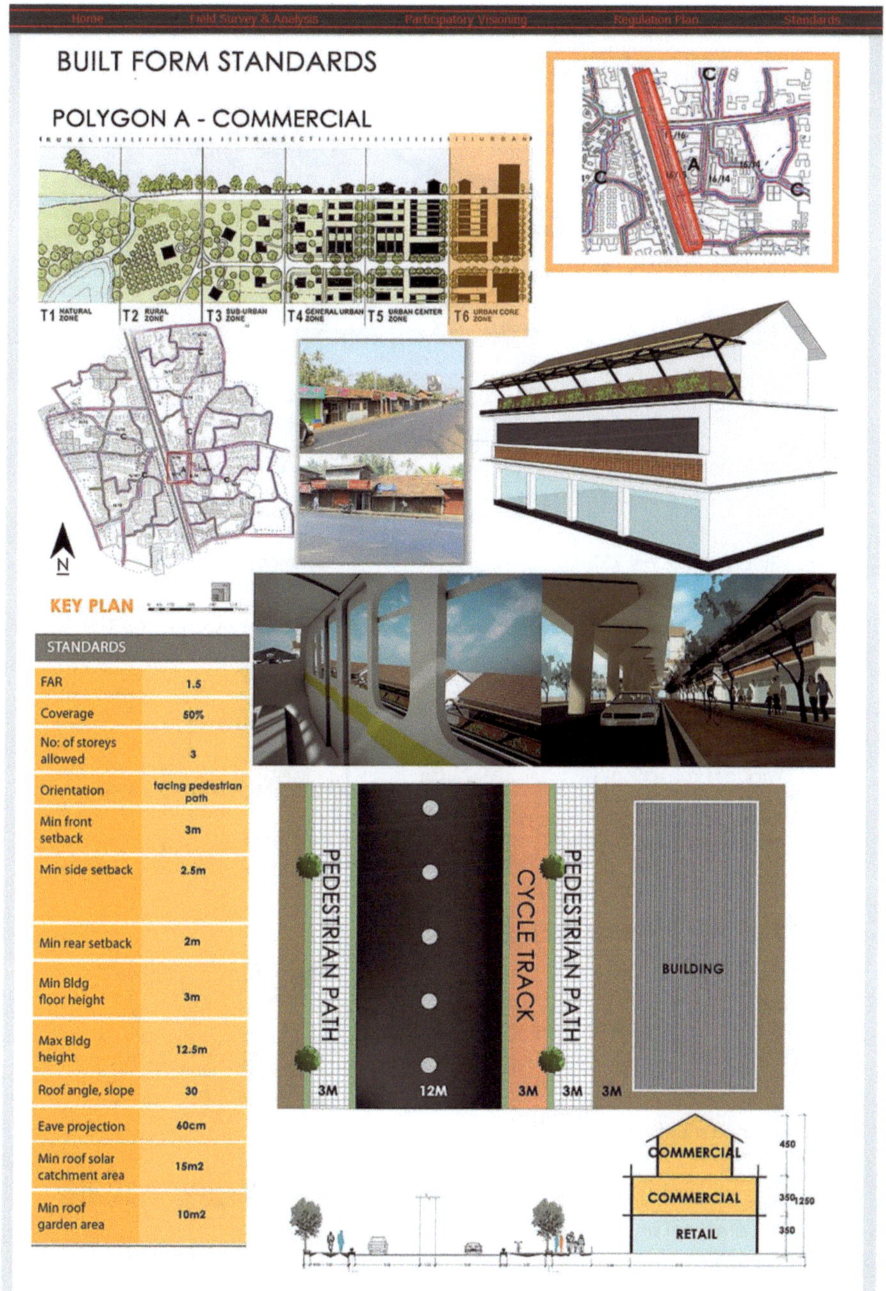

Fig. 21 Built form standard

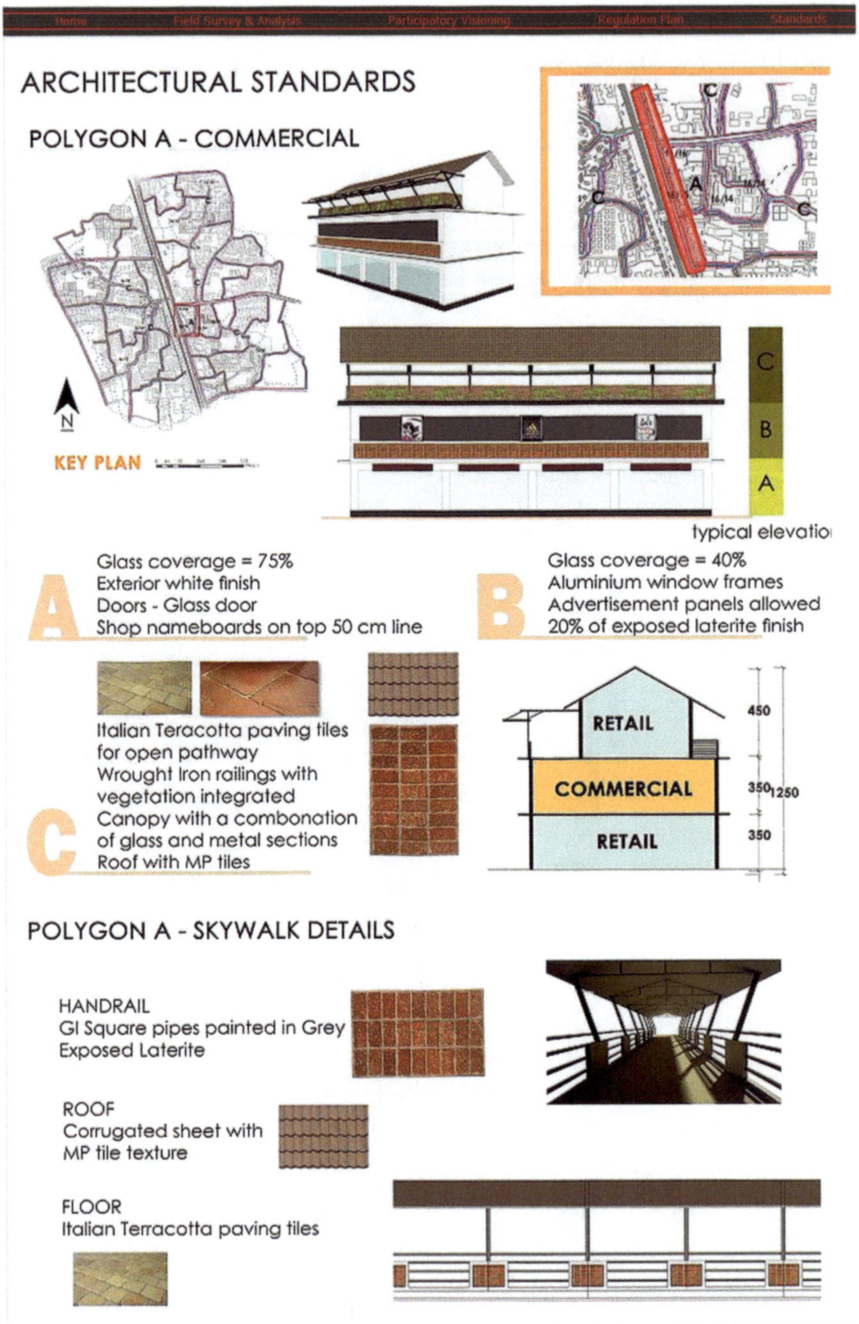

Fig. 22 Architectural standards

Fig. 23 Landscape standards

suggest trees and plants that match the public realm and different transects requiring different plant materials. Such discussions will be initiated in the website, through a set of suggestive examples based on their study. Alternate options can be collected through the E-Governance site, and these options are further evaluated on the basis of their availability, feasibility, intended purpose, and cost of maintenance, etc. Local nursery can make such trees available to the public at a reasonable cost, and other government/business entities can take up the task of maintaining them Fig. 23.

6.11 Signage Standards

Indian public spaces are ill-famed for the uncontrolled use of billboards, each of them trying to catch attention than the remaining. This section of standards attempt to regulate the signage and billboards used in the public realm. Signage includes direction signs, information signs, Identification signs, short term advertisements and large billboards. The placement and location of such signs are initially by initial team entrusted with the task of providing first iteration, which is further discussed and finalized by the public through the website. Signage standards is given in Fig. 24.

Fig. 24 Signage standards

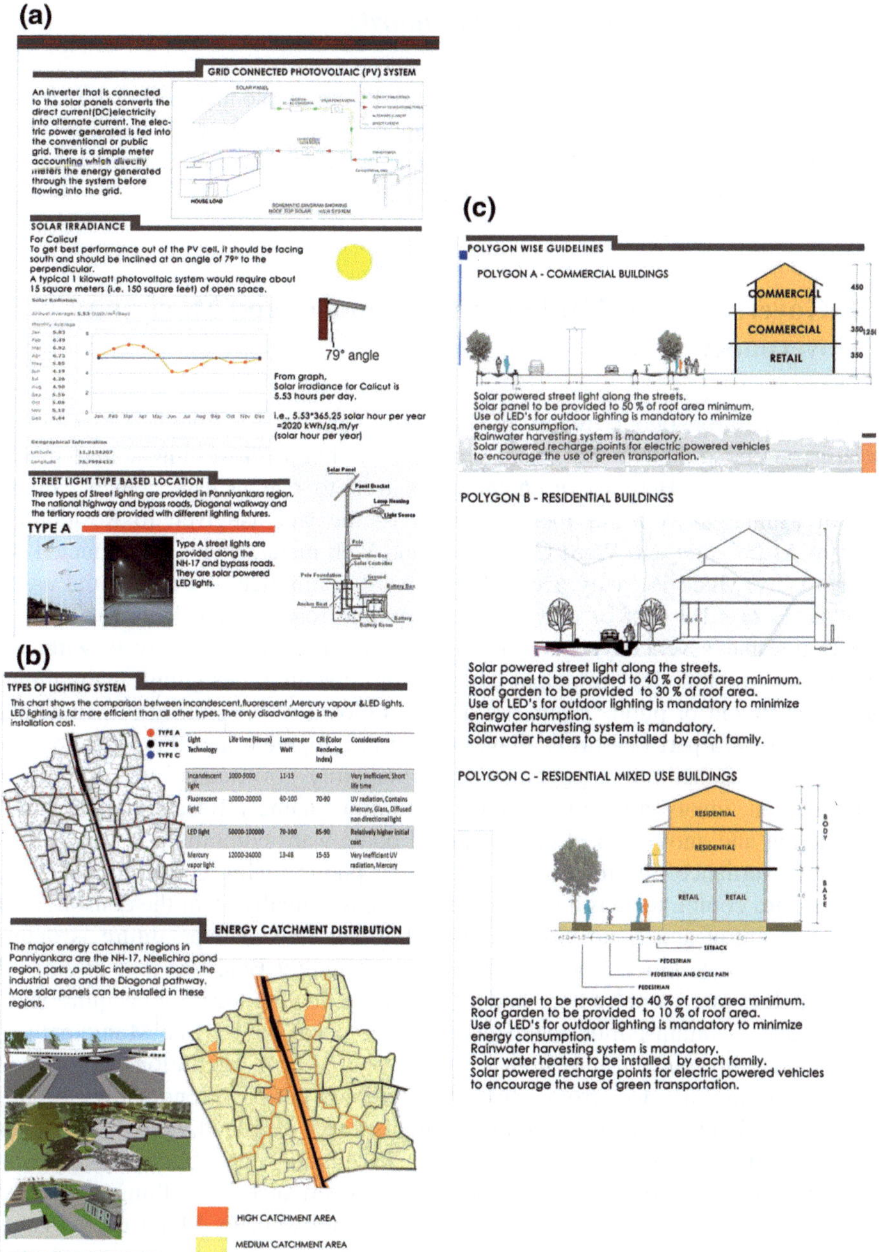

Light Technology	Life time (Hours)	Lumens per Watt	CRI (Color Rendering Index)	Considerations
Incandescent light	1000-5000	11-15	40	Very Inefficient, Short life time
Fluorescent light	10000-20000	60-100	70-90	UV radiation, Contains Mercury, Glass, Diffused non directional light
LED light	50000-100000	70-100	85-90	Relatively higher initial cost
Mercury vapor light	12000-24000	13-48	15-55	Very inefficient UV radiation, Mercury

Fig. 25 Environmental standards

6.12 Environmental Resource Standards

Environmental standards give the details of different elements like the drain, ground water recharging system, sewage management, and waste disposal system. It also provides suggestive solutions for solar power generation, and biogas system. Environment standard is as given in Fig. 25a–c.

7 E-Governance for Minimum Government and Maximum Governance

E-Governance for public realm of Panniyanakara in Kozhikode Corporation is built on the principle that there shall be minimum Government and maximum Governance. Here Government is reduced to one website which can be part of Municipal Website. Field updating of the web is given to Residential Welfare Association and Ward Committee and Web management is conducted by Municipality which is easily accessible to ward committee and residential associations. Here all aspect of Governance of public realm is conducted in twenty four hours and seven days a week. Ownership of public realm rests with the Kozhikode (Kozhikode) Municipal Corporation. Planning, designing, maintaining and management of public realm are a constitutional responsibility of Municipal Corporation as a part of Master Planning and development and this website support it. Since Constitution recommends participatory governance, it is important to create first, involvement of two local institutions namely Residential (welfare) Association and another local constitutional body, ward committee in this participatory Governance. In creating and managing public realm, constitutional body such as ward committee has a right to obtain funds annually from the consolidated funds of Central and State Government. However line of expenditure is determined at the State Government level by State Planning Board as part of Annual Planning. It is rarely one can see budget allocation flowing locally to acquire land, develop and maintain public realm. Autonomous local bodies like Municipalities can include this budget line item on their own. Local institutions mentioned above can obtain finances from citizen and business but past trend shows the full potential of obtaining such source of funding is never exploited. One of the major reasons for this is lack of e-democracy in Kozhikode that effectively supports E-Governance. To activate e-democracy for E-Governance, everything connected with Governance for Public Realm is conducted in a website. Major users of this websites are citizen, business and Government. Citizen needs not necessarily the beneficiary population of ward and residential welfare association. Network of persons living in this area can link with people outside this study area for financial and technical inputs.

7.1 Major E-Governance Functions Performed by Public Realm Website

Since this website is meant to comprehensively govern the public realm, it shall have several functions enumerated as below.

1. Directory function: It shall provide Name, mobile and phone number and email id of all Government Officers who have jurisdiction over this area.
2. Announcement: From time to time all important functions and events locally or in neighbourhood are announced in the website which may have some bearing on the inhabitant of this area.
3. Conduct of various types of surveys online related to public realm: It can publish on line comments and discussion and later can be followed up by a public hearing by Municipal Authorities as part of ward committee meetings.
4. Development of Proposals for public realm: As indicated in the earlier part of this chapter a highly structured approach is developed in the website with data presentation, analysis, SWOT analysis, Strategic Planning, Vision Design, Zonal Plan and requirements, Form-based codes and so on. One or two design examples are given and alternatives designs are solicited from the website. These are conducted based on an online discussion of major issues from interested parties in the website.
5. Online Tenders for various works related to public realm: E-Tendering is mandatory in Government of Kerala so the website need to create proper linkages to such websites.
6. Subscription, Tax and Bill Payments: All bill and Tax payments are linked to selected banks for the citizen and business to pay fees, tax and subscriptions. Bill payments are for Water, Building tax, Land tax, Electricity, Telephone and broadband connections.
7. Maintenance Services.
8. Sanctions and Occupancy certificates for new Construction and Renovation based on item 4 above.

The website performs the function of e-participation, e-service and e-administration. For every citizen, Government then becomes highly accessible entity through his cell phone or computer. This website also provides for Government to Government (G2G), Government to Citizen (G2C), Government to Business (G2B) and vice versa interaction and decision-making . This is because Business can be a funding and maintaining agency using their resources to finance and manage the assets created for public realm.

7.2 E-Proposals and Selection

Proposals are initially developed by in house Municipal planner, urban designer, engineer, landscape architects or architect assuming a function of primer in the website to elicit alternate proposals. Social network within the study area can use their contacts to get design proposals from anywhere to enrich the initial proposals. Proposals can come from Government, Business and Citizen. All of them need not necessarily be local entities. Municipality, Residential Welfare Association and Ward Committee jointly appoint a review committee to review various proposals and finally select the best suited proposal for implementation. The process involved in the website is summarized below. Proposals generally can be from public, Government, or Private Sector Agencies.

1. Sign in for the website by supplying selected inputs on identity.
2. From the pull down menu select the appropriate section of public realm development you are interested in.
3. Select the type of proposal such as Business Proposal, Citizen Proposal and Government Proposal.
4. Proposals are studied by citizen and public and comments can be made on the web.
5. Appointed Committee review all inputs and select the best.
6. Criteria and reason for selection is noted on the website.

Sanctioning procedure is hierarchically structured in the website. Proposals can be for New Projects which can further be structured as Residential, Commercial, Industrial or Mixed Use. In order to conform to various standards developed, they have to follow the regulation plan and indicate the polygon numbers and select. Similarly, item renovations are selected from the website. Renovation can be painting and plastering, structural modifications, changing roofing, changing openings, and landscaping. They then have to upload the necessary documents and also checklist conforming to the form-based codes and deviation if any.

7.3 Proposal Submission and Approval

Proposal submission is effected through website. By studying the website, proposal submitted shall ensure conformity to zonal plans as well as form-based codes. There is provision in the website for filling in on line forms such as Table 2 where a conformity with the form-based code is ensured. Further detailed drawings need to be uploaded. All these uploaded information shall be verified by competent officers or committee. In case of simple residential buildings Municipal Town Planner shall review but if the proposal is complex and of a higher value it may be scrutinized by Urban Arts Commission or specifically appointed standing committee and comments may be provided online. Finally, based on comments, Town Planners approve and online approvals are provided.

7.4 E-Maintenance Services for Public Realm

E-Maintenance involves two aspects with computer date stamps applied. In the first aspect, demanded-maintenance is recorded with time stamp. Since residential welfare association has smaller geographic area than ward, this area is used in

Table 2 Checklist for new construction permit

Polygon A—Panniyankara public realm				
Checklist for new construction permit				
Standards	Component	Compliance (yes/no)	Reason for noncompliance	Proposed alternative
Regulation plan	Tree line			
	Building line			
	Parking line			
Building form standards	Front setback (min-3 m)			
	Side setback (min-2.5 m)			
	Rear setback (2 m)			
	Building coverage (max 50 %)			
	Building height (max 12.5 m)			
	Number of floors (3 m)			
Landscape standards	Distance between trees			
	Tree species			
	Tree location			
	Planting strip			
	Paving material			
	Paving width			
Architectural standards	Wall material			
	Wall type			
	Wall openings			
	Wall finishes			
	Door type—glass door			
	Door size			
	Window type—aluminium			
	Window size			
	Sill level			
	Roof structure			
	Roof material			
	Pitch			
	Overhang			
	Compound wall type			
	Compound wall height			
Signage standards	Sign board size			
	Sign board material			
	Font size			
	Font style			
	Location			

(continued)

Table 2 (continued)

Standards	Component	Compliance (yes/no)	Reason for noncompliance	Proposed alternative
Environmental standards	Rain water harvesting			
	Solar technology			
	Water management			
	Sewage management			
	Biogas plant			

the website to record the requirement of maintenance. Maintenance requirement is further classified in terms of Electrical, drainage, road, waste collection, water, security and other public issues. A reporting form and map is used. Map will help to locate the site with GPS. A form is designed to record the details of demanded maintenance. The process involved in e-maintenance is enumerated below.

1. Using website and date stamp, user report the maintenance issue as classified above.
2. The concerned authorities (say water or electricity department) review the issue from the website and date stamp the review date.
3. Thereafter, repair work is completed by the concerned authorities and completion is reported in the website with date stamp.
4. The supervisor check every week various date stamps for maintenance and evaluate whether maintenance was done as stipulated in the norm for time allocation for service delivery. If not disciplinary action such as fine may be applied to the concerned officers and report it in the website for transparency [20].

Routine periodic maintenance are undertaken (for example cutting trees over electric lines, or cleaning of drains) and reported on the website. The same four step procedure adopted above is used and recorded in the website.

8 Evaluating E-Governance of Public Realm

In this chapter, E-Governance for Public Realm is seen as an exercise in social, political and administrative authority to govern affairs of public realm of Panniyankara. It comprises the administrative mechanisms, legal processes and related institutions such as residential welfare association, ward committee and Municipal Corporation through which citizens, Government and Business groups articulate their interests, exercise their legal rights, meet their obligations and mediate their differences on all aspects of public realm in the study area. Good E-Governance for public realm is participatory, transparent and accountable. It is also effective and equitable, and it promotes the rule of law enshrined in the constitutional amendments 74 of India and various planning legislation and ordinance of State of Kerala. It ensures that political, social and spatial priorities are based on broad consensus in society and that the voices of the poorest and the most vulnerable are heard in decision-making over development of land resources centred on public realm.

For planning, designing, developing, maintaining and managing public realm, there shall be minimum government and maximum governance. The web portal designed for this purpose is the minimal Government accessible from anywhere in 24 h and 7 days a week and various process structured and indicated in this chapter are an attempt to maximize governance processes conforming to best practices in spatial planning, and legal processes of Kerala that is practiced in Master Planning, Zonal Panning (not practiced now but recommended) and Urban Design based on form-based codes. The web portal *de jure* shall eventually be part of Municipal Website but de facto, it is owned jointly by Residential Welfare Associations and Ward Committee who update it and uses it for participative planning and design, administrative sanctions, and linking with probable participants, such as funding agencies, government, non-government organization, and business community. This web is used for identifying and reporting local issues, intimating it to Government for maintenance annual planning and budgeting and as a payment and tendering portal.

There are several characteristics of good E-Governance embedded in this Public Realm E-Governance discussed in this chapter.

They are:

8.1 Better Participation

All, men, women and transgender irrespective of their caste and religion, inclusive of the physically challenged, should have a voice in decision-making for their public realm, directly using web. The aim of these voices is to reach general consensus in an orderly fashion without agitations and strikes prevalent in Kerala. Such broad participation is built on freedom of association and speech, as well as capabilities to participate constructively providing necessary inputs which are their democratic right enshrined in Indian Constitution. Participation often requires certain levels of skills in certain type of activities detailed in the website. Participation is a process whereby policy-making, prioritizing issues, accessibility to public goods and services and also allocating land resources is influenced by key stakeholders. It varies from one context to another, as well as one area to another and subject to different projects and visions. Participatory processes in developing various form-based coding standards discussed above promote information exchange and transparency in decision-making processes. This, in turn, will improve and, as a result, increase the overall governance and social efficiency of development activities.

Generally, public involvement includes three elements:

- Access to information by public of local area say Panniyankara;
- Public participation in decision-making processes involving public realm in Panniyankara;
- Public access to judicial and administrative redress, in case it is required.

Access to information can be passive or active. Passive access is where the public will get information upon request to government institutions based on Right to Information Act or otherwise. Active access is whereby the government is obliged to give and disseminate information. Access to justice involves the procedural rights of the public to information are respected and guaranteed. This is because for rights to be effective there should be a corresponding remedy and cannot be in the absence of full information.

The rationale for public involvement can be discussed from various dimensions. From a human rights dimension, people have the right to know, to be informed and participate in decisions that affect their life as well as seeking redress related to public realm. From a legal, ethical and moral dimension, citizens and government officials are obliged to ensure good governance. It has been argued that government processes are improved through public involvement. This is well provided in E-Public Realm Governance.

8.2 Rule of Law

Public realm is related to use of urban land for public purpose, hence laws, regulations such as Master Plan, zonal development codes and form-based development codes should be fair and enforced impartially. Public realm web is one of the effective ways of tackling Information needs that is critical for the leaders and their constituents to be informed of their problems as well as the solutions. Likewise, it is important to review previous institutional constraints in order to map the future with viable options. There is scope for discussing this in the website. A poor governance system serves private interests at the expense of the poor and they suffer in a multiplicity of ways. A correct diagnosis of poor governance is important in that it determines practical strategies that are sustainable and effective. The governance is to look at the disconnection between institutions within the broader governance environment including the scope of operation of the society in general. The availability of information is critical to good governance and web portal provides for it in all fine details as processed. Since this web portal is comprehensive, it avoids any disconnection of institutions. Access to information and the promotion of procedural rights provide an enabling framework where accountability and improved delivery could enhance institutional changes.

8.3 More Transparency

Transparency is built on the free flow of information on every aspect of public realm provided in the website. Processes, institutions and information are directly accessible to those concerned with them through web, and enough information is provided to understand and monitor them through this web portal. This provides for active participation by eliciting various inputs by those who are interested. It promotes openness of government action, decision-making processes, and consultative processes among public sector and all stakeholders. These processes are subject to scrutiny by other government institutions, civil society and external institutions by studying the website.

Lack of transparency, weaken accountability. Lack of responsiveness and inefficiency in administrative procedures also compromises on good governance. In a corrupt government, public resources such as public land for public realm involving valuable land in strategic location are diverted for other uses from meeting the needs of the poor and benefits do not reach the intended beneficiaries. This happens easily when there is no transparency. Human health and security is compromised by corruption. This is because corruption is both a cause and effect of bad governance. The poor are usually disproportionately affected by poor governance because education and legal and police services are inaccessible to them. Their income is usually eroded through payment of bribes.

Corruption can be classified into two broad categories, state capture and administrative corruption. State capture takes place when a framework of laws and rules has been distorted in Master Planning and Zonal Planning to benefit the rich compromising the needs of the poor. This form of corruption is characterized by enactment of laws, policies and regulations that are influenced through illegal and non-transparent ways, as well as serving private interests. Administrative corruption is about distortion of the implementation of these laws and policies. It refers to the intentional imposition of distortion in the prescribed implementation of existing laws, rules and regulations to provide advantage to either state or non-state actors as a result of illegal transfer or concentration of private gains to public officials. In Public Realm website there is no scope for these two types of corruption since all process are open to public scrutiny and can be questioned by anyone using the website.

8.4 High Responsiveness

In E-Governance for public realm, institutions and processes try to serve all stakeholders within a reasonable timeframe. With Pubic Realm website, it is fully oriented to receive responses from Government, Business, Citizen anywhere in the world establishing C2G, G2G, C2B, C2C and B2B relationship for public realm causes. Time stamp based response system designed in the web helps in this regards.

8.5 General Consensus Orientation

Good governance mediates differing interests to reach a broad consensus on what is in the best interest of the group and, where possible, on policies and procedures related to all aspect of public realm. Underlying this characteristic is the art of consensus and consensus decision-making using a website. Consensus has two common meanings. One is a general agreement among the members of a given group or community. The other is the practice of getting such agreements. The process of achieving consensus involves serious consideration of every group members' or stakeholders' considered opinion. Consensus usually involves collaboration, rather than compromise. Participative public realm website achieves this. Instead of one opinion being adopted by a plurality, stakeholders are brought together until a convergent decision is developed. Appointed experts by community arbitrate in this matter only if it is required. Consensus decision-making is a decision process that not only seeks the agreement of most participants, but also to resolve or mitigate the objections of the minority to achieve the most agreeable decision. Consensus decision-making is intended to de-emphasize the role of factions or parties and promote the expression of individual voices. This method also increases the likelihood of unforeseen or creative solutions by juxtaposing dissimilar ideas. Consensus decision-making involves identifying and addressing concerns, generating new alternatives, combining elements of multiple alternatives and checking that people understand a proposal or an argument. All these activities are performed in a website open to all to understand and interact. This empowers minorities, those with objections that are hard to state quickly, and those who are less skilled in debate. Therefore, consensus decision-making in a website can be seen as a form of grassroots e-democracy. The purpose of E-Public Realm Governance is to arrive at consensus.

8.6 Equity

All men and women irrespective of their caste, caste and or religion have opportunities to improve or maintain their well-being by creating and maintaining all public realms using E-Governance.

8.7 Effectiveness and Efficiency

E-Governance for public realm involves processes and institutions that produce results that meet needs for public realm while making the best use of land resources. In Kozhikode where zonal plans or urban design at micro level for Common man (*aam admi*) is never attempted, E-Public Realm Governance, creates the most efficient system, nowhere to be found in Kozhikode or other parts of Kerala. This characteristic promotes efficient public delivery systems of public spaces and quality public outputs in terms of well designed and well maintained spaces socially acceptable.

8.8 Accountability

In E-Public Realm Governance, decision-makers in municipal government including ward Committee, Residential Welfare Associations, the private sector and civil society organizations are accountable to the public, as well as to institutional stakeholders. This accountability differs depending on the organization and whether the decision is internal or external to an organization. Central to the principle of accountability is information sharing and transparency which should be promoted by E-Governance web. On the contrary, accountability is hard to achieve especially in the absence of access to information. Accountability can be classified in four categories. These are public, financial, horizontal and vertical. Horizontal accountability is the relationship between say for example the executive, legislature and the judiciary. Vertical accountability is whereby one actor reports to another subject to the interpretation of constitutional provisions. Informal checks on these relationships are reinforced by the civil society and the donor community. A web based interface in E-Public Realm Governance provides for high level of accountability.

8.9 Strategic Vision

Leaders and the public have a long-term perspective on good E-Governance for habitat and human development, along with a sense of what is needed for such development. There is also an understanding of the historical, cultural and social complexities in which that perspective is grounded.

Good E-Governance for public realm leads to a number of positive consequences, including:

- People trust Municipal Corporation.
- Local people know where they are going in a continuous manner from website.
- E-Governance connects all public realm to all stakeholders namely citizen, business and government, however far they are located.
- E-Governance for Public Realm web gets better decisions through networking and people value the work evident in the website.
- Public realm web has the ability to weather crises.
- Financial stability.

9 Conclusion

Building codes, land development codes (of zonal plans), and form-based codes are the substantive aspect of participatory spatial development planning and design of public realm. They are arrived based on concensus. A Public Realm website structures the process and sequences of participatory generation of these codes using an interacting website; there by emphasising the best practice in E-Governance. This approach may be considered as still unknown, Spatial planning component of Peoples Planning Movement (*Janakiya Asuthranam*) of Kerala. All aspects of Governance such as spatial issue identification, strategic planning from SWOT analysis, Vision Design that leads to regulation plan and Form-Based Coding standards are incorporated in this website. Maintenance and administrative sanctions for building and payment procedures and e-tendering are all executed in the website. The same website is used by government, business and citizen involved in this process. Sustainable development, social cohesion and environmental management of public realm are dependent on effective E-Governance. E-Governance makes greater and most efficient access to Government than conventional ones. Hence, good governance is successful E-Governance well-articulated that promotes equity and sustainable development. A weak governance system cannot deliver services and benefits to those who need them most. The influence of powerful interest groups biases policies, plans, programs and thereby results in spending away from the poor. Lack of property rights and non-availability of best legal services for the poor, disadvantages the poor and inhibits them from securing their homes and other assets and operating businesses. Poor governance generates and reinforces and subverts efforts to reduce the access to the poor, while bad governance acts as a barrier to economic development to both domestic and foreign direct investment and this may lead to the collapse of the nation state.

Acknowledgments The participatory urban design component of this chapter have a long history of development since 2005 in the Urban Design Studio of Department of Architecture, National Institute of Technology, Kozhikode. Till date several hundred students guided by faculty had worked on it as part of research and development. Many case studies were tested and at least two Manuals were produced. Authors gratefully acknowledge the specific contribution made by Arun, Ashil, Basil, Eapen, Jishnu, Muhsina, Nivin, Dileep, Prasoon, Pratheeksha, Ridha, Rubeena, Safa and Sajna for this chapter.

References

1. Government of Kerala and National Council of Applied Economic Research (2013) The Kerala Perspective Plan 2030 (KPP), Chap. 13 Urban and Rural Development Strategy: State Spatial Strategy, Thiruvananthapuram
2. Vinod Kumar TM (2014) Implementing Smart Cities in India in "Geographic Information System for Smart City (Chap. 154). Copal Publishing Group, New Delhi
3. Bevir M (2013) Governance: a very short introduction. Oxford University Press, Oxford
4. Kooiman J (2003) Governing as governance. Sage Publications, Beverley Hills, p 170

5. Government of Kerala (2013) Law (Legislation –C) Department 23-9-2013, Kerala Gazette Extraordinary, Ordinance No. 51 OF 2013 The Kerala Town and Country Planning Ordinance 2013, Thiruvananthapuram
6. Levi-Faur D (2011) Regulation and regulatory governance. In: Levi-Faur D (ed) Handbook on the politics of regulation. Edward Elgar, Cheltenham, pp 1–20
7. Braithwaite J, Coglianese C, Levi-Faur David (2007) Can regulation and governance make a difference? Regul Gov 1(1):1–7
8. Rosenau JN (1999) Toward an ontology for global governance. In: Hewson M, Sinclair T (eds) Approaches to global governance theory. SUNY Press, Albany
9. World Bank (1991) Managing Development—The Governance Dimension, 1991, Washington DC, p 1
10. http://www.akshaya.kerala.gov.in/
11. Government of Kerala (2012) State Urbanisation Report Kerala, Department of Town and Country Planning, Government of Kerala, Mar 2012
12. Rose R et al (2011) A manual for form based coding regulation in Kerala, Calicut. National Institute of Technology, Calicut
13. Salim AM, Nadarajan KD, Mutunayagam NB (1967) Towards a settlement structure for Kerala, Part II. Retrieved from http://digitalcommons.unl.edu/arch_crp_facultyschol/12/?utm_source=digitalcommons.unl.edu%252Farch_crp_facultyschol%252F12&utm_medium=PDF&utm_campaign=PDFCoverPages
14. Vinod Kumar TM (2001) Annual planning and budgeting system using GIS for Calicut Corporation. School of Planning and Architecture, New Delhi
15. Vinodkumar TM, Bimal P (2011) Participatory urban design for Kerala: the art of community visioning, application of principles of new urbanism and designing form based coding regulation. Spat Dev Rec 19(6):144–156
16. Prasoon G, Nivin D et al (2014) Urban design around monorail stations in Calicut—Panniyankara. National Institute of Technology, Calicut
17. Www.formbasedcodes.org, "what-are-form-based-codes". Available at http://www.formbasedcodes.org/what-are-form-based-codes. Accessed 20 Aug 2011
18. Daniel PG, Karen P, Paul CC (2009) Form based codes-, a guide for planners, Urban Designers, Municipalities, and Developers. Wiley, New York
19. Marye BS, Madden E (2006) Place making with form based codes. Urban Land Institute, New York
20. Government of Kerala (2010) Information technology (B) department, Electronic Delivery of Service Rules Thiruvananthapuram

Smart Systems and Smart Grids for Effective Governance of Electricity Supply in India

Veena Aggarwal and Parimita Mohanty

Abstract Despite several reform measures, the electricity sector in India is grappling with multiple challenges including significant system losses, supply shortages, demand management and integrating renewable energy into the system network. At the same time institutional arrangements and relations have also transformed and consumers are asserting their demand for improved quality and services through the regulatory process. Both utilities and policy makers see Information and Communication Technology (ICT) as an important tool in meeting growing aspirations in the electricity sector. Many state utilities have improved their interaction with consumers through IT enabled web-based systems. Some are providing billing, payment and grievance redressal facilities online. Utilities also see significant use of ICT in improving operational efficiencies, theft detection, mapping of assets, managing load, outage management, etc. Integration of all urban basic services as envisaged under a smart cities, requires collection and analysis of electricity consumption data on real-time basis. Most utilities have deployed, automated meter reading (AMR) systems, prepaid meters and time of day meters for large revenue industrial consumers. In future, advanced Metering Infrastructure (AMI) compatible smart meters are expected to facilitate two-way communication between utilities and consumers. A Smart Grid Vision for India has been drafted to 'transform the Indian power sector into a secure, adaptive, sustainable and digitally enabled eco-system that provides reliable and quality energy for all with active participation of stakeholders'. With part funding from the central government, state utilities are experimenting with smart pilot projects and if successful and effective, these pilots would be rolled out in a bigger way. The Energy

V. Aggarwal (✉) · P. Mohanty
The Energy and Resources Institute (TERI), Darbari Seth Block, India Habitat Centre, Lodhi Road, New Delhi 110003, India
e-mail: veenaa@teri.res.in

P. Mohanty
e-mail: parimita@teri.res.in

T.M. Vinod Kumar (ed.), *E-Governance for Smart Cities*, Advances in 21st Century Human Settlements, DOI 10.1007/978-981-287-287-6_6

and Resources Institutes (TERI's) recent initiative on making a renewable-based mini-grid smart showcases the possibility and utility of smart technologies in load management and reduced need for human intervention. The drive for smart grids and smart cities would certainly improve quality of civic life. However, it would require huge investments, resources and greater co-ordination between state level electricity providers and city governments. The investments envisaged will result in greater stress on state and city governments and ultimately on consumers. Hence, policy makers and regulators will have tread cautiously and weigh the costs and benefits of the smart initiatives in light of the equity objectives of the country.

Keywords Smart cities · Smart grids · ICT · Electricity · Smart meters · Mini-Grid · Renewable

1 Introduction

E-Governance is a means of enhancing quality of public service and public life through the use of information and communication technology (ICT). According to the UNESCO 'E-Governance is the public sector's use of ICT with the aim of improving information and service delivery, encouraging citizen participation in the decision-making process and making government more accountable, transparent and effective'. Governance today is considered a process of decision-making that involves multiple actors besides the government. Hence E-Governance would imply the use of ICT by various actors in the governance process for improving public service delivery and making decision-making more participatory.

"The 'Smart' concept originally comes from the integration of ICT in products and systems with the proclaimed aim of optimising their efficiency" [1]. In the context of 'Smart Cities', E-Governance is the key and Smart Cities essentially focus on use of IT for greater integration in the management of urban public services including real-time two-way communication between citizen and government and between citizen and service provider. "In a smart city, sensors will provide real-time inputs to a control centre to provide real time inputs to a control centre on clean water, energy, public transport, public safety, education and health care. Intelligent communication tools will let administrators manage and respond to emergencies quickly as well as provide residents with constant real-time inputs" [2]. Smart cities focus on optimal utilization of resources and effective management of effluents [1].

A smart efficient and sustainable energy infrastructure is one of the most important building blocks of a smart city. Smart cities of the future are expected to consume lesser energy, use more renewable energy and will have low-carbon footprint. The European Commission's Smart Cities programme, with its focus on active building, decentralized renewable energy technologies, smart energy grids, low-carbon mobility and urban energy planning, has also identified its end objective as 'meeting its 2,020 and 2,050 energy and carbon reduction targets' [3].

2 Electricity Sector in India and Institutional Arrangements for Electricity Supply

The Electricity Sector in India has seen a series of reforms beginning with measures to increase private sector participation in capacity addition in the early 1990s to the reforms in distribution in early 2000s with the enactment of the Electricity Act, 2003 and setting up independent regulatory commissions. The establishment of independent electricity regulatory commissions was seen as a major move in the Indian power sector towards better accountability and transparency. It was assumed that crucial issues of pricing, competition, and consumer interest could be best taken by a statutory authority that would not be dependent on the state government [4].

As part of the reform process, state electricity boards in most states were unbundled into separate entities looking at generation, transmission and distribution. Some attempts have been made at privatization of distribution first in Orissa and subsequently in Delhi. There are also some private companies serving cities and industrial hubs such as CESC in Kolkatta, Torrent Power in Ahmedabad and Surat, Reliance Energy and Tata Power in Mumbai. More recently, some utilities have appointed franchisees in certain urban cities and conglomeration such as in Bhiwandi and Nagpur.

Institutionally, electricity distribution is not arranged at a city level but at a state level with commonly three to four electricity distribution utilities per state and each utility serving multiple districts comprising both urban and rural consumers. Hence a state utility will serve many cities coming under several separate Municipal jurisdictions. Overall Planning for electricity is undertaken vis-a vis the area served and may not necessarily be in tandem with administrative boundaries of cities. For electricity utilities, Municipal bodies are large consumers of electricity and offer prospects for trying out new initiatives on energy efficiency (for instance in water pumping) and renewable energy (for instance solar energy-based traffic lights and street lights). However, from a planning perspective, there is not enough interaction between the two. The city level electricity supply companies mentioned earlier are an exception and would be suited for trying out an integration of municipal services with electricity supply services in India.

3 Challenges in Electricity Distribution Sector

There are several challenges in the distribution end of the electricity sector. Distribution losses, comprising both technical and commercial losses, including theft remain a major concern affecting the financial health of utilities. On an average, the Aggregate Technical and Commercial (AT&C) losses across the country in 2011–2012 stood at 27 % [5]. Notably, while AT&C losses in 2011–2012 were around 42.6 % in the eastern states, the southern states have managed to bring down losses to around 18.6 % (ibid). This shows that there is considerable scope for improvement in many states.

Electricity deficits have come down over the years with all India average energy deficits reported at 4.2 % for 2013–2014 [6]. However, there are significant differences from state to state and even in 2013–2014, states like Himachal and Punjab had average deficits in the range of 10–15 % (ibid). Managing electricity demand, particularly during peak summer and winter months and during particular hours in a day is another major challenge. In the month of May 2014, Uttar Pradesh reported peak shortages of 22 % (ibid). It is now recognized that it is more economical and efficient to better manage such peak demand rather than setup additional capacity.

Connecting more and more people to the grid and meeting their electricity requirement is another concern. States are now exploring off-grid solutions for a large section of people in rural and remote areas. Concerns of climate change and energy security have become compelling reasons for regulators and governments to push for renewable energy.

Another major challenge today, which perhaps was missing earlier, is a more demanding and informed consumer. With increasing tariffs, consumers are no longer ‘recipient of public services’ as made available by the state but are more active and involved consumers who pay and demand better services. The regulatory setup is encouraging, building capacity and providing consumers more avenues to participate in decision making.

4 Role of Information Technology in Electricity Distribution

Both utilities and policy makers see ICT as an important tool in managing these multiple challenges. The National Electricity Policy, 2005 under Sect. 4.10 ‘emphasizes on use of modern IT systems, to facilitate creation of network information and customer data base which will help in management of load, improvement in quality, detection of theft and tampering, consumer indexing and mapping, customer information and prompt and correct billing and collection’ [7].

The Restructured Accelerated Power Development and Reforms Programme (R-APDRP) is a major IT focused scheme of the central government, primarily focused at cities and towns, intended to help utilities reduce losses and strengthen the distribution network. It is an ambitious scheme with an intended investment of about INR 515,770 million and work under this scheme was originally planned for only the XIth Plan but has subsequently been taken forward into the XII Plan. However, most of the funding will be first provided to utilities as loans and will be converted into grants subject to utilities meeting certain laid objectives within a stipulated time. MoP [8] For certain projects, the central government will provide loans for only 25 % of the total amount and the rest will have to raise from financial institutions. To begin with, R-APDRP funds were to not be made available for private distribution companies. Subsequently during the XIIth Plan, the Central

Electricity Supply Utility of Orissa (CESU) of Orissa has been included. However, other private Distribution Companies (Discoms), including Delhi's Discoms do not receive funds under APDRP.

Projects that are being funded under APDRP include those for establishment of baseline data and IT applications for energy accounting/auditing- and IT-based consumer service centres. MoP [8] Funds are being made available for asset mapping of the distribution network using GIS. The funding scheme is also consumer focused and provides funds for adoption of IT applications for meter reading, billing and collection; energy accounting and auditing; MIS; redressal of consumer grievances; establishment of IT enabled consumer service centres etc., [8]. It can be said that this R-APDRP scheme is the major driver of 'smartness' in electricity.

4.1 IT Enabled Online Utility—Consumer–Regulator Interface for Improved Governance

Supported by funds from the government, many state utilities are working on ways to improve their interaction with consumers, through call centres, customer care centres and increasing through IT enabled web-based systems. Many utilities provide consumer online facilities on their websites to apply for a connection, receive and pay bills, track consumption and also register and track electricity-related complaints. All these requests are managed in an IT enabled environment at the utility end. Box 1 briefly describes the online services provided by one of Delhi's distribution company—Tata Power Delhi Distribution Limited (TPDDL). Utility websites are also more informative and provide advanced notice on load shedding, suggestions for energy conservation, etc.

In the provision of service, utilities need to abide by a 'Standard of Performance (SoP)' which is set out by the regulator. SoPs are essentially benchmarks for parameters such as reliability and quality of electricity supply, metering, billing, new connections, disconnections, etc. The utilities are required to submit quarterly and annual reports on their performance on these benchmarks to the regulator. Many utilities and regulators put out these reports on their website. Such information, made available in the public domain helps improve transparency and accountability in the functioning of the utility.

Box 1: Tata Power Delhi Distribution Limited: IT Enabled Facilities for Consumer

TPDDL provides facility for consumers to apply for a connection online on the company's website. Consumers can download their bill and make payments online. More recently the utility has commenced a 'pull sms' service whereby consumers can apply for connection, register complaint and pay bill through SMS. Consumers can register complaints regarding supply services online, besides other venues such as consumer care centre, a call centre and a zonal complaint centre.

IT is the backbone for registering all supply request and complaints and the system is supported by software called 'Sambandh'. All the complaints registered through various sources are fed into the software by the executives. Once the complaint is registered, the system generates a unique complaint number and the executives can locate the status of the complaint at any stage of the redressal process and it can be accordingly communicated to the customer. Till the complaint is closed and issue is resolved, concerned officials, at each stage of the process, have to update the software on the status of a particular complaint. At any stage of the redressal process, if the action is not taken within the stipulated time frame or the status is not updated, the complaint gets automatically escalated to the senior official for an appropriate action.

Source Excerpts, [22] and website for TPDDL (www.ndpl.com).

Websites of regulatory commissions are an important point of interaction with both the utility and consumers. All consultation papers and draft regulation are put up by the regulators on their website seeking suggestions from the public and various stakeholders. Utilities put out their annual tariff proposal on their websites, sometimes with detailed excel sheets explaining various cost estimate. Regulators upload the final tariff orders for regulated utilities in their jurisdiction. These orders detail out reasons and data based on which decisions regarding the performance of utilities are made. From a consumer perspective, the tariff orders, give insight into how, electricity prices are set every year for the concerned utilities. Increasing regulators and regulated entities are exploring ways in which the regulated entities can directly upload data and information in standardized formats on the website of regulators.

4.2 IT for Efficient Utility Operations and Future Smart Grids

Utilities are increasingly using ICT to improve operational efficiencies, for theft detection, mapping of assets, managing load, outage management, etc. "Key systems such as enterprise resource planning (ERP) for streamlining business processes, supervisory control and data acquisition (SCADA) for energy accounting and customer relationship management for consumer interface have been deployed" [9]. Utilities are exploring more advanced IT options such as Outage management systems (OMS) for managing outages. "OMS can process outage reports from various operational systems, including SCADA and phone calls from consumers" (ibid). Overall, all these initiatives will help utilities to have accurate and real-time control over their assets and systems and also help in understanding customer needs.

Collection and analysing of data is a crucial requirement for a smart city. Somewhere in the future it is anticipated that all basic urban services, including electricity, will be provided in an integrated manner. A large amount of data will be collected in the process, which will help in more integrated planning. Urban planners will try to organize city life so as to reduce consumption of resources such as water and different forms of energy. Smart meters with focus on collection of real-time data are thus a building block for future smart cities. In electricity distribution, particularly for industrial consumers, most utilities have deployed, AMR systems, prepaid meters and time of day meters [10]. Smart meters help to monitor electricity consumption remotely, load profiling and for identifying leakages in real time. Inconsistencies in consumption pattern over a time period can help detect faults in the system and even the possibility of theft [11]. While most utilities focus on smart metering for high end industrial consumers, gradually smart meters are being tried out on a pilot basis for LT domestic and commercial consumers too. Study of load data collected through smart meters help utilities to analyse demand and manage their load better. Both the private Discoms in Delhi, TPDDL and BSES (Yamuna and Rajdhani Power Limited) have utilized AMR technology (amongst other measures) for effectively curtailing distribution losses which were in the range of 50–60 % when these private companies were handed over the distribution network from erstwhile Delhi Vidyut Board [11].

In future AMI compatible smart meters are expected to facilitate two-way communication between utilities and consumers (ibid). 'AMI can help utilities remotely connect or disconnect services, record waveforms and support time of use and real time rate structures' [11]. This kind of communication is a key ingredient of smart grids for smart cities. Smart grids in the future would allow for influencing consumer behaviour to reduce overall consumption and encourage use of electricity when excess capacity is available. If price of electricity is available on an hourly basis, consumers can switch of consumption when prices are high. They can also invest in control technologies that automatically react to price changes. Pilots on demand response at a smaller scale are being tested out by utilities for select consumers with the intent of larger roll outs based on the outcomes and benefits accrued from the pilots. The Tata Power Delhi Distribution Company, for instances has commenced in January 2014 an INR 125 million pilot for implementing auto demand response for high end industrial and commercial consumers. Details on this pilot are given in Box 2.

Box 2: Pilot on Auto-Demand Response

The Tata Power Delhi Distribution Company's auto demand response pilot combines ADR with smart meters and AMI- includes 250 industrial and commercial consumers with a load greater than 100 kW and collective shed potential of 20 kW. There will be focus on peak load saving. A dedicated customer portal will allow consumers to set threshold values for load and power factor, they will receive alerts either through the portal or via SMS if these thresholds are breached. Smart meters will also provide instant information on outages, helping the company manage peak power demand and avoid buying expensive power. The savings is in terms of not setting up additional power generation capacity. The smart meters installed at the consumers' premises integrated with the Control Centre will provide Tata Power Delhi Distribution with instant information on outages and other factors related to quality of power. This will help Tata Power Delhi Distribution to improve the reliability of power and to manage peak power demand more optimally and avoid expensive power purchase during the peak hours.

(*Source* TPDDL's press release dated 1/28/2014).

There is however no standard definition of a smart grid. Multiple factors would decide how 'smart' the grid could be. It would depend on technologies available, supporting infrastructure, financial support from government, besides capacity of utility to invest and for consumer to pay. Accordingly different countries, based on their current level of development and peculiar challenges in their electricity supply system have devised their smart grid programmes. Smart Grid Vision for India is to 'transform the Indian power sector into a secure, adaptive, sustainable and digitally enabled eco-system that provides reliable and quality energy for all with active participation of stakeholders' [12]. While the drivers on smart grids in western country have been reliability, reducing cost of manpower, lower emissions, in India, the drivers are—reduction in losses, reducing human error through greater automation, peak load management and integrating renewable energy into the grid [13]. Policy makers see multiple uses of smart grids. 'AMI can help locate major sources of theft, leading to administrative action. It can help consumers make wiser choices about how they use power, for what purpose and at what time. Grid automation technologies help power grids to heal themselves in case of outages—allowing utilities to restore supply to citizens much faster' [14]. A budget of INR 2,000 million has been set aside from the RAPDRP budget with matching funds from state for smart grid-related pilots in 14 cities [14]. Individual projects are estimated cost about INR 500–600 million [15]. Each pilot smart grid project will trial a different combination of seven technologies—AMI for residential applications, AMI for industrial applications, outage management, peak load management, power quality, micro-grid and distributed generation [14]. The pilot projects were identified recently in 2013 and are still in the preliminary stages with feasibility studies being undertaken and work orders being placed. Some progress has been achieved in the pilots in Panipat and Mysore where work has been awarded and project implementation has commenced.[1] Most of the utilities are testing a combination of all categories of consumers-industry, commercial, residential; and

[1] This is progress noted on the smart grid pilots from the official smart grid website http://indiasmartgrid.org/en/Pages/pilot.aspx as on 28 June 2014.

some even agricultural consumers. Typically any pilot comprises of about 20,000–30,000 consumers and around 100–300 million units of electricity annually. Most pilots are experimenting primarily with peak load management and outage management using AMI. The success, benefits and scalability of these pilots will determine the future implementation of smart grids in India.

The smart grid vision of India also envisages various other policy and programmes which will have to be adopted in tandem. This includes establishing micro-grids in 1,000 villages/industrial parks/commercial hubs and policies for mandatory rooftop solar power generation for large establishments (with connected load of more than 20 kW). Renewable and Micro-grids have been identified as an important component by the Indian Smart Grid Forum. This Forum is a public-private partnership that provides inputs to government on smart grid implementation, is divided into several groups focused on specific areas and one of the working groups (WG), WG-9 is on 'renewable and micro-grids'. This group is responsible for developing an architecture and methodology for micro-grids—both off-grid and grid connected; and also develop related set of standards [16]. Since in cities and towns, the consumers are categorized based on their electricity consumption and type of establishments such as industrial, commercial, residential etc., various different sizes and configurations of micro-grids can be established for those consumers with different financial and institutional arrangements.

5 Micro/Mini Grids and TERI's Smart Mini Grid

A mini-grid is defined as an electricity distribution network operating below 11 kV (in India, generally with 415 volts 50 Hz three-phase AC electricity supply), providing electricity to a community. The mini-grid has small power plant (it could be as small as 2 kW or as big as several kW), and spread within 2–3 km radius from the power plant. It supplies power through low-tension distribution networks to households for domestic power, commercial (e.g. shops, cycle repair shops, flour mills) activities and community requirements.

A mini-grid may or may not be connected to the wider grid.

Isolated mini-grid is defined as a mini-grid that is not connected to the utility grid in any way. It is a distinct island for which no point of common coupling (PCC) exists.

Connected mini-grid is defined as a mini-grid which may or may be connected to the utility grid. It may operate as a distinct island but features a point of common coupling that allows interaction with the utility grid.

A smart mini-grid is a mini-grid, which delivers electricity from suppliers to the consumers using digital technology. Smart mini-grid is a subset of smart grid concept. Smart mini-grid could be defined as the application of digital information technology to optimize electrical power generation, and delivery and ultimately its end-use within the domain of the mini-grid. In smart mini-grid model, at every instance, the load as well as the DG resources are optimally managed and distributed through advanced controls and interfaces. Hence, smart mini-grid is an integrated energy system that consists of (i) variable loads which are connected to the distribution grid (ii) diverse range of small, local generators based on distributed energy resources with (or without) storage system and (iii) control and power conditioning systems. The distributed energy resources can meet the local energy requirement or feed power to the conventional grid. The smart controllers can disconnect or reconnect the distributed energy resources to the conventional grid with minimal disruptions.

The mini-grid system may or may not be intelligent enough to manage the resources and load smartly through automation and digitization. Table 1 discusses the key characteristics which make a regular mini-grid smart.

Table 1 Essential of smart mini-grid

Features	
1. Decentralization of power generation and integration of multiple distributed generation (DG) resources into the same mini-grid	Smart mini-grid allows individual consumers to generate power through suitable DG-based power generating technology and feed it to their mini-grid. Multiple DG-based small generators are integrated into the same mini-grid system. A smart mini-grid manages the situations, when the distributed generation is more than the locally connected load, by feeding the excess generation into the main grid and vice versa for making the grid more reliable
2. Intelligent Load and resource management	The total load connected to the network, which is the sum of individual loads, is not stable and varies significantly with time. In cases of sudden increase of loads, the smart mini-grid takes care of load reduction temporarily by either asking consumers to reduce their loads or drawing instant additional supply from online standby DG generators so as to allow the time required for increasing the generation to match with load. Here at every instant, the load and DG resources are managed simultaneously in order to provide the most efficient, reliable power supply
3. Automatic response to network problems	A smart mini-grid, within its mini-grid network should automatically anticipate, detect, and respond to system network problems and avoid or mitigate power outages, power quality problems and service disruptions using real-time information from embedded sensors and automated controls
4. Resist network disruptions	Smart grid technologies identify and respond to man-made or natural disruptions in the network. The real-time information enables mini-grid operators to isolate affected areas and redirect power flows around damaged facilities
5. Greater resilience to loading	The power flow through multiple routes is also a feature of smart mini-grid as in the conventional grid system. However, the smart mini-grid avoids cascade tripping of network elements. A smart mini-grid has greater resilience to loading as compared to conventional grid system

In other word, a smart mini-grid is a technologically advanced version of distributed generation-based micro-mini grid. "Smart Mini-Grid is an application of digital ICT and uses advanced sensing, communication and control technologies to optimise electrical power generation and delivery" [17]. A smart mini-grid allows for a two-way communication with consumers, corrects supply demand imbalances and detect faults in a 'self-healing process'. While micro-grids can be used for any localized load (hospitals, institutions, offices) and for electrifying remote areas, they are also now seen as means of providing reliable electricity supply in the wake of natural disasters [18].

Box 3 describes TERI's recent initiative on setting up a Smart Mini Grid (SMG) project at its retreat at Gual Pahari, Gurgaon.

Box 3: TERI's Smart Mini Grid

The TERI has setup a SMG at its research facility at TERI retreat in Gual Pahari, Gurgaon. This initiative commenced in 2009 under the Asia-Pacific Partnership on Clean Development and Climate and with support from the Ministry of New & Renewable Energy (MNRE), Government of India.

Initially the retreat had a renewable based mini-grid with the distributed generation resources comprising of 12.5 kW, crystalline silicon solar PV, 1 kWp thin-film solar PV, 3.3 kWe wind generator, 100 kWe biomass gasifier, 48 V, 600 Ah storage battery and a diesel generator. All these energy resources were used separately or individually to cater to the dedicated load of the Retreat building. There was no intelligence incorporated into the system and interventions such as switching over from one energy resources to other and switching off some of the non-essential loads in times of lesser energy availability were done manually. There were some instances, when due to a sudden increase in loads and less electricity available, the entire mini grid collapsed due to the instability of the grid. This made the system vulnerable, inflexible and heavily dependent upon human intervention.

Against this backdrop, the smart mini-grid project was designed with the objective of converting the existing in-efficient mini-grid system into a SMG system that would allow smarter control of existing distributed energy sources and intelligent management of loads. Figure 1 gives a schematic diagram of the SMG. In the modified smart mini-grid system, the load of the Retreat building is categorized based on its priority (i.e. critical, non- essential and essential load). Each energy resource has its own local controller to condition the power quality of individual energy generating resources.

A communication line was added to the distribution system for real-time data acquisition and monitoring of various electrical, weather and physical data. Similarly, in addition to the local controller, a central controller known as intelligent dispatch controller was added, which uses real-time data and decides the load dispatch strategies based on the type of load and load profile, criticality of the load, resource availability etc.

The use of digital switches has reduced the transition time (from one resource to other) and as result the number of occurrence of brownout in the Retreat building has reduced by 40–50 %. This system now requires less human interventions as the Intelligent Dispatch Controller decides the dispatch strategies automatically based on the pre-set criteria. Diesel consumption has also declined due to overall efficiency in the system.

Such smart, reliable systems can be setup at any campus or facility and will go a long way in decreasing reliance on the main grid while increasing the consumption of cheaply and locally generated renewable energy. Through the implementation of SMG technology, communities and complexes will be empowered to better manage and utilize their energy resources while also decreasing costs across

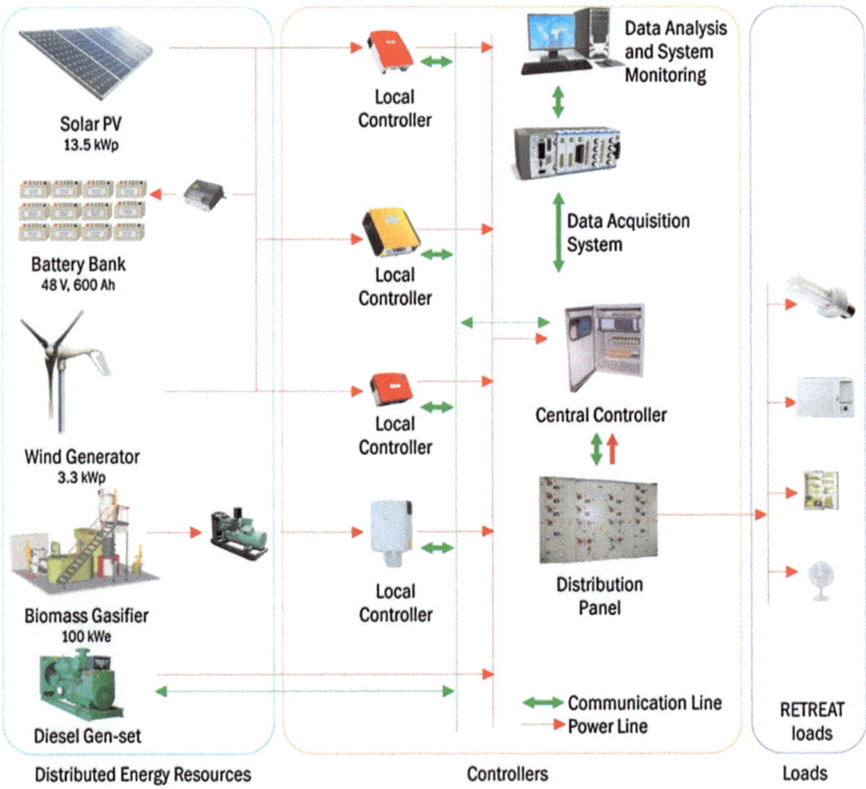

Fig. 1 Schematic diagram of the SMG at TERI. *Source* TERI Compilation

the board. Many such SMGs setup across the country will eventually lead to the grounds-up smartening of the main grid while simultaneously reducing the need for larger, expensive, and environmentally damaging centralized power plants.

While looking to retrofit or setup a campus or building with a SMG system, there are several available government subsidies that can be considered. These initiatives from state and central governments provide capital relief and can help lower the upfront costs of such an SMG undertaking. Applicable government grants include capital costs subsidies on RE projects from MNRE and various state Energy Development Agencies; property tax rebates from certain municipal corporations; and Demand Side Management based rebates from local utilities. In addition, a rooftop solar generation program providing capital benefits is currently being planned by MNRE and various state governments, which will further help with project costs.

6 Smart Energy Systems and Smart Cities—Challenges on the Way

Smart grid is not one single initiative but a series of initiatives in automation backed by various policies and programmes which will eventually connect consumer with utilities and various utility devices with each other so as to enable better demand management, effective outage management, enhanced energy efficiency and use of renewable energy. Utilities have already adopted some technologies while some are being tested out on a pilot bases. The Smart Grid programme as envisaged by the Government is expansive and will require substantive support in terms of enabling policies and programmes, say on differential and dynamic pricing, net metering etc.

The smart grid programme will also have huge financial implications. To begin with, the pilots are being funded through government support shared equally between central government and state utilities. But in future, during full roll out it is expected that distribution utilities will have to bear these costs. The current literature and policy documents do not indicate what these costs would be. Funds under R-APDRP will be first disbursed as loans and only converted into grants on achievement of certain loss-related targets and other objectives specified within a stipulated. Until and unless the loans get converted into grants, utilities will have to shell out huge interest outflows. These interest outflows, on completion of work, will be capitalized. In both cases, consumers will have to pay for this interest amount through pass through in tariffs. Overall, across states, there have been substantial delays in execution of projects under R-APDRP for various reasons including issues in procurement, management of funds, inadequate supply of devices, delays in appointing consultants [19]. Distribution utilities are finding it difficult to raise loans in the markets owning to their financial ill-health. The aggregate book losses (on accrual basis) of all the utilities increased from INR 304,300 million in 2009–2010 to INR 516,020 million in the year 2010–2011 and to INR 625,810 million in 2011–2012 [5].

Another major concern is security and maintenance of the IT systems created. Experience with another capital subsidy scheme—the Rajiv Gandhi Grameen

Vidhyutikaran Yojna (RGGVY) shows that some of the states are finding it difficult to maintain the network created through the central government capital subsidy scheme and instances of theft of network have increased [20]. The theft of components, from copper wires to transformers, makes the network redundant and villages continue to remain without electricity. Since, smart grids, involve substantial investments in smart meters and other equipment, additional investments will have to be made in ensuring their security.

Training and capacity building of the utility staff is emerging as another major challenge. Under R-APDRP, however, this need has been identified and a sum of INR 2,000 million has been set aside for capacity building. Notably, a sum of INR 4,000 million has been set aside towards 'incentive schemes' whereby state utilities would reward and incentives staff for reducing losses below 15 % [8].

Automation is expected to bring in efficiency and help in reducing losses in electricity supply, but it would be difficult to quantify the impacts of ICT intervention [11], and analyse the benefits vs the costs. Some benefits may also not be tangible and others may accrue over a period of time. As stated earlier, R-APDRP funds will be treated as grants only on meeting of certain objectives. Till then utilities will have to bear the interest on this loan and utilities will try to pass these expenditures to consumers through the annual tariff revisions. Private utilities such as in Delhi, so far have not been covered under the R-APDRP funds, and hence any expenditure they incur on their IT interventions will have to recovered through tariff hikes. There have already been concerns regarding the high capital expenditure of the Delhi Discoms and political drives for auditing the accounts of the Discoms. It will be for the regulators to ensure that all capital expenditure is closely scrutinized in the interest of consumers and that consumers are not unjustly loaded with heavy tariffs. Amongst consumers, smart grids, as envisaged offer more saving potential for large consumers and hence utilities are correctly targeting them for pilots. While smart grids may benefit smaller domestic consumers, it will have to be examined what would be the cost and benefits.

Collection and analyses consumption of data is one of the key elements of a smart grid and a smart city. IT-based applications can be developed that analyse the data and share the interpretations with consumers, utilities, and policy makers in a user friendly manner. However, use of such data would also raise concerns of privacy. Utilities that invest significantly in data generation and collection technology may want to sell such data to third party for commercial consideration. Issues on what can and cannot be shared will have to be examined vis-à-vis the needs for citizen privacy.

With growing emphasis on sustainable cities, reduced carbon footprints and greater intake of renewable energy, electricity service providers will have to start engaging more with city governments and consumers. A major challenge in transforming existing cities into 'smart cities' will be the integration of electricity services with other basic municipal services. 'The legislative policies and acts governing the power sector in India assign no particular roles, responsibilities and authority to the urban local bodies (ULBs) with regard to electricity services except the provision of street-lighting, the only component in electricity services under their purview' [21]. City planning done by Municipal bodies does not typically engage much with the electricity utilities and ULBs have no important role in electricity. A start perhaps has been made with utilities focusing on how to get local bodies to use energy more efficiently in water pumping and in street lights. Utilities are also partnering with local bodies for harnessing rooftop solar PV for local electricity generation and consumption as well as for feeding additional power, if any, to the utility grid. In future, utilities will need to engage even more with local bodies, for instance for promoting energy efficiency and energy conservation measures, demand side management and green buildings. It will require the utility to work with the local body in developing the latter's capacity to operate and manage such systems.

Developing capacity of Municipal bodies around IT can prove to be a major challenge. A study by TERI in 2009–2010 showed that even larger Municipal Corporations such as in Ahmedabad are not able to effectively use IT enabled systems for simpler work such as addressing consumer grievances. "In Ahmedabad it was seen, that at a lot of centres, the staff continue to take down complaints in registers and at the end of day feed all the complaints into the system" [22]. Discussions with the Municipal officials brought out that "resistance to new technology was observed not only at the level of the staff but also at the senior level" (ibid). In such a set up more advanced ICT systems, bought incurring huge expenditure, may ultimately be wasted, for lack of capacities.

If the level of automation, integration and resource efficiency, as envisaged under smart cities become a reality, urban dwellers will see a tremendous improvement in quality of life. However, these benefits will have to be weighed with existing ground realities, particularly the health of utilities and municipal bodies in the country. In an ideal scenario, with an e-democracy framework perfectly in place in the electricity sector, consumers of a smart city would decide what mix of conventional and non-conventional energy they want, what levels of automation they require and in what areas and more importantly what prices they would be willing to pay for these services. Currently, India is far behind such an ideal scenario. Electricity regulatory commissions have tried and struggled in many states, to provide an enabling environment where consumers can participate in important decisions that affect them. Competition is very limited in the sector and consumers rarely have the options to even choose their supplier. It is often left to the regulator to work in public interest and at times faces stiff resistance from the government and even from within. It therefore becomes imperative for the regulator to use its wisdom in allowing costs on capital expenditure that it thinks are necessary. It will also have to ensure that investments are made such that those who can bear the costs, should pay.

As regard the municipal bodies, it is generally agreed that our 'third tier of governance' remains weak in capacities with limited 'actual devolution' of powers and functions. Providing or urging them to purchase expensive technologies and equipment, without adequate and trained man-power and without adequate user-charges will only further add to their burden. Hence, we should ride the 'smart cities' drive rather carefully and not be swayed by international experiences where the drivers of technology are different.

References

1. Ramamurthy A, Devadas MD (2013). Smart sustainable cities: an integrated planning approach towards sustainable urban energy systems, India. Int J Soc Hum Sci Eng 7(1):136–156
2. Spizzichino T (2013, Jan/Feb) Smart cities-the critical need of the hour in India. Energetica India 30:69–70
3. Morthorst PE, Page J, Schmid RR (2011) Planning the smart city. In: Hans Larsen LS (ed) RIS Energy Report 10: energy for smart cities in an urbanized world. Technical University of Denmark, Denmark, pp 13–16
4. Aggarwal V (2014) Electricity governance in India: case of regulatory governance of Maharashtra power sector. TERI, Delhi
5. PFC (2013) Performance of state power utilities for the years 2009–2012. GoI, Delhi
6. CEA (2014) CEA monthly report. Retrieved Jun 2014, from CEA: http://www.cea.nic.in/monthly_power_sup.html
7. MoP (2005) National Electricity Policy 2005 Ministry of Power, GoI, retrieved in April 2014, from www.powermin.nic.in/whats_new/national_electricity_policy.htm
8. MoP (2008, September) R-APDRP retrieved Apr 2014, from R-APDRP: http://www.apdrp.gov.in/OrderSGuidelines/Mop%20Order(19.09.08)%20-%20Presidential%20approval%20of%20R-APDRP.pdf
9. Powerline (2012, August) IT thrust: focus on customisation for efficient operations. Powerline 76–77 (August 2012), India Infrastructure Publishing Pvt Ltd
10. Powerline (2013, August) Smart power. Poweline 54–55
11. TERI & NASCOM (2011) Sustainble tomorrow: harnessing ICT potential. TERI, Delhi
12. MoP (2013) Smart grid vision and roadmap for India. Ministry of Power, Government of India
13. USAID (2010) The smart grid vision for India's power sector—a Whilte Paper. USAID, Delhi
14. Pitroda SG (2012, July) Interview. Governance Now available at http://www.governancenow.com/views/interview/we-are-awaiting-cabinet-nod-financingsmart-grid-pilots
15. Nexant (2012) Smart grids electric systems—inception report. USAID, Delhi
16. ISGF (2014) India smart grid forum retrieved Jun 2014. http://indiasmartgrid.org/en/isgtf/wg/wg1/Pages/default.aspx
17. CSIRO (2013) Technical report on mini-grids—scope and definition under the project accelerating the deployment of smart mini-grids in APP country, 2011, CSIRO, Australia
18. Abey et al (2014, June) Powering through the storm. IEEE Power Energ Mag 12(3):67–76
19. Mint (2013) State power boards seek extension of reform programme deadline. Mint, Jan 30, 2014
20. TERI (2013) Evaluation of RGGVY. TERI, Dehi
21. TERI (2009) An exploration of sustainability in the provision of basic urban services in Indian cities. TERI, Delhi
22. Aggarwal V, Krithika P, Garg D (2010) Strengthening municipal governance for effective consumer. Int J Regul Gov 10(2):63–76

Bibliography

1. AF Mercados EMI (2013) India smart grid day. India electricity. FICCI & Ministry of Power, Delhi
2. Depura W (2011) Electricity theft: overview, issues, prevention and a smart meter based apprach to control theft. Energy Policy 39(2):1007–1015
3. Ganesh D (2013) Das Ganesh, The ever evolving smarter grid. Presentation at the India International Cleantech Summit 2013, FICCI, New Delhi, 8–10 October 2013

4. Gupta N, Jain A (2011, August, Volume 5, issue I) Smart grids in India. Akshay Urja.
5. Haney A, Jamasb T, Pollitt M (2009) Smart metering and electricity demand:technology, economics and international experience. Cambridge Working Papers
6. Nexant Inc. (2014) Smart grids: An approach to dynamic pricing in India. USAID, Delhi
7. Sreedharan P et al (2011) Electricity grid evolution in India. Energy and Environment Economics Inc, Sansfrancisco
8. TERI (2010) Framework for effective consumer grievance redressal system. TERI, Delhi
9. Vijayapriya, Bapna, Kothari (2010) Smart tariff for smart meters in smart grid. Int J Eng Technol 2(5):310–315

E-Governance for Solar Photo Voltaic Powergrid: Solar City Gandhinagar, Gujarat, India

Jignesh G. Bhatt and Omkar K. Jani

Abstract Energy generation, distribution and consumption include different personnel at various hierarchical levels such as Production, Transmission and Distribution, Billing, Corporate, including the End User or Consumer. Enabling simple, user-friendly, reliable, efficient, timely and seamless connectivity to critical information to all such personnel remains the most desirable feature and benchmark for Smart City. Reliable availability of critical data on timely basis in desired formats for different decision-making authorities plays key role for successful e-Governance of energy distribution systems. Emerging approaches based on hybrid communication technologies possess capability to transform the present ones into smart, intelligent and adaptive sites and could become basis of exploring more options including *anytime–anywhere monitoring* in future. Welcome to be the part of transition of epoch and total paradigm shift in the domain of electricity utility services! E-Governance system for Smart City could be well defined as *System of Systems*, wherein E-Governance of Energy Generation and Distribution functions as vital source of power. Solar photovoltaic (SPV) is the proven method of electricity generation with significant sustainable future potential. Grid-tied PV rooftop system is gaining worldwide acceptance as one of the most widely accepted models for wide-scale deployment due to their simplicity, ease of installation, operation, maintenance, scalability, etc. E-Governance of power-grid implies the responsible and accountable delivery of electrical power assisted or supported by communication devices providing efficient and effective communication between Government to Government (G2G), Business to Employee (B2E), Business to Government (B2G), Business to Citizen (B2C) and finally, Government to Citizen (G2C). Globally, today many cities are adapting e-Governance for enhancement of

J.G. Bhatt (✉)
Department of Instrumentation and Control Engineering, Faculty of Technology,
Dharmsinh Desai University, Nadiad 387 001, Gujarat, India
e-mail: jigneshgbhatt@gmail.com

O.K. Jani
Solar Energy Division, Gujarat Energy Research and Management Institute,
Gandhinagar 382 007, Gujarat, India
e-mail: omkar.j@germi.res.in

T.M. Vinod Kumar (ed.), *E-Governance for Smart Cities*, Advances in 21st Century Human Settlements, DOI 10.1007/978-981-287-287-6_7

utility services and improved citizen convenience. Gandhinagar has emerged as the city in India taking leadership in application of advanced hybrid communication technologies and thereby transforming itself into Smart City. Earlier Gandhinagar has been facing frequent power shortages as well as demand–supply fluctuations causing not only inconvenience, but also resulting in large work losses and production losses as well as chaotic situations. With the untiring efforts of visionary and inspirational leadership of Gujarat-India on political-diplomatic-administrative fronts and initiatives such as Gandhinagar Rooftop Photovoltaic Programme, Solar City—Smart Grid Project, etc., have been implemented with efficient, effective and citizen-friendly objectives, wherein Solar Policy of Gujarat state has been of immense support. For sustainable growth and enhanced lifestyle of citizens, reliable and affordable power is a mandatory requirement and Gandhinagar city administration has been doing its best to approach every citizen through hybrid communication network, especially for power distribution and thereby resulting in strong trustworthy relationship of citizens with the government. Today, Gandhinagar is considered to be one of the most advanced and world-class self-reliant city having most resilient power infrastructure in India. Well-planned, systematically designed, and customer-oriented power distribution network encouraging participation of citizens has played key role in making Gandhinagar more liveable and a place to stay today. Complete Gandhinagar has been geographically distributed into separate zones, interested citizens are invited to express their willingness to participate in the programmes and after completion of formalities including registration, document verification, etc., scope and size of SPV installations are fixed up. Commercial aspects such as subsidy, revenue sharing and billing details are worked out on actual basis. Using seamless data connectivity using GUIs based on GPS, GIS and Virtual Instrumentation, real-time data of generation and consumption have been availed to the end user for determination and implementation of his real-time choice to run important equipments, if any. This chapter also includes descriptions on hybrid communication technologies deployed to timely serve need-based information. *Remote Energy Parameter Monitoring System using Hybrid Communication Technologies for Solar Photovoltaic Energy Generation System for Gandhinagar Solar City* has also been presented. The chapter ends with discussion over transformation of Solar City into Smart City.

Keywords E-Governance · Smart city · Solar city · Energy generation and distribution · Rooftop photovoltaic system · Solar power · Hybrid communication networking · Virtual instrumentation · Remote monitoring

Abbreviations

AC	Alternating current
ADB	Asian Development Bank
AEC	Ahmedabad Electricity Company
AMR	Automated Meter Reading
APERC	Andhra Pradesh Electricity Regulatory Commission

AUDA	Ahmedabad urban development authority
BAS	Building automation system
BG	Biomass Gasifier
BP	Biomass Power
BU	Building utilization
C-WET	Centre for wind energy technology
CERC	Central Electric Regulatory Commission
CSR	Corporate social responsibility
DC	Direct current
DDU	Dharmsinh Desai University
DISCOM or DisCom	Distribution company
DG	Diesel generator
DPR	Detailed project report
DR	Demand Response
GBI	Generation-based incentive
GEDA	Gujarat energy development authority
GERC	Gujarat energy regulatory commission
GERMI	Gujarat Energy Research and Management Institute, Gandhinagar
GIS	Geographic information system
GHG	Green house gas
GoI	Government of India
GoG	Government of Gujarat
GPCL	Gujarat Power Corporation Limited
GPRS	General packet radio service
GPS	Global positioning system
GSM	Global system for mobile communication
GUDA	Gandhinagar Urban Development Authority
GUI	Graphical user interface
HMI	Human machine interface
HVAC	Heating, Ventilation and Air Conditioning
IEMU	Intelligent energy measurement unit
IFC	International Finance Corporation
JNNSM	Jawaharlal Nehru National Solar Mission
LabVIEW	Laboratory Virtual Instrumentation Engineering Workbench
MGVCL	Madhya Gujarat Vij Company Limited
MNRE	Ministry of New and Renewable Energy
PDPU	Pandit Deendayal Petroleum University, Gandhinagar
PLC or PLCC	Power line communication or power line carrier communication
PPA	Power Purchase Agreement
PPP	Public–private partnership
PV	Photovoltaic
RES	Renewable Energy Sources
RPSSGP	Rooftop PV and small solar generation programme
RTP	Real-time pricing

RTU	Remote terminal unit
SECI	Solar Energy Corporation of India
SG	Smart grid
SHP	Small Hydro Project
SPV	Solar photovoltaic
SRRA	Solar radiation resource assessment
SSGP	Small Solar Generation Programme
U & I	Urban and Industrial
UGVCL	Uttar Gujarat Vij Company Limited
WLAN	Wireless local area network
WSN	Wireless Sensor Network

1 Introduction

This chapter is an outcome of the contribution of authors from their experiences and learnings from 'Solar City—Smart Grid Project—Gandhinagar, Gujarat State, India'. The chapter starts with providing overviews on power scenario in the country and in the state and thereafter includes conceptual descriptions on Solar Energy, Solar City, Smart Grid (SG) and Rooftop PV Systems.

SPV and more specifically, PV rooftop systems are these days considered proven method of power generation with promising-reliable capabilities, and hence, rapidly emerging as popular models for massive installations on account of their simplicity, ease of installation, operation and maintenance.

Reliable availability of critical data on timely basis for different decision-making authorities plays key role for successful e-Governance of energy distribution systems. This chapter presents, as a case study, how application of hybrid communication technologies has been deployed to serve need-based data for Gandhinagar Rooftop PV Programme. *Development of Remote Energy Parameter Monitoring System using Hybrid Networking Technologies* with necessary details have been reported.

Emerging approaches, such as presented in this chapter, based on hybrid communication technologies possess capability to completely revamp the current power generation locations into smart, intelligent and adaptive sites and could result in becoming basis for exploring more options in future. The chapter concludes with acknowledgments and list of references utilized.

2 Country's Power Scenario and Opportunity in Gujarat

2.1 Power Scenario in India [1, 2]

Information presented in both the Table 1a, b, sketches overall power scenario of the country quite clearly. It is quite clear from the data of Table 1a that the overall installed capacity has been almost equally divided between the energy utilities

Table 1 Power sector at a glance 'All India' (as on 30 September 2013)

Sector	MW	Percentage
(a) Total installed capacity		
State sector	90,062.14	39.37
Central sector	65,732.94	28.73
Private sector	72,926.66	31.88
Total	**2,28,721.73**	
Fuel	MW	Percentage
(b) Generation contribution based on fuel type		
Total thermal	**1,55,968.99**	**68.19**
Coal	1,34,388.39	58.75
Gas	20,380.85	8.91
Oil	1,199.75	0.52
Hydro (renewable)	39,788.40	17.39
Nuclear	4,780.00	2.08
RES (MNRE)	28,184.35	12.32
Total	**2,28,721.73**	**100**

Source Om Section [1]
Renewable energy sources (RES) include SHP, BG, BP, U & I and Wind Energy
SHP small hydro project, *BG* biomass gasifier, *BP* biomass power
U & I urban and industrial waste power, *RES* renewable energy sources

managed by state governments, central government and private sector companies. Although the combined stack of states and central governments run utilities is more than double to that of managed by private sector companies of the total capacity.

Table 1b indicates heavy dependency of Indian power sector upon Thermal (mostly on Coal)-based power production, which has quite recently became a cause of everyone's sincere concern. Among the non-thermal-based units, hydro type units lead with largest contribution, while, combined contribution from solar, wind, biomass, etc., type renewables contribute less than hydro alone. Much-debated nuclear contributes miniscule. These figures indicate the scope and hence opportunities for renewables especially solar has the highest compatibility to country's demographics and geography.

The figures of Table 2a clearly indicate steady improvements in hydro and thermal power with small, but noteworthy contribution from nuclear power, resulting in generating subdued aggregate overall growth rate figure. Significant reduction (36 %) in imports from Bhutan is a welcome change. Table 2b shows that except for the last year of 2012–2013, the 11th plan has been successful in steering the growth achievements incrementally. The overall generation in the country has been increased from 877 BU during 2011–2012 to 911.652 BU during the year 2012–2013. The growth in electricity generation during 2008–2009 was constrained due to delay in commissioning of new units during 2008–2009, long outages, shortage of coal/gas/nuclear fuel, poor hydrology, etc. It is important to note here that the improvements in coal- and gas-based thermal capacities have been steady, while that of in diesel-based thermal capacities has been reduced to almost nil sequentially this year. No significant

Table 2 Category-wise generation performance and annual growth of installed capacity [2]

Category	Achievement	Targets	Achievement	Trend of change	% change
	May-2013	May-2014	May-2014		w.r.t. 2013
(a) Category-wise generation performance					
Thermal	70.3	72.713	74.882	Improved	+6.52
Hydro	11.025	11.096	11.77	Improved	+6.76
Nuclear	2.601	2.494	2.738	Improved	+5.27
Bhutan import	0.425	0.279	0.272	Reduced	−36.00
All India	84.351	86.582	89.662	Improved	+6.30

Plan/Year	Thermal								Nuclear		Hydro		RES (MNRE)		Total	
	Coal		Gas		Diesel		Total									
	MW	Percentage	MW	Percentage	MW	Percentage	MW	Percentage	MW	Percentage	MW	Percentage	MW	Percentage	MW	Percentage
(b) Annual growth of installed capacity (since 6th plan)																
End of 6th plan	26,310.83	–	541.50	–	177.37	–	27,029.70	–	1,095	–	14,460.02	–	–	–	42,584.72	–
End of 7th plan	41,237.48	56.73	2,343.00	332.69	165.09	−6.92	43,745.57	61.84	1,565	42.92	18,307.63	26.61	18.14	–	63,636.34	49.43
End of 8th plan	54,154.48	31.32	6,561.90	180.06	293.9	78.02	61,010.28	39.47	2,225	42.17	21,658.08	18.30	902.01	4,872.49	85,795.37	34.82
End of 9th plan	62,130.88	14.73	11,163.10	70.12	1,134.83	286.13	74,428.81	21.99	2,720	22.25	26,268.76	21.29	1,628.39	80.53	105,045.96	22.44
End of 10th plan	71,121.38	14.47	13,691.71	22.65	1,201.75	5.90	86,014.84	15.57	3,900	43.38	34,653.77	31.92	7,760.6	376.58	132,329.21	25.97
End of 11th plan	112,022.38	57.51	18,381.05	34.25	1,199.75	−0.17	131,603.18	53.00	4,780	22.56	38,990.4	12.51	24,503.45	215.74	199,877.03	51.05

Table 2 (continued)

Plan/Year	Thermal								Nuclear		Hydro		RES (MNRE)		Total	
	Coal		Gas		Diesel		Total									
	MW	Percentage	MW	Percentage	MW	Percentage	MW	Percentage	MW	Percentage	MW	Percentage	MW	Percentage	MW	Percentage
End of May 2014	**147,568.39**	**31.73**	**22,607.95**	**23.00**	**1,199.75**	**0.00**	**171,376.09**	**30.22**	**4,780**	**0.00**	**40,661.41**	**4.29**	**31,692.14**	**29.34**	**248,509.64**	**24.33**
Since 6th plan or inception		**460.87**		**4,075.06**		**576.41**		**534.03**		**336.53**		**181.20**		**174,608.60**		**483.57**

Plan/year	Thermal				Nuclear	Hydro	RES (MNRE)	Total
	Coal	Gas	Diesel	Total				
End of 6th plan	–	–	–	–	–	–	–	–
End of 7th plan	56.73	332.69	−6.92	61.84	42.92	26.61	–	49.43
End of 8th plan	31.32	180.06	78.02	39.47	42.17	18.30	4,872.49	34.82
End of 9th plan	14.73	70.12	286.13	21.99	22.25	21.29	80.53	22.44
End of 10th plan	14.47	22.65	5.90	15.57	43.38	31.92	376.58	25.97
End of 11th plan	57.51	34.25	−0.17	53.00	22.56	12.51	215.74	51.05
End of May 2014	**31.73**	**23.00**	**0.00**	**30.22**	**0.00**	**4.29**	**29.34**	**24.33**
% growth since inception	**460.87**	**4,075.06**	**576.41**	**534.03**	**336.53**	**181.20**	**174,608.60**	**483.57**

capacity growth has been observed in Nuclear as well this year. Interestingly, earlier leading contributor hydro capacities have also been observed shrinking and renewables are observed contributing phenomenal capacity additions (almost equal to thermals) this year. The data is indicative of shifting of power generation paradigm from thermal-hydro combination to thermal-renewables combination with noteworthy changes in trends in diesel- and nuclear-based capacities.

Recent expert opinions [3] suggest that renewable energy can be an important part of India's plan, not only to add new capacity but also to strengthen energy security since more than ¾th of India's electricity production depends on fossil fuels such as coal and natural gas. Development of renewable energy sources, which are indigenous-distributed and have low marginal costs of generation, can increase energy security by diversifying supply, reducing import dependence, and mitigating fuel price volatility. Accelerating the use of renewable energy is also indispensable, if India is to meet its commitments to reduce its carbon intensity. Renewable energy development can also be an important tool for spurring regional economic development, particularly for many underdeveloped states, which have the greatest potential for developing such resources.

India's renewable energy installed capacity has grown impressively to 32,911.37 MW as on 31 May 2014. Wind energy dominates India's renewable energy industry, accounting for 66.79 % (21,262.23 MW). It is followed by small hydropower (11.95 %—3,803.65 MW), biomass power, including biogas cogeneration (4,761.22 MW—14.47 %) and solar power (2,829.50 MW—8.60 %).

Changing the Energy Landscape [3, 4]

Solar panel installations on just 1.6 % of Delhi's roof space is sufficient to generate 2 GW of electricity by 2020, says a recent study. The report, titled 'Rooftop Revolution: Unleashing Delhi's Solar Potential', not only emphasizes the immense potential of power generation from rooftop solar in India by highlighting its possible role and utility in addressing the power crisis in a city like Delhi, but also triggers a fresh debate over its scope and viability factor. Brought out by Greenpeace India in collaboration with Bridge to India, the report has tried to draw the attention of policymakers, industry and other stakeholders and urged them to seriously weigh the potential, capability and viability of harnessing power through solar panels installed on rooftops.

The launch of the Jawaharlal Nehru Solar Mission (JNNSM) in 2010 propelled the solar energy sector in India. The utility-scale solar sector has posted a significant growth over the last three years, as the focus has been primarily on large-scale grid-connected power plants. However, in the context of falling prices of SPV modules, grid-connected as well as off-grid installation of solar panels on rooftop presents a viable and workable solution to addressing the country's energy concerns. Despite a slow start, the rooftop segment has been gaining prominence in recent times. Several European countries, most notably Germany, have proved the merit of grid-connected power from rooftop solar that has added significantly to their overall solar power capacities.

India is endowed with vast solar energy potential. About 5,000 trillion kWh/year energy is incident over India's land area with most parts receiving 4–7 kWh/

m^2/day. Hence, both technology routes for conversion of solar radiation into heat and electricity, namely, solar thermal and SPVs, could effectively be harnessed providing huge scalability for solar in India. Solar energy also provides the ability to generate power on distributed basis and enables rapid capacity addition with short lead times. Off-grid decentralized and low-temperature applications will be advantageous from a rural electrification perspective and meeting other energy needs for power and heating-cooling in both rural and urban areas. From an energy security perspective, solar energy could be the most secure of all sources, since it is abundantly available. Theoretically, a small fraction of the total incident solar energy (if captured effectively) could meet the entire country's power requirements. It is also clear that given the large proportion of poor and energy unserved population in the country, every effort needs to be made to exploit the relatively abundant sources of energy available to the country. While, today, domestic coal-based power generation is the cheapest electricity source, future scenarios suggest that this could well change.

Under First Phase of Jawaharlal Nehru National Solar Mission (JNNSM) implemented between 1 April 2010 and 31 March 2013, 200 MW capacity equivalent off-grid SPV systems and 7 million m^2 solar thermal collector area have been installed in the country. During first year of first phase (2010–2011), a target of 32 MW SPV off-grid systems and 5 Lakh m^2 solar thermal collector area was set. The targets are to be achieved through various channel partners in the scheme. As on February 2011, the achievements' figures were 38.5 MW for off-grid PV systems and 1.2 Lakh m^2 solar thermal collector area.

India is densely populated and has high scope for using solar power. Much of the country does not have an electrical grid, so one of the first applications of solar power has been used for water pumping; to begin replacing India's four to five million diesel-powered water pumps, each consuming about 3.5 kW, and off-grid lighting. Some large projects have been proposed, and a 35,000 km^2 area of the desert has been set aside for solar power projects, sufficient to generate 700–2,100 GW [4].

Announced in November-2009, the Government of India (GoI) proposed to launch its JNNSM under the National Action Plan on Climate Change with plans to generate 1,000 MW of power by 2013 and up to 20,000 MW grid-based solar power, 2,000 MW of off-grid solar power and cover 20 million m^2 with collectors by the end of the final phase of the mission in 2020 [4].

2.2 Opportunity in Gujarat [3–6]

Supported by proactive national and state policies, including the National Action Plan on Climate Change, the country is spearheading a massive deployment of renewable energy, predominantly from solar and wind. 29 % wind and 7 % solar potential utilized in the state of Gujarat so far, indicates the sheer scale of opportunity for further deployment of the energy from these sources.

The state of Gujarat is blessed with abundant solar potential. If only 0.1 % land is used in the state, the realizable potential could be above 10,000 MW. The state has been a pioneer in solar capacity installations, with 690 MW installed in the state as on July-2012 amounting to 67 % of total installed solar capacity in the country.

For example, Charanka Solar Park of Gujarat is the Asia's largest concentration of SPV power plants co-located with centralized facilities for developers. Asian Development Bank (ADB) contributed to the Park's development by funding the US$ 100 million transmission line for evacuating solar power, and optimizing its utilization by using SG technologies. Since then, ADB has worked closely with the Government of Gujarat (GoG) on various innovative initiatives around topics of sustainable development and technological breakthroughs, including rooftop SPV installations and strengthening the high voltage distribution system and introducing the pilot concept of smart villages—using SG and clean energy interventions to make rural Gujarat—a model for responsible energy choices.

Gujarat is endowed with huge solar energy potential with most of the state having about 300 sunny days/year with annual mean daily global solar radiation in the range of 5–7 kWh/m^2/day. In this context, Gujarat aimed to pioneer a movement towards adoption and promotion of cleaner sources of power as a potential solution to the mounting global energy crisis and took a proactive approach by formulating and announcing Solar Power Generation Policy in January-2009 and appointing Gujarat Power Corporation Limited (GPCL) as the nodal agency for the development of solar projects. The Gujarat Solar Policy promises to suffice the growing power demand with an objective to promote generation of green and clean power in the state using solar energy and promoting investment in the state for solar component manufacturing and solar power plants. The encouraging Solar Energy Policies of Gujarat have attracted global investors and brought Gujarat on every investor's radar which in turn set Gujarat to emerge as the Global Solar Energy Hub.

It needs imperative mentioning that owing to proactive approach adopted by the State Government, out of 1,045 MW of grid-connected solar projects in India, 690 MW of the capacity is commissioned in Gujarat.

Milestones Achieved [6]:

- Solar Power Projects Installed: Total capacity of 852.31 MW Solar Power Projects installed at various locations in Gujarat till 31 March, 2013
- Gujarat aims to be the Solar Capital of the world.
- First state in the country to announce Solar Power Policy 2009, with targets and implementation mechanism. The state today leads in the implementation solar power projects.
- Asia's first 500 MW Solar Power Park at Charanaka, District, Patan.
- 85 investors signed PPA for setting up 971.50 MW Solar Power Projects.
- LED Village Amrapura, Gandhinagar
- Energy Audit in small-scale industries.
- Replacement of energy inefficient pump sets with efficient pump sets in agriculture sector.
- Investment Grade Energy Audit in government buildings.
- Replacement of incandescent bulbs with T5 Tube lights in government buildings

- Canal-top Solar Power Plant: 1 MW grid-connected Solar Power Plant at village, Chandrasan, T: Kadi, Dist: Mehsana on Narmada canal is an engineering innovation as it serves the triple purpose—generation of clean energy, annually reducing 90,000 L of water evaporation from the canal as well as avoiding the use of precious land for the solar power plant.
- Gujarat has the largest base of solar thermal system manufacturers, nearly 40. Many of these manufacturers also service other regions of the country.
- Solar Rooftop Systems 5 MW SPV Rooftop Scheme launched in 2012 in the cities of Gandhinagar, Mehsana, Vadodara, Surat, Rajkot and Bhavnagar.

3 Solar Energy [7]

The Sun is the ultimate source for all forms of energy. The Sun is over a hundred times larger than the earth in size and is located at a distance of 150 million kms from the earth.

Solar Energy Utilization
The solar energy can be broadly classified in two categories on the basis of its use—Solar Active (Direct Use) and Solar Passive (Indirect Use).

Solar Active
In Solar Active category, the solar energy is directly converted in the application form and can be further divided into two forms—Solar Thermal (Heating Application) and Solar Photo Voltaic (SPV) (Electricity Generation).

Solar Thermal
Solar Thermal technology is employed for collecting and converting the sun energy to heat energy for applications such as water and air heating, cooking and drying, steam generation, distillation, etc. Basically a solar thermal device consists of a solar energy collector—'the absorber', a heating or heat transferring medium and a heat storage or heat tank. Solar thermal technology employs an elaborated use of a black body, good heat conducting materials, insulation and reflectors. Solar geyser, solar concentrators, solar cookers, solar still are some example of such devices based on solar thermal technology.

Solar Photovoltaic (SPV)
SPV technology is employed for directly converting solar energy to electrical energy by the using 'solar silicon cell'. The electricity generated could be utilized for different applications directly or through battery storage system. SPV has found wide application in rural areas for various important activities besides rural home lighting. Remote villages deprived of grid power could easily be powered using the SPV technology. The economics of rural electrification could be attractive considering the high cost of power transmission and erratic power supply in the rural areas.

Solar Passive for Buildings
In Solar Passive, the solar energy can be put into use by incorporating appropriate designs in buildings itself to maximize utilization of solar energy for various purposes such as lighting, seasonal air conditioning, water and space heating/cooling and thereby reducing external energy inputs.

Solar Energy is generated when photovoltaic (PV) cells convert heat from the sun directly into electricity.

Pros

1. Non-polluting
2. Most abundant energy source available
3. Systems last 15–30 years

Cons

1. High initial investment
2. Dependent on sunny weather
3. Supplemental energy may be needed in low sunlight areas
4. Requires large physical space for PV cell panels.

4 Solar PV Cell: Construction and Power Generation [7–9]

SPV Cell is a solid state device that converts the energy of sunlight directly into electricity by the PV effect. Typical SPV cell, hierarchy of the PVs and construction of solar cell have been shown in Fig. 1.

Materials presently used for construction of PV solar cells include monocrystalline silicon, polycrystalline silicon, amorphous silicon, cadmium telluride and copper indium selenide or sulphide. Many currently available solar cells are made from bulk materials that are cut into wafers between 70 and 240 μm thick that are then processed like other semiconductors. Other materials are made as thin-film layers, organic dyes and organic polymers that are deposited on supporting substrates.

A solar cell or PV cell is constructed of semiconducting material. The most common semiconductor material employed in the construction of a solar cell is known as silicon. When sunlight falls upon a solar cell, part of the light energy is absorbed by the semiconductor material, which bangs electrons loose, causing them to flow freely. However, pure Silicon is a poor conductor of electricity, so impurities such as Phosphorus and Boron have been added to the Silicon to create its semiconductor. This not only allows the Silicon to conduct electricity, but also forces electrons freed by the absorption of light to flow in a specific direction. This directional flow of electrons is also referred to as an electric current. Due to the special composition of solar cells, the electrons are only allowed to move in a single direction. An array of solar cells converts solar energy into a usable amount of direct current (DC) electricity. Because this electric current flows in single

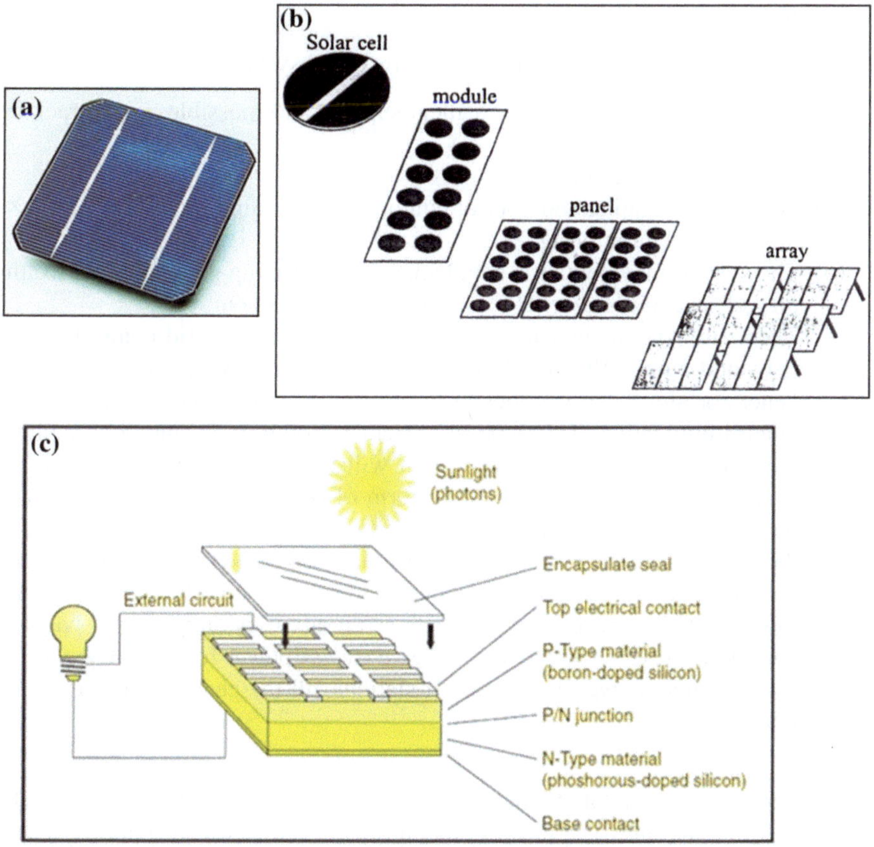

Fig. 1 **a** Solar PV cell [8], **b** the hierarchy of the photovoltaics [21], and **c** construction of solar cell [20]

direction only, hence it is called DC. This current can be drawn off externally by placing metal contacts on the top and bottom of the solar cell.

Solar cells are not the only components of complete PV system. Many other parts are usually required to provide satisfactory electricity supply. PV systems contain a provision for energy storage to supply electricity at night and during periods of severe weather. Solar cells generate DC. Since most available appliances work with alternating current (AC), some form of power conditioning is usually required. Other power conditioning or control elements are needed to interface different parts of the system, and to allow for the variable nature of the converted solar energy. All these components have to be properly interconnected, sized and specified for given expected PV operation. The size of the system and, indeed, of the PV generator and storage subsystems, depend on the geographical location and on the application for which the system is intended. Thus, solar energy could be a very reliable power source in isolated locations, with the minimum attention and maintenance.

1. *Structure of a PV system*

   The PV system consists of a number of parts or subsystems:

   (a) The PV generator, with mechanical support and, possible, a sun-tracking system
   (b) Batteries (storage subsystem)
   (c) Power conditioning and control equipment, including provision for measurement and monitoring
   (d) The system may contain a supplementary or backup generator [for example, a Diesel Generator (DG)] to form a hybrid system.
   There are two main categories of PV systems, (i) grid-connected and (ii) stand-alone. The simplest form of the latter consists simply of a PV generator alone that supplies DC power to a load whenever there is adequate illumination. This type of system is common in pumping applications. In other instances, the system will usually contain a provision for energy storage by batteries. Some form of power conditioning could also be included, as is the case when AC current is required to be generated at the output from the system. In some situations, the system could contain a backup generator as well. Grid-connected system could be subdivided into those in which the grid merely acts as an auxiliary supply (grid backup) and those in which it may also receive excess power form the PV generator (grid interactive). In PV power stations, all the generated power is fed into the grid.

2. *The PV generator*
   PV generator consists of PV modules which are interconnected to form a DC power-producing unit. The physical assembly of modules with supports is usually called an array. The structure of a PV module is based on crystalline or semi-crystalline silicon cells. Sequential development from solar cell to an array has been shown in Fig. 1.
   Most frequently, the cells in a module are interconnected in series. A typical 4-inch crystalline silicon solar cell, or 10 × 10 cm multi-crystalline cell, could provide between 1 and 1.5 W under standard conditions, depending on the cell efficiency. This power is usually supplied at a voltage 0.5–0.6 V. Since there are few appliances that work at this voltage, the immediate solution is to connect the solar cells in series. Schematic diagrams of such PV system and PV generator consisting of several modules have been shown in Fig. 2(a,b). Additionally, the generator contains bypass and blocking diodes to protect the modules and prevent the generator acting as a load in the dark.
3. *Energy Storage (Battery)*
   The solar energy supply is intrinsically variable with time; stand-alone PV systems usually make a provision for energy storage. The majority of stand-alone PV systems use battery storage. The batteries in most common use are lead–acid batteries, because of their good availability and cost-effectiveness. Nickel-Cadmium batteries are used in some smaller applications, where their

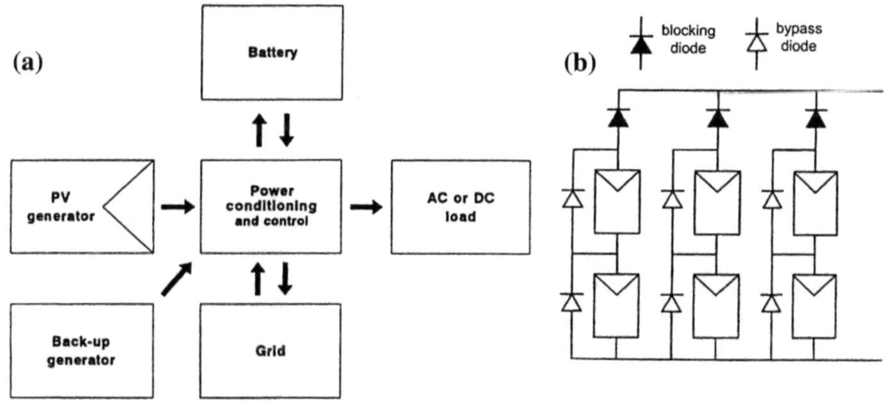

Fig. 2 Schematic diagram of **a** Photo Voltaic system [21] and **b** Photo Voltaic (PV) generator [21]

ruggedness, both mechanical and electrical, is considered essential. However, their high cost per amount of energy stored has prevented their wider use in PVs.

4. *Power conditioning and control*

   Various electrical devices are used to accommodate the variable nature of power output from the PV generator, to avoid the malfunction of the system or to convert the DC power produced by the PV generator into output. DC/DC convertors (also known as switching power convertors) are used to transform DC power between one voltage and another. The buck converter reduces the voltage, while the boost converter increases the voltage. In both cases, the voltage transformation is performed with only a small loss of power. DC/AC convertor (Inverter) is used to convert the input DC power from the PV generator or battery to the output is AC power, which can be used to run AC appliances or fed into the utility grid.

5 Solar City [10–14]

Needs

Urbanization and economic development are leading to a rapid rise in energy demand in urban areas in our country leading to enhanced green house gas (GHG) emissions. Many cities around the world are setting targets and introducing polices for promoting renewable energy and reducing GHG emissions and the countries like Australia and USA are developing the solar cities.

Several Indian cities and towns are experiencing rapid growth with the peak electricity demand. The local governments and the electricity utilities are finding it difficult to cope with this rapid rise in demand and as a result, most of the cities/towns are facing electricity shortages. In this context, the 'Solar City Programmes' in India are designed to support/encourage urban local bodies to prepare road maps to guide their cities in becoming 'renewable energy cities' or 'solar cities'.

Objectives

The Solar City aims at minimum 10 % reduction in projected demand of conventional energy at the end of 5 years, through a combination of enhancing supply from renewable energy sources in the city and energy efficiency measures. The basic aim is to motivate the local governments for adopting renewable energy technologies and energy efficiency measures. Although, the major focus for Indian Solar Cities remains solar energy due to its obvious merits (and hence the name *Solar City*), different types of renewable energy-based projects like solar, wind, biomass, small hydro, waste to energy, etc., have also been commissioned along with possible energy efficiency measures depending on the need and resource availability.

The Solar City programme aims

- To enable and empower urban local governments to address energy challenges at city level.
- To provide a framework and support to prepare a master plan including assessment of current energy situation, future demand and action plans.
- To build capacity in the urban local bodies and create awareness among all sections of civil society.
- To involve various stakeholders in the planning process.
- To oversee the implementation of sustainable energy options through Public–Private Partnership (PPP).

Eligibility to be a Solar City

The city is identified based on city population, potential and commitment for adoption of renewable energy and energy conservation in the city activities, initiatives already taken by City Council/Administration/Private Developers/Industry/ General Public in promoting renewable energy and energy conservation, regulatory measures taken on deployment of renewable energy technologies and their willingness to provide resources and sustenance of activities initiated under the programme. The cities may have population between 0.50 and 50 Lakhs, however, relaxation could be considered for special category states including North Eastern states and hilly states, Islands and Union Territories.

Prior to taking a final call on investments over solar power units, systematic Solar Radiation Resource Assessment (SRRA) is carried out. The core idea behind such an assessment is to scientifically calculate and verify the capacity of geographic location to receive sufficient sun light and generate electricity from the same.

Solar Radiation Resource Assessment (SRRA) [12]

Ministry of New and Renewable Energy (MNRE) has initiated a major project on SRRA across the nation to assess and quantify the solar radiation availability along with weather parameters with a view to develop Solar Atlas. Centre for Wind Energy Technology (C-WET), Chennai is implementing the project by installing a network of 51 SRRA stations in the first phase in different states using high quality, high resolution equipment/instruments.

Each SRRA station consists of two towers of 1.5 and 6 m tall each. The 1.5-m tall tower houses a Solar Tracker equipped with Pyranometer, Pyranometer with Shaded Ring and Pyrheliometer to measure solar parameters, such as, global, diffused and direct radiation. The 6 m tall tower houses instruments measuring rainfall, ambient temperature, atmospheric pressure, relative humidity, wind speed and direction. Each SRRA station is totally powered by 160 W SPV panels and consists of 13 equipments/instruments and records 37 parameters inclusive of both measured and derived. The data from each SRRA station averaged to 10 min will be transmitted to a Central Receiving Station established at C-WET, Chennai through GPRS mode. The implementation of the project has started from February, 2011 and by now, almost all stations have already been installed, completed and commissioned. The monthly average (daily)-wise data received from each SRRA station is available with C-WET. The quality checking process of the data is also carried out.

Table 3 also proves preparedness of Gujarat as the second highest number of SRRA stations that have been proposed and completed in state-wise allocation.

Number of cities to be developed as Solar Cities

A total of 60 cities/towns are proposed to be supported for development as Solar Cities during the 11th Plan period. At least one city in each state to a maximum of five cities in a state may be supported by the Ministry.

Assistance to Urban Local Governments

The programme assists urban local governments by providing financial assistance and technical help for:

- Preparation of a master plan for increasing renewable energy supply and energy efficiency measures in the city.

Table 3 State-wise number of SRRA stations [12]

Sr. no.	State	Proposed	Completed
1	Rajasthan	12	12
2	Gujarat	11	11
3	Tamil Nadu	7	7
4	Andhra Pradesh	6	6
5	Karnataka	5	5
6	Maharashtra	3	3
7	Madhya Pradesh	3	3
8	Jammu and Kashmir	1	1
9	Chhattisgarh	1	1
10	Pondicherry	1	1
11	Haryana	1	1
Total		**51**	**51**

- Setting up institutional arrangements for the implementation of the master plan.
- Awareness generation and capacity building activities.
- Implementation of projects as per financial incentives under various programmes of MNRE.

Financial Assistance under Solar City Programme

Up to Rs. 50.00 Lakhs per city/town is provided depending upon population and initiatives decided to be taken by the City Council/Administration as per following details:

- Up to Rs 10.00 Lakhs for preparation of a master plan within a year along with few implementable Detailed Project Reports (DPR).
- Up to Rs. 10.00 Lakhs for setting up of Solar City Cell and it's functioning for a period of 3 years.
- Up to Rs. 10.00 Lakhs for oversight of implementation during 3 years.
- Up to Rs. 20 Lakhs for capacity building and other promotional activities to be utilized in 3 years.

In addition, the financial and fiscal incentives available under various programmes of the Ministry could also be applicable on the Solar Cities for installation of renewable energy projects, systems and devices.

Cities Identified for Developing as Solar Cities

In-principle approval is given to those cities, which fulfil the requirements as per criteria, have commitment for undertaking renewable energy installations and energy efficiency measures and have submitted their proposals in the prescribed format. So far, based on the proposals received and the cities identified by some of the State Governments, in-principle approval has been given to **48 Cities** which are:

Agra, Moradabad, Rajkot, **Gandhinagar**, Surat, Nagpur, Kalyan-Dombiwali, Thane, Nanded, Aurangabad, Indore, Gwalior, Bhopal, Imphal, Kohima, Dimapur, Dehradun, Haridwar-Rishikesh, Chamoli-Gopeshwar, Chandigarh, Gurgaon, Faridabad, Coimbatore, Vijayawada, Bilaspur, Raipur, Agartala, Guwahati, Jorhat, Hubli, Mysore, Thiruananthapuram, Amritsar, Ludhiana, Ajmer, Jaipur, Jodhpur, Bhubaneswar, Aizawl, Panjim City and Environs, Itanagar, Hamirpur, Shimla, Kochi, Howrah, Rewa, Shirdi and SAS Nagar Mohali.

Cities Sanctioned

Sanctions have been given to **31 cities** which have received in-principle approvals and they have started working upon preparation of master plan. Those cities are:

Agra, Moradabad, Rajkot, **Gandhinagar**, Nagpur, Kalyan-Dombiwali, Kohima, Dehradun, Chandigarh, Gurgaon, Faridabad, Thane, Panjim City and Environs, Bilaspur, Raipur, Imphal, Itanagar, Jodhpur, Jorhat, Guwahati, Agartala, Ludhiana, Amritsar, Shimla, Hamirpur, Haridwar and Rishikesh, Vijayawada, Aizawl, Mysore, Hubli and Gwalior.

Funds towards 50 % cost of preparation of master plans have also been released to each city. Draft master plans have been prepared for 11 cities. The setting up of solar city cell in these cities is in progress.

The Master Plan of Solar City
The master plan of the city was to include the baseline for energy consumption during the year 2008, demand forecasting for the years 2013 and 2017, sector-wise strategies and action plan for implementation of renewable energy projects so as to mitigate the fossil fuel consumption in the city. This will contain details of identified project sites so that, the project proposals could be developed.

Implementation of Master Plan
After preparation of master plan, the Municipal Corporations and the State Nodal Agencies will prepare the implementable proposals on various sectors of renewable energy for implementation under particular schemes of the Ministry.

Actions to be Taken by the Municipal Corporations and City Councils/ District Administrations for Developing a Solar City

- To create a '*Solar City Cell*'.
- To constitute a '*Solar City Stakeholders' Committee*'.
- To promote National Rating System for construction of energy efficient Green Buildings in particular to commercial and institutional buildings.
- To amend building bye-laws for making the use of solar water heating systems, mandatory in certain category of buildings.
- To provide rebate in property tax through Municipal Corporations/ Municipalities and in electricity tariff though Utilities/Electricity Boards to the users of solar water heaters especially in domestic sector.
- To comply with MSW Rules 2,000 notified by the Ministry of Environment.
- To organize rigorous publicity, and also the training programmes/business meets for various stakeholders, e.g. architects, engineers, builders and developers, financial institutions, NGOs, technical institutions, manufactures/suppliers, RWAs, etc., so as to involve them actively in meeting the objective of solar city.
- To generate necessary funds from State Governments and other funding organizations for achieving the objective of making the city as '*Solar City*'.

6 The Solar City Gandhinagar [7–15]

Just like any potentially massive programme, even the rooftop PV model for India also needed pilot implementation. The learning from this pilot was aimed to provide accurate technical and administrative insights for the next phase of full scale deployment.

In a vast and diverse country like India, rooftop PV programmes are more suited for local initiatives instead of central initiatives. Due to their diversified requirements and demographic situations, different regions of the country would require different recipes.

Gandhinagar, the capital of Gujarat state, is already earmarked by the Government of India as '*The Solar City*' and is also one of the most planned cities of India, thus was making itself an ideal candidate for pilot implementation for rooftop PV programme. Figure 3 depicts conceptual understanding of implementation of Gandhinagar in form of Solar City. As shown, housing communities, individual houses, parkings, large buildings, gardens, etc., have been chosen to implement PV cells to generate solar power, hence bidirectional power flow takes place between the grid (shown as distribution towers) and them, on the other hand, the utility power generation units always function in supply mode, hence unidirectional power flow has been shown for them.

Typically, Gandhinagar's peak summer season demand has been 40 MW, non-peak demand has been 24 MW and planned distributed grid-connected solar power installation has been 5 MW, approximately. Under-mentioned core functions have been planned to realize:

1. Smart Generation: Real-time energy monitoring and forecasting of solar power generation for effective performance and fault detection
2. Smart Maintenance: Proactive fault detection and repairs
3. Smart Consumption: Real-time energy consumption and other applications making the grid smarter
4. Demand Side Management/Demand Response
5. Automated Meter Reading (AMR)

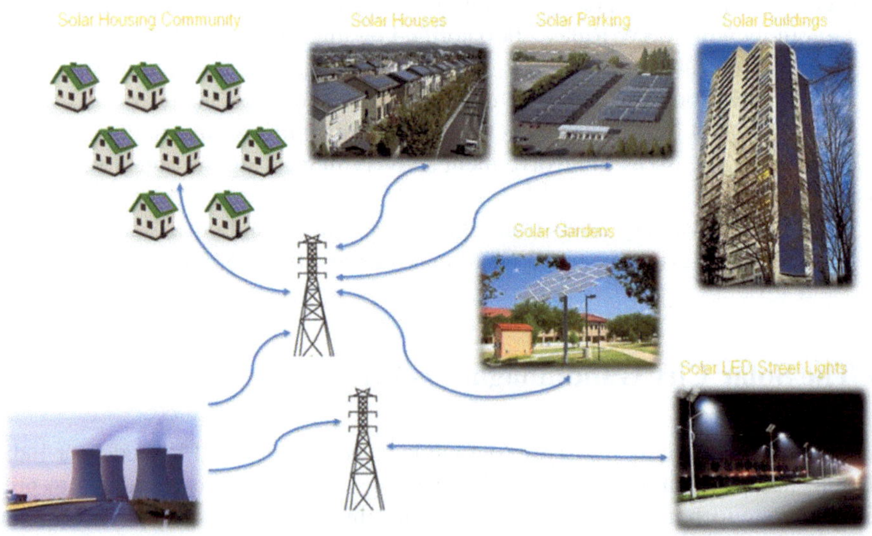

Fig. 3 Gandhinagar as Solar City [7]

The Gandhinagar PV Rooftop Programme under has been jointly proposed by the Gujarat Energy Research and Management Institute (GERMI) and the Gujarat Energy Development Agency (GEDA). The programme is aimed towards exploiting maximum potential of solar energy by installing distributed grid-tied PV systems on available roofs and terrace spaces of domestic, commercial, institutional and government buildings in Gandhinagar. The programme envisages demonstrating a successful pilot of *Solar City*, which can be then replicated in similar other cities at a larger scale all over India. The initial steps in this work involve sampling of buildings to estimate the potential to generate power in case of rooftops commissioned as shown in Fig. 4. For this, GIS software has been utilized and potentially feasible buildings for rooftop installations have been identified as shown. The buildings have

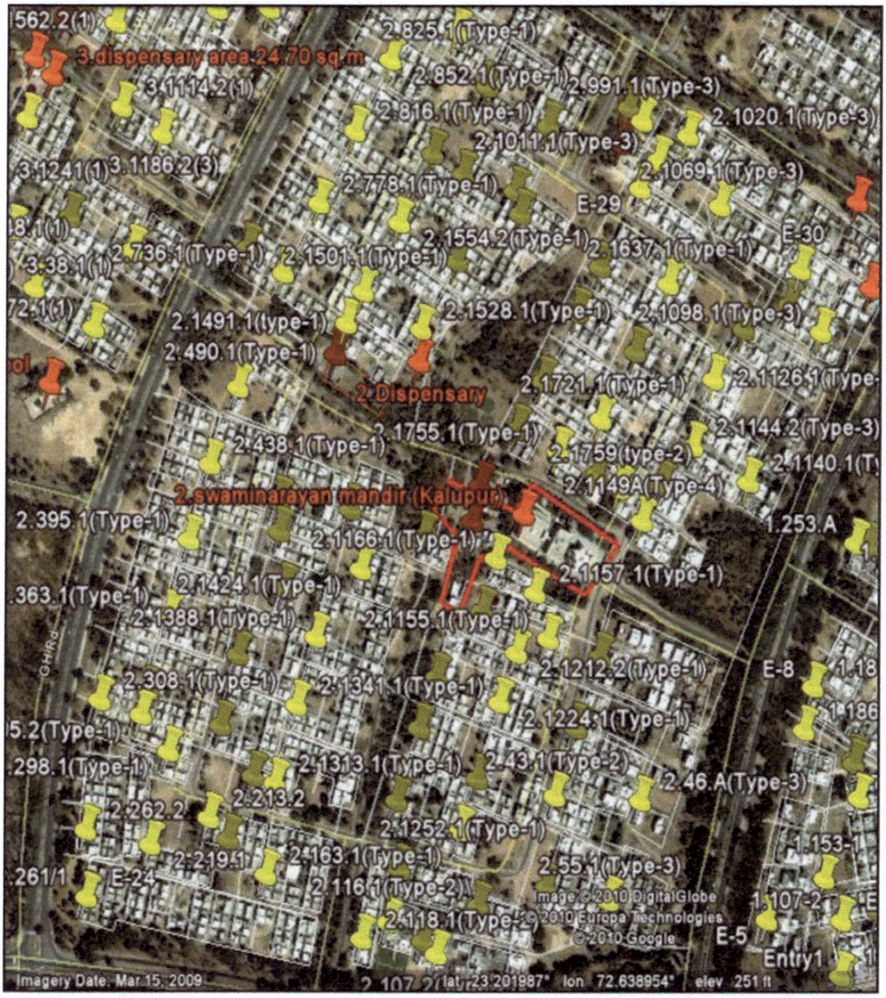

Fig. 4 Sampling of buildings for calculation of Gandhinagar rooftop photovoltaic potential [7]

also been marked with type and quantity of PV panels to be installed. At later stage, the Graphical User Interfaces (GUIs) of the respective PV panels are linked with these marked locations, so that real-time information of energy generation and utilization with other important details could be availed via Internet.

The major objective for implementing the programme has been to evolve a 'methodology' that could be implemented without too many modifications in other cities all over India.

The city of Gandhinagar is systematically divided into 30 Sectors, each measuring about 1 km × ¾ km, i.e. 75 acres each. The total footprint of buildings in the entire city is about 285 acres, which can be classified into two categories: (i) Residential, and (ii) Non-residential buildings.

The residential plots are further classified into eight official categories by the Gandhinagar Town Planner, ranging from a minimum of 50 to 1,600 m^2, out of which 40 % is typically used for building construction.

The net capacity of the proposed programme has been proposed at 5 MW, out of which 3 MW have been dedicated to residential buildings, and the remaining 2 MW for non-residential buildings.

Out of the net 5 MW of targeted solar installations, government buildings will be available for installation of 4 MW, while 1 MW will have to be installed on private rooftops and terraces. As each individual PV system on a rooftop will be of kilowatt-size, it is estimated that approximately 45 government buildings and 400 private homes will be involved in the programme.

While the Central and State Governments in India are acclimatizing themselves with the modalities of deploying megawatt-sized solar power plants, another attractive segment, which is the rooftop solar power, offers a substantial potential for harnessing this green and clean energy. In fact, majority of the solar energy installations globally consist of distributed rooftop systems. GERMI is currently heading the structuring of the 5 MW Gandhinagar Photo Voltaic Rooftop Programme, which aims at deploying net 5 MW of distributed rooftop PV systems in Gandhinagar through a Public Private Partnership (PPP) model. This programme is promoted by the Energy and Petrochemicals Department, Government of Gujarat, and funded by the Gujarat Energy Development Agency (GEDA). Gujarat Power Corporation Ltd. (GPCL) has been appointed as the implementing agency for this programme, while Torrent Power Ltd., a private utility company operating the electricity grid in Gandhinagar, will be off-taking the solar electricity generated from this programme. Complete functional structure of the programme described above has been depicted graphically in Fig. 5.

This programme is multidisciplinary and involves various stakeholders including the administrators for permissions and funding, utility company for buying the solar power, various government departments for their terraces, individuals from the society for their terraces, regulators for approval of the transaction structure, project developers and an appropriate implementing agency on behalf of the government for ensuring successful execution of the programme for 25 years. As on date, the transaction structure has been finalized, draft request for proposal

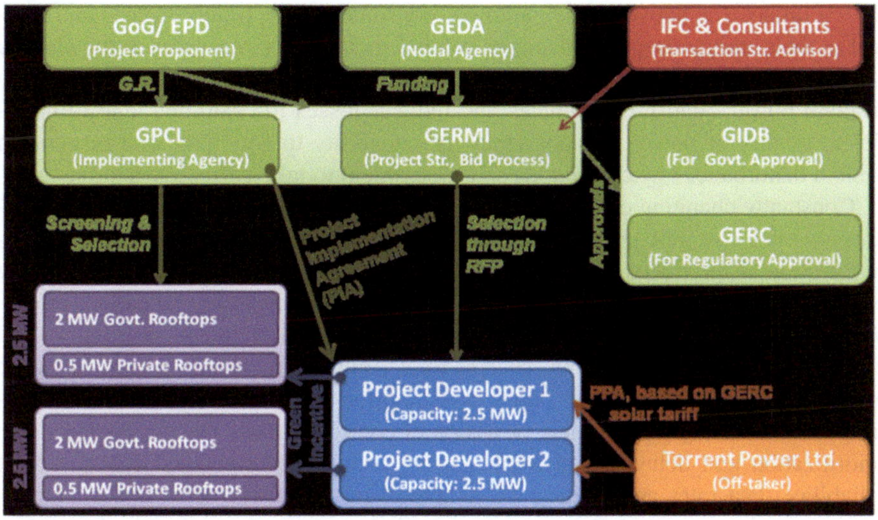

Fig. 5 Structure of the Gandhinagar photovoltaic rooftop programme [15]

has been floated and a pre-bid meeting with interested developers has been conducted. Further, a meeting with government stakeholders involved in leasing the terraces of the respective department buildings has also been conducted under the chairmanship of the Chief Secretary, Government of Gujarat, and the 'in-principle' approvals for most of these departments for participating in this programme has been received. GERMI is now in the process of filing the transaction structure of the programme to GERC for their approval, upon which the proposal from bidders will be invited. The successful bidders will be selected based on quoted tariff, and then they will commence with installation of the PV systems. This programme has also been instrumental in supporting Master's and a Ph.D. Thesis of the research scholars.

7 Smart Grid [7, 16–18]

The SG has started attracting attention of global research community and demonstrating rapid growth potential. SG is transformation of the legacy stand-alone unidirectional non-intelligent electric grids into automatic intelligent adaptive system of systems for bidirectional exchange of electric power and information. Reliability of critical power infrastructures has been the area of major focus today, wherein SGs are expected to play game changing role. SG, the modernization of conventional powergrid using technological advancements, is digital automation of electric power system from power generation to consumer appliances for improvements of quality, reliability, efficiency and environmental friendliness.

SG facilitates for active participation of consumers with timely access and control to their energy usage. Consumers can bid their energy resources at the electric market. SG supports real-time power quality monitoring and active diagnostics to respond power quality deficiencies and reduces losses to customers due to insufficient quality of power. SG possesses the capability to anticipate and respond to system disturbances by continuous self-assessment to take corrective action.

Constantly changing customer choices behaviour, increased integration of renewables, varieties of DR programmes, etc., all are likely to increase the fluctuations in the ratio of produced and consumed power. Hence, utilities must make strong efforts to deal with the increasing volatility and vibrancy in classification, affordability, feasibility and final choices of power production, demand, distribution and consumption. Important features, such as Real Time Pricing (RTP), require intensive monitoring of the consumers' power consumption patterns along with close real-time asset monitoring and timely provision of control actions. This necessitates data prioritization and delay-responsiveness using communication links with sufficient reliability, data rates and latency. SG is a modernized power network for addressing independence, global and emergency resilience issues, including the key aspect of efficient integration of distributed sources of renewable energy.

A SG transmits and distributes electricity using information and communication technology. It helps control electricity use by making consumption patterns more visible. Each solar panel has been fitted with a device. The devices have been also placed at the points of consumption of electricity. The device, about the size of a matchbox, converts the information of electricity flow into digital format and encrypts it to save it from hacking. It uses electricity to send out signals that carry data. This data gives information on when and how the consumption of electricity is increasing or decreasing. SG at home means seeing how much electricity the refrigerator, television and air conditioner are using through smart metres. Depending on the information, users can switch appliances OFF or ON. SG uses technologies such as Power Line Communication (PLC), Wireless Sensor Network (WSN), Wireless Local Area Network (WLAN), General Packet Radio Service (GPRS), ZigBee, etc. WLAN and GPRS are services available on computers and cell phones to provide access to the Internet. An interesting illustration of SG has been shown below in Fig. 6, wherein smart houses-buildings-offices with rooftop solar panels, wind farms, industrial plants with generators, utility power stations, energy storages—all are seamlessly connected by bidirectional data highway with central data centre for data acquisition and processing. The central data centre commands and controls the SG, thus formed as well as gets the desired functions and estimates calculated. Each building with rooftop solar mounted is locally controlled by Intelligent Home Power Monitoring System.

1. SG incorporates the following as its important subsystems:
   (a) Intelligent Home Power Monitoring System [or Building Automation System (BAS)]—provides consumers with essential data to make smart energy choices

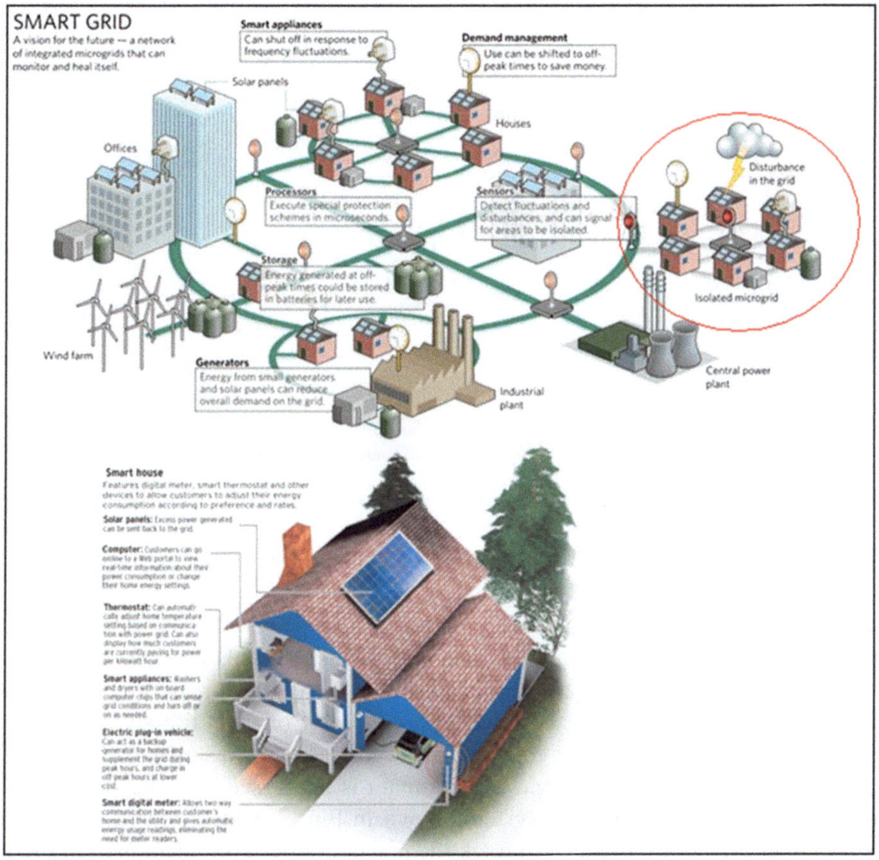

Fig. 6 Smart grid [17]

(b) Integrated Intelligent Grid Monitoring System—Integrating and tracking energy from various energy sources including grid power
(c) Analysis and energy forecasting—Provides valuable information to utility companies in efficient generation and transmission of power

Intelligent Home Power Monitoring System [7]
Conceptual system architecture of Intelligent Home Power Monitoring System or BAS has been presented in Fig. 7. Herein, all the electrical home appliances have their exclusive wired (PLC based) or wireless (ZigBee based) interfaces via which they communicate with central control and monitoring host systems via local home level aggregator. Using Internet, the access could be extended to local/remote user as well as to the system technician.

Ideally, any energy management solution should have multiple levels of energy management and conservation to make it a viable and proactive solution. While the SG solution provides the utility company with accurate information of power

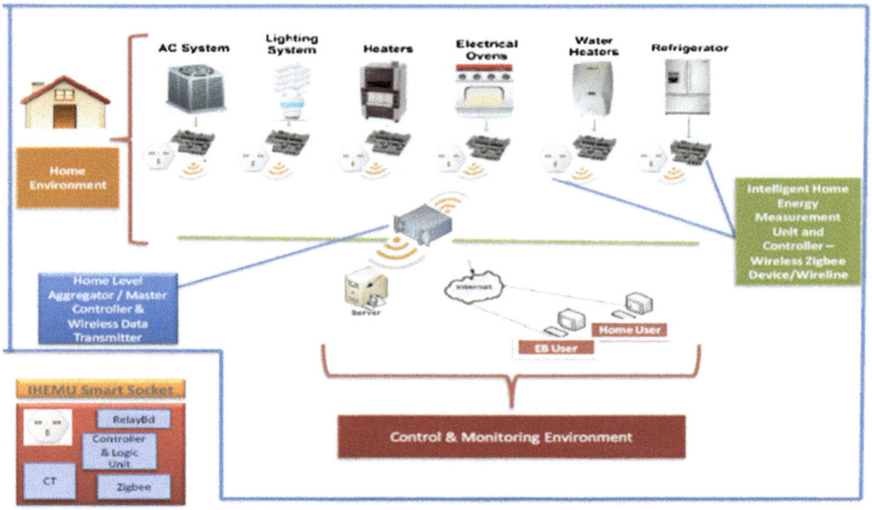

Fig. 7 System architecture intelligent home power monitoring system [7]

consumption for a particular home, and, access to power the information, and unaccounted for power information. The end user should also be empowered to know the areas she or he is consuming power in a rightful or incorrect manner, for example, a consumer will be interested what part of their life style or activity is causing very high power consumption, what are the wastages involved and tracking faulty electric equipment. This domain is addressed by smart home grid power consumption monitoring system shown in Fig. 7, which consists of smart sockets with inbuilt intelligence to track the power consumption details of particular equipment and relay it to an information database linked to a command and control dashboard. In future, this system will have a command control feature that enables a consumer to program, schedule and optimize usage of a particular device, remotely power ON or OFF the device. The system can also be programmed to automatically switch off in case of violation of any parameters set by the user.

The system could be designed to suit the aesthetics of a home/office environment; a compact form factor and can also make a lifestyle statement for conscious consumers. The system can monitor and control all home equipments such as:

1. HVAC System
2. Lighting System
3. Television
4. Water Heaters
5. Washing Machines, etc.

Advanced version of Intelligent Home Power Monitoring System is BAS [7] resulted post-transformation from the legacy stand-alone security and safety systems to intelligent computerized network-based solutions. It is a data acquisition and control system that incorporates various functionalities provided by the control system of a building. Modern BAS is a computerized, intelligent network of electronic devices, designed to monitor and control the lighting, internal climate and other systems in a building for creating optimized energy usage, safety, security, information, communication and entertainment facilities. It presents the design of a three-layer BAS validated by lab implementation for residential building using sensors, MODBUS compliant digital input/output data acquisition modules wired to form an RS-485 network, and PC-based main controller running LabVIEW GUIs with local and/or remote connectivity using intranet or Internet, respectively.

Integrated Intelligent Grid Monitoring System [7]

Proactive energy boards are seeking better and cost-effective ways to intelligently track and trace electricity consumption behaviour of customer, identify misuse of power and ability to effectively monitor and regulate the usage of power. System architecture of integrated intelligent grid monitoring system has been shown in Fig. 8. Intelligent Energy Measurement Unit (IEMU) attached to each smart home has been bidirectional wireless communication data acquisition device for communicating information related to power consumption and other details to local aggregator via ZigBee platform. This information can then be relayed to the Remote Terminal Unit (RTU), which aggregates data from multiple such IEMUs and reports the data to central server. Software-based dashboard provides real-time parameter display along with database and querying functionalities. Software could be scaled up to feature custom analytics, trending, predictive modelling and

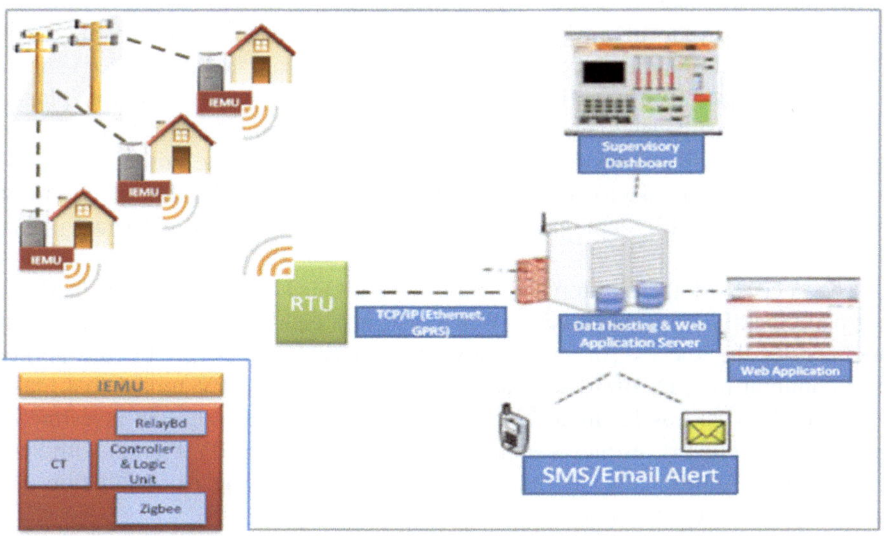

Fig. 8 System architecture of integrated intelligent grid monitoring system [7]

could offer wide range of live, graphical interpretation of parameters and alarm conditions; the dashboard could also provide custom data in the form of reports as required for supervisory control, facility managers, and for the management. The information could be displayed live and provide controllability feature through the custom GUI software interface. This could also include a large size LED/ LCD screen, which should constantly display the amount of solar energy, and thus amount of CO_2 emissions saved by the rooftop solar at the current time.

Future SG solutions could include sophisticated analytic engine that takes as its input real-time data from each end user, pole, transformer, substation, etc., and combines the load at each level with current local weather data to produce accurate load forecasts for next 24 and 48 hrs or even longer timeframes. Detailed, on-demand reporting and control could offer utilities unprecedented information and control over their distribution networks.

8 Rooftop Solar PV System [4, 7, 19–26]

PV is the direct method of converting light into electricity using a device known as a solar cell. When semiconductors such as silicon are exposed to sunlight, they produce small amounts of electric charge (electrons and holes). A well-designed solar cell separates this charge to form a positive and negative terminal. Hence, these terminals produce a voltage, and when connected to an external circuit, cause a current flow. In this way, a solar cell in the sunlight works just like a battery. PV systems can be designed for various applications and varying electricity requirements. Rooftop systems are designed in the range of kilowatts, while SPV power plants are installed in the range of megawatts.

More than 30 years ago, when solar energy experts debated about which type of PV system would be more prevalent, there were two main competitors: (i) stand-alone PV systems with batteries, and (ii) large megawatt-sized PV power plants. Ironically, the last decade witnessed an unexpected growth of a third type of a PV system—the distributed rooftop PV system, which has led the entire PV industry into an exponential growth. Today, the rooftop PV model is one of the most popular models of PV deployments and thereby of solar based electrical power generation.

The success of the rooftop PV model is attributed to its ability in facilitating participation from the masses. Whether it is a cell phone, a computer or an automobile, the success of technology in each case is marked by the ability of a common man to afford it. The attributes of a rooftop PV model that make it a potentially successful model are its simplicity and relatively low investments (compared to huge investments required for megawatt-sized PV Plants) with economic sense. Additionally, its relatively small size is making its installation and subsequent maintenance simple.

In recent years, India has emerged as one of the fastest economies in the world; however, it is a sad reality that around 400 million of its population is still deprived of electricity. Here comes the role of renewable energy in bridging that power gap, and the opportunity presented by the small-scale solar segment, especially rooftop solar, can have a compelling effect in addressing that issue to a great extent.

However, the opportunity presented by the rooftop segment can be converted into success only if the persistent challenges are overcome through adequate policy, regulatory measures and subsidy schemes. Cost and lack of awareness are often cited as the reasons for the rooftop segment not picking up its well deserved growth momentum.

The government needs to play a facilitative role in promoting rooftop solar. There is a need to start with grid-connected system in large cities, where the grid is more stable. Net metering is the way to go for promoting rooftop PV in India according to our present analysis. It allows for better system sizing and a two-way flow of electricity in the grid system. Bringing in greater clarity in net metering rules, grid interconnection and metering issues at the regulatory and policy level emerge as the needs of the hour.

A renewable energy law would be the most ideal thing, taking into consideration the direction in which the country's economy is currently headed. The government should have a favourable policy, backed by easily available subsidies. Net metering should be introduced soon with a policy framework suitable for the common man, and thrust should be given to local supply by not levying taxes on solar products for at least 10 years, so that it can be viable for the common man.

Several cities and towns in the country are experiencing substantial growth in their peak electricity demands. Municipal corporations and the electricity utilities are finding it difficult to cope with this rapid rise in demand and as a result, most of the cities/towns are facing severe electricity shortages. Various industries and commercial establishments, e.g. malls, hotels, hospitals, nursing homes, housing complexes, etc., in cities and towns use DGs for backup power even during the daytime. These generators capacities vary from few kilowatts to couple of MWs. Generally, in a single establishment more than one generator is installed; one to cater the minimum load required for lighting and the other emergency operations during load shedding; the others are for running ACs and other operations such as lifts and other higher power consuming applications. With an objective to reduce dependency on diesel gensets, a scheme to replace them with SPV could be proposed. Further, in order to utilize the existing roof space of buildings, the scheme to include rooftop SPV systems on buildings to replace DG sets installed for minimum load requirement for operation during load shedding. These loads could generally be varying between 25 and 100 kW or so.

A rooftop SPV system could be with or without grid interaction. In grid interaction system, the DC power generated from SPV panels is converted to AC power using power conditioning unit and is fed to the grid either of 11 kV three-phase lines or of 220 V single-phase line depending on the system installed at institutional/commercial establishment or at residential complex. They generate power during daytime, which is utilized fully by powering the captive loads and feeding excess power to the grid as long as grid is available. In cases, where solar power is not sufficient due to cloud covers, etc., the captive loads are served by drawing power from the grid. The grid-interactive rooftop SPV systems thus work on net metering basis, wherein the beneficiary pays to the utility on net meter reading basis only. Ideally, grid-interactive systems do not require battery backup as the grid acts as the backup for feeding excess solar power and vice versa. However, to enhance the performance reliability of the overall systems, a minimum battery-backup of one hour of load capacity is strongly recommended. In grid-interactive systems, it has, however, to be ensured that in case the grid fails, the solar power has to be fully utilized or stopped immediately feeding to the grid (if any in excess), so as to safeguard any grid person/technician from getting shock (electrocuted), while working on the grid for maintenance etc. This safety precaution is termed as 'Islanding Protection'.

Non-grid-interactive systems ideally require a full load capacity battery power backup system. However, with the introduction of advanced load management and power conditioning systems, and safety mechanisms, it is possible to segregate the daytime loads to be served directly by solar power without necessarily going through the battery backup. As in the previous case of grid-interactive systems, minimum one hour of battery backup is, however, strongly recommended for these systems also to enhance the performance reliability of the systems. The non-grid-interactive system with minimum battery backup are viable only at places where normal power is not available during daytime. In case the SPV power is to be used after sunshine hours, it would require full load capacity battery backup which will increase the cost of system which may not be economically viable even with support from Government.

Potential, Scope, Issues, Challenges [22]
Going by the 2011 census, India has around 330 million houses, with 140 million houses having proper roofs, which can accommodate on an average 1–3 kW SPV systems. Among the 140 million houses, around 130 million are having two rooms. Besides, there are other commercial buildings, shopping complexes and offices that can accommodate larger SPV capacities. Taking into account the houses with two rooms, approximately 25,000 MW of solar capacity can be installed by putting solar panels on 20 % of the roofs.

At the same time, rooftop solar can help in reducing carbon footprint by curbing dependence on fossil fuels, and save huge amount of money in imports of gas and oil. With all the benefits of lower investment, no requirement of land, as a sustainable alternative to frequent power cuts in cities and a solution to remote rural areas in the absence of grid, rooftop solar could be the next big thing in the energy domain.

The solar energy industry is still at an evolving stage, and with programmes of Government of India such as JNNSM, it is believed that the sector is heading in the right direction. However, when we look at the extent of implementation, the focus seems to be primarily on-grid connected utility-scale projects. Quite evidently, small-scale off-grid or grid-connected projects are often overlooked. The rooftop solar segment has huge potential in the country, yet it lags behind.

Solar rooftop PV has huge potential in India. The rising consumer tariffs are already making solar rooftop viable for various consumer categories (commercial, high use residential and certain industrial) in certain areas in India. This viability exists without the support of any subsidies, only on the basis of avoided energy and other variable costs. Certain studies have predicted rooftop market of close to 4,000 MW by 2017.

One of the primary and much-talked-about issue is the high upfront cost involved in putting up solar panels on rooftops. The price range in India is much stiffer and higher compared to developed markets such as Germany and Japan. The major cost of owning a solar power unit is involved primarily with the equipment such as solar panels, housing components, inverter, metering equipment, cables and wiring accessories. Battery storage has its own additional cost, which could be saved in case of grid-connected systems, as the utility pays for excess electricity that is fed into the grid. The cost of solar panels now falls around Rs. 55/W. Over and above the installation costs, there are costs linked with operating and maintaining a PV solar array. The total investment, taking into account all the costs, would come to anywhere around Rs. 1.25 Lakhs/kW. Therefore, the upfront investment for switching to a solar-powered system for a 5-kWh household will be close to Rs. 6.5 Lakhs.

The quality of rooftops also plays an important role in assessing the potential of solar installations. For example, there are a huge number of roofs in the country which are not suitable for installing solar panels. Roofs of relatively older buildings which require major refurbishment are also not deemed suitable for installing solar panels with an expected lifespan of 20–25 years. Also, a home's orientation is as important as other factors. Since solar panels on roofs should be mounted south facing, it should be made clear that there are no shadows of trees or adjoining buildings, etc.

Besides the cost factor, the other major issues that are seen as roadblocks in the proliferation of rooftop solar in India are connectivity to the grid and metering. The existing regulations do not recognize connectivity of rooftop solar systems to the grid at low voltage. On the other hand, lack of clarity on the location, type, accuracy, sealing and reading of the metres makes it a tedious proposition for both utilities and owners alike.

The government provides capital subsidies and tax benefits in the form of accelerated depreciation, but that is applicable to SPV systems only up to 500 kW. At the same time, the long delay in getting the approval of projects and subsidy is acting as a deterrent. There is a demand for factoring in the labour cost and the engineering and project management costs while deciding the subsidy structure for solar projects.

Rooftop the answer to power woes [4]
India is a power-starved nation, and during the peak summer months, even the metros like Delhi and Mumbai have to deal with long hours of power cuts. There is definitely a need to search for alternatives, and that too, viable and sustainable alternatives. With quality sunshine around the year, and huge availability of rooftop space in the country, India can kick out the power woes by installing solar panels on them. As of now, there is clear lack of awareness and the will to harness the potential that could help the country address the issue of energy security to a greater extent.

Rooftop solar plants offer several advantages over other forms of power units which include a lower electricity bill, decrease in transmission and distribution losses, low gestation time and improvement of tail-end grid voltages. However, there are several other factors that would play decisive role in the magnitude of the benefit. These include the available conditions and the tariff at which the utilities will charge. It is also believed that implementing a feed-in tariff system for rooftop solar on the lines of major European countries could be a landmark movement in stimulating the growth of the segment in India.

Central initiatives [4, 23–26]

The flagship programme of Government of India, JNNSM, aims to install 20,000 MW of grid-connected solar power by 2022; however, there has been a proper incentive scheme for setting up solar power systems on rooftops. To meet the JNNSM objectives, the MNRE, Government of India, feels that large-scale rooftop solar development could be an ideal alternative, where the generated power may be self-consumed by households, while the excess power could be fed to the grid. Small grid-connected rooftop PV plants of 2 MW or less capacity and grid connection at <33 kV are supported under the Rooftop PV and Small-Scale Generation Programme. The best suitable long-term design solution for India would be a highly distributed set of individual rooftop power generation systems connected through a local grid. However, the major issues to be resolved are still linked to the grid and getting credit for the power that is fed to the grid.

Initially, the JNNSM supported only small-scale and off-grid rooftop projects in Phase-I, but now it has taken up pilot projects to assess the feasibility of large-scale rooftop grid-connected projects. To study the various aspects of grid-connected rooftop solar in India, the MNRE has launched a pilot scheme which allows a system size ranging from 100 to 500 kW. The Solar Energy Corporation of India (SECI) is the nodal agency for the programme, which aims to generate feedback and further promotion of the concept in the country. Under this programme, rooftop solar systems are to be connected to the grid without battery backup. The surplus power, after consumption in the building, will be exported to the grid. Under the scheme, 30 % of the cost would be provided as subsidy and 70 % is to be met by the consumer.

With the Companies Bill, 2011 finally becoming law after receiving the approval of the Upper House of Parliament, it is expected that solar rooftop may get a fillip as Corporate Social Responsibility (CSR) activities. The bill mandates large-sized corporations to spend 2 % of their net profits on CSR activities, and since solar qualifies in ensuring environmental sustainability (one of the nine activities that quantify as CSR initiatives), the solar rooftop market is likely to get the necessary boost. Being clean and sustainable, solar rooftop installations also make financial sense for the corporate.

State initiatives [4, 26, 27]

Besides central schemes, there are several states that have announced their own initiatives for the rooftop solar segment. Different states are coming out with separate plans and the MNRE is supporting in implementation and making them a success.

The state of Gujarat, which is the leader in installed capacity of solar power in the country, is also looking towards harnessing power by making use of roof spaces. The capital city of Gandhinagar has seen an installation of 1.39 MW of solar power generation capacity through rooftop installation. Another scheme to develop 25 MW of power from rooftop in five other cities has already been launched. Gujarat thought its 'Gandhinagar Solar Photo Voltaic Rooftop Programme' targets to set up 5 MW capacity grid-interactive rooftop SPV systems. Under this pilot project, SPV systems of various sizes ranging from 500, 100, 50, 20, 10, 5, 1 kW or more are expected to be set up on rooftops. However, the size of an individual SPV system would depend on the shadow-free rooftop space available. Total capacity is to be divided into two subprojects having 2.5 MW capacities each and out of 2.5, 2 MW capacities SPV systems will be installed on Government Rooftops and 0.5 MW on private rooftops. The proposed metering system for these projects is the feed-in metering system.

Kerala launched its 10,000 rooftop power plants programme for 2012–2013 targeting a capacity of 10 MW. Since the per capita limit is small, this programme targeted households and small cottage industries. Each applicant is eligible to apply for 1 kW system. Also apart from MNRE subsidy of 30 % project cost or Rs. 81,000 per system, the state offers a discount of Rs. 39,000 per system.

Haryana on the other hand is targeting commercial and industrial units for setting up rooftop solar systems to overcome the shortage of power in the state. Haryana's solar policy focuses upon commercial and industrial entities and it has approved two pilot projects of 100 kW providing financial assistance of Rs. 75 Lakhs each.

Tamil Nadu in its State solar policy announced a target of 350 MW capacities to be set up through solar rooftop systems by 2015. Of this capacity, it envisions 50 MW to be set up via Generation-Based Incentive (GBI) route. GBI is proposed to be provided for a period of 6 years with the break up for solar or solar-wind hybrid rooftops installed before 31 March 2014 as—first 2 years, next 2 years and last 2 years, at different costing per unit in each phase. Consumers who wish to avail GBI benefits are required to install separate metres to measure rooftop generation. Net metering mechanism has also been proposed to set up solar rooftop systems. Net metering will be allowed for commercial establishments and individual homes to feed excess power generated back to the grid with 'power credits' accruing to the power producer.

Karnataka has a target of installing rooftop solar systems of 5–10 kW capacity on 25,000 roofs across the state. Karnataka this year in January is targeting a capacity 1.3 MW to be set up through 0.5–1 kW household solar systems across 1943 houses in several cities. The total cost was specified as Rs. 34 crores. Tariff of Rs. 3.4/kWh along with net metering facility has been proposed for such SPV systems.

Andhra Pradesh launched a campaign to encourage house owners to install solar panels on their rooftops and sell surplus power to the state. The state government has announced a net metering incentive programme for both grid-connected and off-grid installations. The owners of rooftop solar systems will have the option of connecting to the grid or to remain off-grid. Following the AP Solar policy, the AP Government released an order for setting up net metering facility in the state. This was done to encourage setting up SPV plants on rooftops. As per recent notification, for excess energy imported from the DISCOM, the consumer is expected to pay the DISCOM as per applicable retail supply tariff decided by the regulator commission. This settlement is carried out on monthly basis. For excess energy exported to the DISCOM by the consumer, no payment was to be made by the DISCOM earlier and the generator was also required to bear the entire cost of metering arrangement provided including its accessories. Later, changes were made to this order, wherein the state Government decided to make payment for the surplus energy injected into the grid by a generator. This payment will be made at pooled cost as may be decided by APERC for that year. This cost will be made effective for a period of 7 years from the date of establishment of such SPV plant and the settlement of this surplus energy will be carried out on half yearly basis. Along with this, the Government will provide 20 % subsidy from the state budget to be paid for installation of rooftop system up to 3 kW capacity in domestic sector only. This subsidy will be provided in addition to the eligible Central subsidy.

Uttarakhand has come out with a new policy to harness solar energy by installing solar panels on rooftops and wasteland around buildings to meet the power demands at a time when the state's hydropower plants are going through a difficult phase. Under this scheme, the MNRE will be providing one-time subsidy of 30 % on the benchmark cost of the project.

Rajasthan through its state Solar Policy announced in 2011 aims to promote the setting up of rooftop plants connected to LT/11 kV grid as per the MNRE guidelines under Rooftop PV and Small Solar Generation Programme (SSGP) of the National Solar Mission. Accordingly, PPA shall be executed with the concerned DISCOM of Rajasthan with metering arrangements made as per the CERC regulations.

Chhattisgarh in its solar policy (2012–2017) mentioned that solar rooftop plants can be set up under supported types of power generation plants. Under this policy, the total grid-connected capacity target is 500–1,000 MW by 2017 to be set up through grid-connected plants including solar rooftop plants.

India's PPP model for rooftop solar programmes [26]

Rooftop solar projects are a known phenomenon nowadays. From governments to citizens, everybody agrees that the combination of one infinite resource (solar energy) and a finite resource (land) can produce beneficial results. However what is socially desirable is not always economically achievable. Rooftop solar projects till now have not been economically viable and face other issues such as ownership disputes. Governments worldwide have tried to tackle this problem through providing subsidies to either the developers—who would then set up installations on rooftops; or to the citizens—who can then invest in the huge upfront costs required to procure and install a rooftop solar system. Recently, the Indian State of Gujarat has tried a novel Public Private Partnership (PPP) concept for providing boost to the rooftop solar programme in order to achieve the targets set under the JNNSM. The Gujarat Government aims to set up 5 MW rooftop solar projects in the state capital city of Gandhinagar—promoted as the model solar city of India.

The way forward [28]

After the success of the Gandhinagar Solar Rooftop SPV project in 2010–2011, the Government of Gujarat has decided to replicate and scale up the rooftop SPV initiative across the state through the development of similar pilot projects across five large cities in the state—Vadodara, Rajkot, Mehsana, Bhavnagar and Surat.

The project shall use a model similar to that used in the case of Gandhinagar, i.e. a PPP-based model which facilitates development of rooftop installations through investments in rooftop projects by private players. The private rooftop project developers shall be selected through a competitive process. These developers shall be responsible for taking the rooftops from private individual residential, commercial and industrial rooftop owners on long-term lease.

Private rooftop owners shall be invited to participate in the programme and earn an income from leasing out their rooftops to the private project developers. The benefit accruing from lease of rooftops to be known as the green incentive.

This is a grid-connected programme, wherein SPV systems will be installed on rooftop and connected to electricity grid. Under this Project, the Gujarat Government has targeted to set up PV installations on various types of rooftops aggregating to around 25 MW (i.e. 25,000 kW) in five cities in Gujarat as shown in Table 4.

The Gujarat Government will, through transparent selection procedure, select the developer(s) to install the SPV systems on building rooftops such as residential, industrial, commercial as well as Government/Public buildings; the selected Project Developer(s) will also maintain and operate the SPV installations. Project developer/s will execute the Green Incentive Agreement with property owners, whose rooftop or terrace will be utilized for the SPV system. Rooftop owner will earn by way of Green Incentive.

Table 4 Details of projects and capacity additions [28]

Sr. no.	City	DisCom	Capacity addition (MW)
1	Bhavnagar	PGVCL	3.5
2	Rajkot	PGVCL	6.5
3	Mehsana	UGVCL	5.0
4	Surat	DGVCL	5.0
5	Vadodara	MGVCL	5.0
Total			**25.0**

9 Gandhinagar Solar Photo Voltaic Rooftop Programme [5, 13, 14]

In order to promote clean energy and self-sustainability, the Gujarat Government has promoted the Gandhinagar Photo Voltaic Rooftop Programme. This programme aims at maximizing installations of SPV systems on rooftops and terraces of private homes; commercial and institutional buildings; and government-public buildings. The SPV systems installed on rooftops and terraces will convert sunlight directly into electricity and feed into the electrical grid. This programme gives an opportunity to property owners in Gandhinagar city for participation by offering their rooftop or terrace for installation of SPV System for solar power generation. In this programme, the developer is selected through a transparent competitive bidding process. The selected developer will finance/install/own/operate and maintain PV rooftop systems on owners roofs. The energy generated from this system will be sold to a utility or DISCOM and the DISCOM will pay duly approved tariffs to developers. The owner of the property will be paid a 'Green Incentive' which shall be based on the units (kWh) electricity generated by the SPV system installed on their property. This programme is the first of its kind in India and aims to be a benchmark for green energy generation at a household level. Such successful programmes will directly result in generation of clean energy without any greenhouse gas emissions, preserving our environment. Such programmes will also reduce our dependency on foreign oil as well as restrict the price of electricity in the future. For this, some developers have already been selected for carrying out the implementation process under the Gandhinagar Solar Rooftop Programme. Similar concept has also been conceived and proliferated in other cities of Surat, Rajkot, Mehsana, Bhavnagar and Vadodara which receives ample solar radiation and are suitable for installing solar rooftop installations (Fig. 9).

About Programme Structure

In this programme, the Gujarat Government will select a limited number of Solar Project Developers through a transparent competitive bidding process.

Fig. 9 Official logo of Gandhinagar rooftop programme [5]

Each Developer will install PV systems on various buildings and houses. The Developers will invest in the capital for the PV systems; further, they will own and maintain the PV systems. They will earn their revenue by generating solar electricity from the PV systems and selling it back into the electricity grid to the private DISCOM Torrent Power Limited.

Preamble
Gujarat Government has launched the ambitious 'Gandhinagar Solar Photo Voltaic Rooftop Programme'.

Objective
To encourage utilization of Green Energy sources like Solar Energy to reduce GHG emissions caused by use of conventional sources.

Aim
To encourage public participation for generation and utilization of Green Energy sources like Solar Energy.

Goal
To establish and generate Solar Power by setting up grid-interactive rooftop SPV systems in Gandhinagar.

Capacity Target—5 MW (For Phase1).

The Project in brief
The programme envisages selection of two developers, who would install solar capacity of 2.5 MW each, power from which would be supplied to Torrent Power, which is the incumbent Distribution Licensee in Gandhinagar. Selection of project developer is on the basis of a tariff-based competitive bidding process, which requires the bidders to quote tariff for a period of 25 years from the scheduled commercial operation date.

To each of the selected developer, the Gujarat Government shall provide access to around 25 state governments owned building rooftops which should be sufficient to generate about 2 MW, while the balance 0.5 MW would be generated by installing rooftop systems on around 250 households. As per a government notification the city residents shall be encouraged to come forward to participate and shall be offered '***green incentive***'—which is electricity generated by the SPV system installed on their property. The project developer shall get paid by the procurer, i.e. Torrent Power according to his submitted bid.

Results of the bid process and lessons
While about 40 companies had expressed interest in the project, finally only four companies submitted their bids. Out of these SunEdison and Azure Power were selected and have been awarded for individual phases of 2.5 MW each. The IFC was the Lead Transaction Advisor for designing and implementing the transaction. Depending upon the results of this pilot programme, the Gujarat Government wishes to launch similar projects in the cities of Surat and Rajkot.

This unique PPP model can serve as a boost for developers who are increasingly finding it ***difficult to acquire land*** for their projects. Further such projects which ensure capacity addition within city limits also ease the burden of creating ***incremental transmission capacities*** required to wheel electricity generated from remote places to demand centres. However, from the public participation, there could be some concerns – primarily the benefit of allowing the ***right of use to one's land*** (in this case rooftop), which could be a source of income in future to the owner. Even with a very modest assumption of solar energy reception of about 200 KWh/sq ft per year and a conversion factor of 7 %, the land owner would at best receive about INR 8,878 per month for a 1,000 sq. ft. of leased area. This is almost equal to an average rent value of about INR 8–10/ sq. ft. of undeveloped land in Gandhinagar. Under such a scenario, the terms and conditions binding on the land owner would play a major role.

Benefits for the Property Owner and 'Green Incentive'

The Project developers, while utilizing the rooftops and terraces of property owners, will pay a Green Incentive to the property owner. This Green Incentive will be a monetary amount paid monthly or a bimonthly to the property owner, which will directly depend upon the solar electricity generated by the PV system and fed into the electricity grid.

The property owner can avail generation-based rent from the developer based on total PV electricity generated. Since the typical tariff for conventional Electricity is around Rs. 3–5/kwh, the tariff for selling the PV electricity is around Rs. 15/kwh which implies 'Pay less for conventional electricity consumption and Gain more from solar electricity generation.

Eligibility

The property should be in city limits of Gandhinagar and shall be a permanent structure with rooftop/terrace having easy access for system developer. The applicant must be the genuine owner of the property which can be either a Residential or Commercial or Educational Building.

Accounting for PV generation

The metering technology stands out to be the key advantage of this programme. For the ease of accounting for both conventional electricity consumption and SPV electricity generation, Dual metres would be installed as shown below in the PV system architecture.

Solar Power Technology [29]

SPV system generates electric power by converting solar radiation into electrical energy. This consists of solar panels installed on rooftop made out of crystalline silicon. When sun hits the panels, radiation gets converted to DC electricity. An inverter converts the DC electricity into alternate current. The electric power generated is fed into the conventional or public grid. There is a simple metre accounting which directly metres the energy generated through the system before flowing into the grid. SPV rooftop system is relatively simple to install, operate and maintain and it is most popular and globally accepted configuration (Fig. 10).

Working of Solar Rooftop PV

PV is the direct method of converting sunlight into electricity. This direct conversion of sunlight into electricity is carried out through devices known as 'Solar Cells' which are packed (encapsulated) in a 'PV Module' (also generally known as a 'Solar Module'). PVs is the 'solar electric' technology. As shown in Fig. 11, the dc power produced by solar panel is converted to ac power using convertor and the connected to serve the domestic load. Thus generated solar power could be an excellent option of electricity supplied via grid by the utility company.

The Gandhinagar Photovoltaic Rooftop Programme is focusing only on the PV technology for installations on rooftops and terraces.

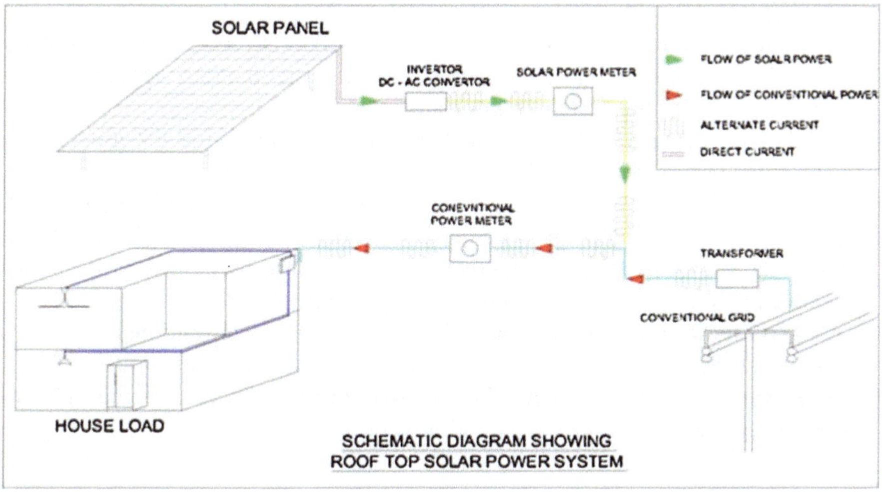

Fig. 10 Schematic diagram of solar rooftop PV system [29]

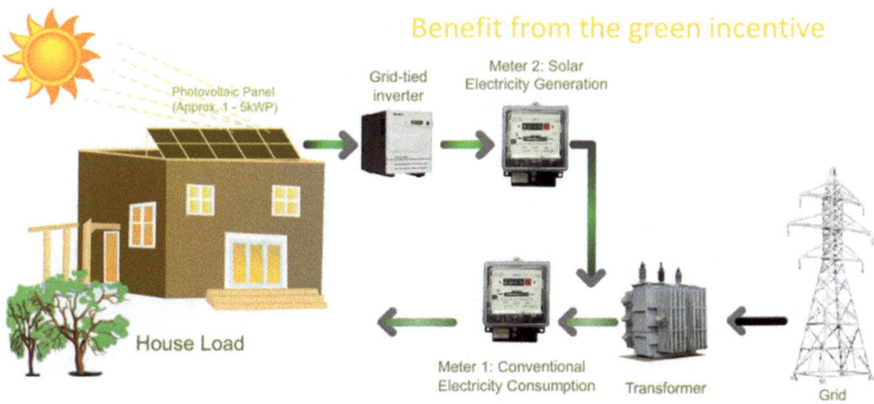

Fig. 11 Graphical representation of rooftop solar PV system [29]

A PV system can be physically divided into two parts:

- **Photovoltaic Array**, which consists of arrays of PV modules placed on the rooftop or terrace to capture sunlight, and
- **Balance of system**, which consists of electronics and other hardware such as inverters, metres, junction boxes, wiring, switches/MCBs, mountings, etc., which may be connected to the PV array and placed indoors very near to the conventional energy metre (Fig. 12).

The PV systems for the Gandhinagar Photovoltaic Rooftop Programme will not use any batteries because storage of electricity is not intended here. The electricity generated by the system will directly be injected into the electricity grid.

Components of a PV System

- PV Modules, which directly convert sunlight into DC electricity. Many PV modules are connected in series and parallel combinations to achieve the desired capacity of the system in kilowatts.
- **Module Mounting Structures**, which are used to mount the PV modules on the rooftops or terraces (Fig. 13).
- **Inverter**, which converts the DC electricity coming from the PV modules AC electricity. In the Gandhinagar Photovoltaic Rooftop Programme, a special type of inverter known as 'Grid-Interactive' inverter will be used; such grid-interactive inverters can feed the AC electricity directly into the electricity grid.
- **Energy (Electricity) Meter**, which is the same as a conventional electricity meter, to measure the solar energy (electricity) injected into the grid (Fig. 14).
- **Wires, junction boxes, switches/MCBs, surge arrestors, etc.**, to make electrical connections, and ensure safety and protection.

Space requirements

A typical 1 kW PV system would require about 15 m^2 (i.e. 150 square feet) of open space. Larger PV systems would require proportionally more space to absorb more sunlight. Hence, for example, a 5 kW system would require 75 m^2.

Energy generation capacity

A typical 1 kW PV system would generate an average of 4.5–5 units (i.e. kilowatt-hours, kWh) per day in Gandhinagar. Larger PV systems would generate

Fig. 12 **a** Photovoltaic array on a rooftop, **b** balance of system (BoS) including an inverter, and **c** energy meter, junction boxes, wiring, etc., mounted on a wall [29]

Fig. 13 **a** Different types of photovoltaic modules and **b** mounting structures [29]

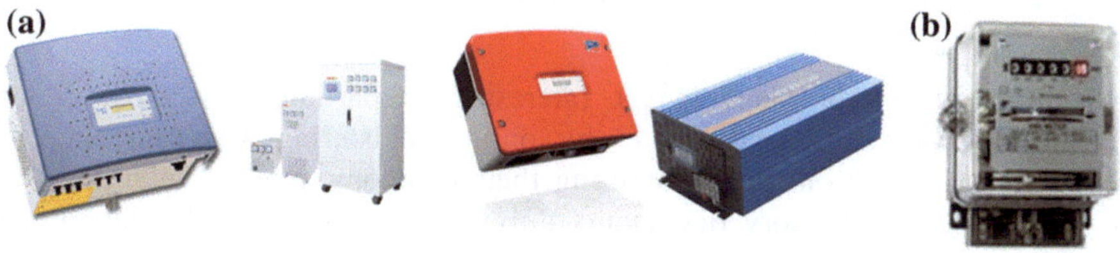

Fig. 14 **a** Different types of grid-interactive inverters [7] and **b** typical energy meter [29]

proportionally more energy. Hence, for example, a 5 kW system would generate around 22.5–25 units/day.

Advantages

1. PV systems offer one of the cleanest, most practical and scalable methods of generating clean energy.
2. The fuel, which is sunlight in this case, is free. Moreover, the earth receives enough sunlight in 1 h to provide the world's energy needs for an entire year.
3. PV systems are very easy to install, operate and maintain; further they are very easy to understand. The main activity in maintenance of such systems involves cleaning of the PV modules with water.
4. PV modules, which are the main cost of the PV system are very robust and, now come with a 25 years warranty.
5. There are no moving parts in typical rooftop or terrace PV systems. This further makes the maintenance very easy and ensures a long life of the system.

Participation Procedure [30]

Step 1
Being a rooftop owner the first step is to register the property through **REGISTRATION PROCESS** called under **EXPRESSION OF INTEREST**.

Step 2
Once the registration process is completed, the request will be technically and legally scrutinized, legally it will be verified whether the documents and various statutory approvals are valid of not.

Step 3
A technical team will visit the property to verify and assess the potential of maximum solar power installation. Technical team will also review the various aspects like accessibility, Strength of the roof, connectivity with meter, etc.

Step 4
Technical team will recommend the maximum capacity of solar power generation to project developer, who installs, operates and maintains such systems. Subsequently project developer decides the capacity to install on rooftop and he will execute the *Green Incentive* agreement with Rooftop owner.

Registration
Rooftop owners of Gandhinagar interested to participate in 'The Solar Photo Voltaic Rooftop programme' by online registering after agreeing to the terms and conditions and submitting their hardcopy application form with the self attested mandatory documents duly attached at the under-mentioned address:

GPCL (Website: http://www.egujarat.net/gg/index.html)
Block No. 8, Sixth Floor,
Udyog Bhavan, Sector 11,
Gandhinagar—382 011, Gujarat, India.

10 E-Governance for Photo Voltaic Powergrid: Gandhinagar [7]

10.1 Introduction

E-Governance of rooftop PV powergrid in Gandhinagar is based on GIS-based remote monitoring. GIS-based geographical maps are accessible locally or remotely using intranet or internet, respectively. By clicking the area of locality and location of installation in the corresponding map, the installation site information is served. By clicking the marked rooftop PV installation of building, required information in various formats is served using customized GUIs. Using proper GUI with right level access authority using password, configuration and settings could be modified for given installation as well as installed electrical appliances could also be switched on/off remotely.

10.2 GIS in E-Governance of PV Powergrid

Geographic Information System (GIS) integrates hardware, software and data for measurement, acquisition, management, analysis and real-time display of all forms of information. It allows viewing, querying and studying data in various customizable formats. Information could be retrieved and analysed in form of relationships, patterns and trends using GIS-based maps, globes, reports and charts. Due to geographical context, the data could be quickly understood and easily shared as well as integrated into other enterprise information system framework. Major GIS enabled functionalities for utilities are data management, planning and analysis, workforce automation and situational awareness.

On account of restrictions in getting suitable site for new transmission lines because of development of rural areas and the growing concern over environmental issues, power industry consequently has to keep track of numerous poles, circuits, power lines, and transformers based on location information, voltage, and distribution of electricity. Hence, GIS can effectively manage information on the distribution of electricity to customers and information describing the attributes of each customer such as location and electricity use.

10.3 Requirements for Remote E-Governance System

1. It helps to quickly debug faulty conditions or error occurrences.
2. If web enabled, it provides facility to access the system from anywhere in the world using internet.
3. Remote access helps energy auditors to remotely audit the site without visiting the site physically.
4. Crucial data of measurement and analysis are stored centrally and accessed centrally to provide protection to data from being stolen and getting corrupted.
5. Real-time remote I/O monitoring and/or control possible.
6. Local as well as remote display of different event messages, alarms, events, notifications, ON/OFF operations of devices, building-wise energy information, report generation in various customizable formats, adjustments and settings, etc.
7. 24×7 remote monitoring.
8. Protection from intrusion, theft, eavesdropping and hacking.

9. Identification of exact point of alarm, i.e. diagnosis of root cause of failure.
10. Easy system expansion of upgradation.
11. Provides better protection to operators and associated equipments.
12. Enhances communication and coordination between various units and work forces, which results in better and quicker response to abnormal occurrences.
13. Provides cleaner, distraction free work environment, finally results into focused operator free of the distractions and congregation from staff and equipments.
14. Results into reduction of unskilled/labour-type manpower requirements, but emphasizes on rise of skilled, highly trained and knowledgeable workforce.
15. Storage of huge amount of critical process related data can be possible on associated peripherals, such as hard discs, etc., that are subsequently utilized for analysis purposes to improve the system performance.
16. Reduces the costs of expansion as compared to decentralized systems.
17. Provides accurate and timely updated real-time information related to the process being monitored.
18. Provides benefits such as redundancy and other fail-safe techniques to help promote high system reliability.
19. Promotes Centralized Processing and Database Administration at single location based server(s).
20. Results in to high security, low cost, scalable system with ease of restoring backup to quickly recover faulty system.

10.4 Critical Parameters

Critical parameters identified have been listed in Table 5. Comprehensive list of parameters has been split into parameters for generation, transmission–distribution and consumption respectively for simplicity and better understanding.

10.5 Equipments, Instruments and Software

Set of equipments for implementation of typical 3 kW rooftop PV systems include Energy metres (with wired/wireless connectivity), Rooftop SPV Modules, PV Module Frames, Junction boxes, DC distribution panel, Power conditioning

Table 5 List of critical parameters

Energy generation parameters	Energy transmission and distribution parameters	Energy consumption parameters
Voltage (V)	Voltage (V)	Voltage (V)
Current (I)	Current (I)	Current (I)
Kilo Watt Hour (kWH)	Frequency (F)	Frequency (F)
Power fed into grid (P)	Power factor (ø)	Power (P)

unit and AC distribution panel. Instruments include convertors for wired RS485 to wireless ZigBee protocols, convertor for ZigBee to GSM/GPRS technology, Internet modem or router and workstation PC as host for GUIs for remote monitoring and control to suit typical site level requirements. Software requirements include windows operating system with drivers for connected peripherals, LabVIEW platform for GUI programming and GIS for integration. Remote connectivity could be provided via internet/intranet by converting PC to web server.

10.6 System Description

The system described in this chapter has been installed as pilot version for testing and validation purposes at the Vidhansabha (Assembly) and Sachivalay (Secretariat) buildings, Gandhinagar, Gujarat. This system monitors electric energy generated by the solar panels mounted on building roofs. The produced power is distributed among Sachivalay buildings to run various types of electrical appliances.

Via junction boxes installed at necessary locations, the power generated from solar panels is first supplied to DC Distribution panel, where power is filtered to remove harmonics from DC power. Then this rectified DC power is supplied to Power Conditioning Unit to convert DC power in to AC Power, for this purpose Inverter has been used in the unit. Thus, reconverted, conditioned and filtered AC power is then transmitted and distributed to the consumer buildings through AC distribution panel. An energy metre (Energy Metre-1) has been installed just after the AC distribution panel to monitor the energy parameters at the generation side. Similar Energy metre (Energy Metre-2) has also been installed for each consumer building that consumes energy for recording the actual energy usage.

All the energy metres are equipped with RS-485 wired network communication facility. Data packets of energy parameters measured are first formed in form of RS-485 packets, which are later converted to corresponding wireless ZigBee packets using RS-485 to ZigBee convertors. These wireless convertors send data wirelessly using ZigBee network to the main wireless ZigBee router. It contains ZigBee to GPRS/GSM convertor that converts wireless ZigBee packets in to wireless GSM/GPRS packets. Thus, converted data are then transmitted through the Internet connectivity via cellular network of the corresponding service provider to the remote centralized data monitoring and analysis centre located at Ahmedabad via web client–server technique. Data acquired by the web server in Ahmedabad is then grabbed by LabVIEW GUIs. GUIs open the packets supplied, interpret them and display the values of the monitored parameters. The GUIs are also equipped with facility to plot energy data in to various formats such as bar plot, pi chart, trends, etc., for detailed diagnosis and analysis purposes. Similar type of remote data acquisition and monitoring has been developed for the energy consumption data.

The network diagram shown in Fig. 15 depicts the graphical overview of the system with necessary details of communication technologies and protocols deployed at various levels.

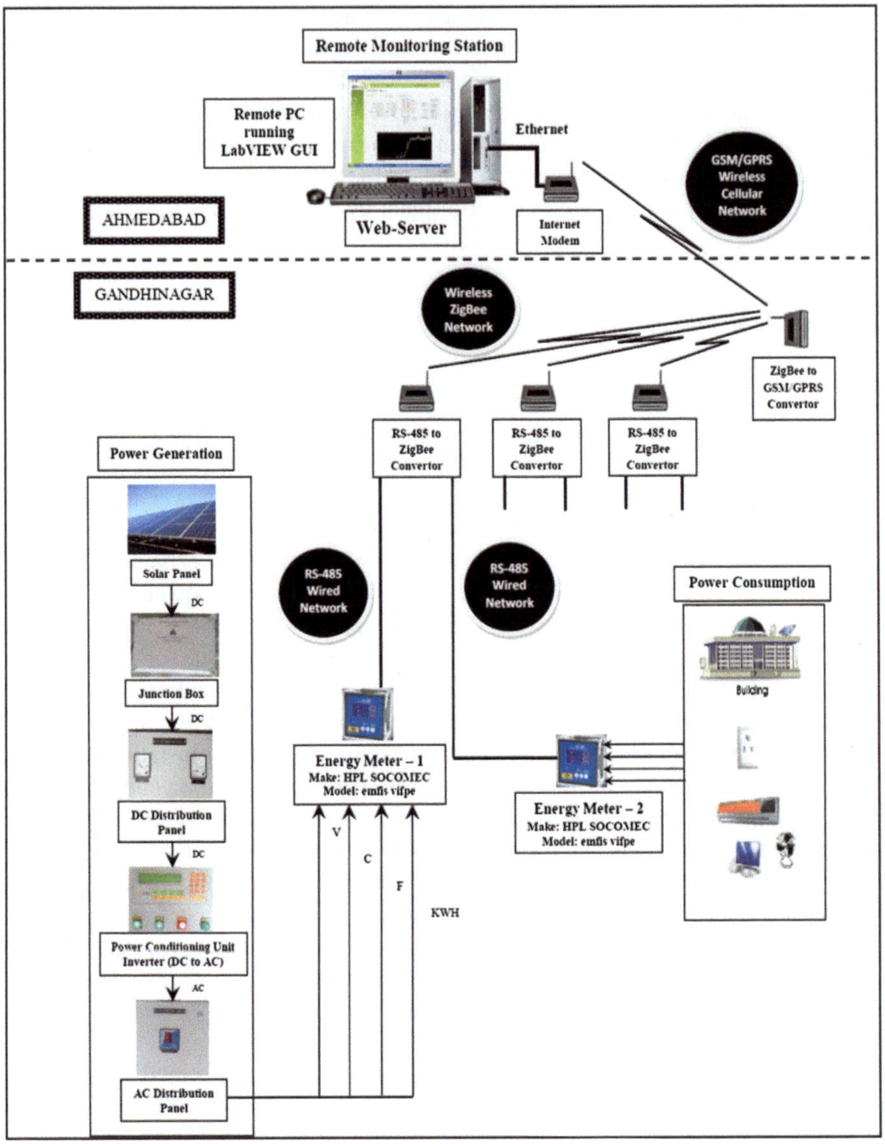

Fig. 15 Network diagram of remote energy parameter monitoring system using RS-485/ZigBee/GPRS-based hybrid networking [7]

10.7 System Implementation

The system has been installed at Sachivalay (Secretariat) and Vidhansabha (Assembly) buildings. Figure 16 shows the photographic maps of the installation sites. These installation sites are divided into two zones A and B. Block No. 1–7

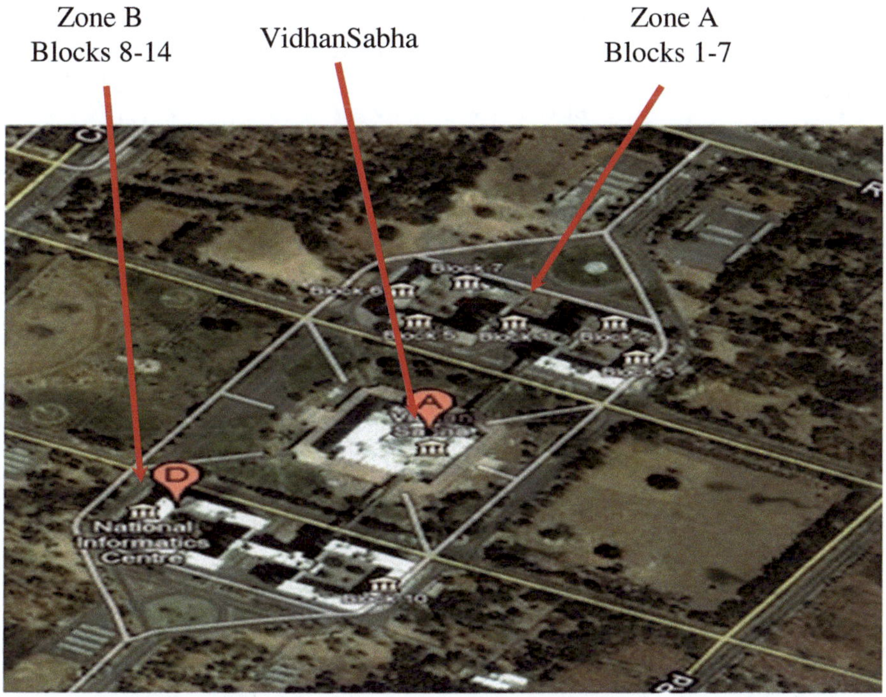

Fig. 16 Photographic maps installation sites of the proposed system on Sachivalay and Vidhansabha buildings [7]

are included in Zone A and Block No. 8–14 are included in Zone B. 10/20 KW Solar panels are installed on each building with total 130 KW generation capacities. Entire generation has been monitored by Energy Metre-1, while the entire consumption has been monitored by Energy Metre-2. Overall, this pilot system tracks the energy generation and consumption of total nine blocks of the Sachivalay buildings, which has solar cells installations on its rooftop. Each block is separately tracked with showcasing feature to track real-time energy parameters along with device level control.

The details of data being tracked block-wise are as shown in Table 6.

Notes

1. All Solar installations are 10 kwp each, with a grid connection
2. The Physical location of all the IEMU-C is in the Vidhansabha
3. Blocks 9, 11, 14 have 2 separate metres for light and power (L+P)
4. There are 13 IHEMU's installed on floor 5 of Block 7

Besides the equipments listed above, there are 4 repeaters (ZigBee signal boosters) kept in the area, 3 in Zone A (Blocks 1–7) and 1 in Zone B (Blocks 8–14).

The design has been validated by implementation at Sachivalay and Vidhansabha buildings in form of a WSN comprised of ZigBee and PC with intranet/internet

Table 6 Block-wise details of devices installed for implementation of the proposed system [7]

Block no.	Installed devices				
	IEMU-G	IEMU-C	IHEMU	RTU	LCD display
Block 1	✓	✓		✓	✓
Block 2	✓✓	✓			
Block 4	✓	✓			
Block 5	✓	✓			
Block 7	✓	✓	✓ (13)		
Block 9	✓✓	✓✓ (L+P)			
Block 11	✓	✓✓ (L+P)			
Block 12	✓✓	✓		✓	
Block 14	✓✓	✓✓ (L+P)			
Total	**13**	**12**	**13**	**2**	**1**

N.B
IEMU-G Intelligent Energy Metre Unit-Generation
IEMU-C Intelligent Energy Metre Unit-Consumption
IHEMU Intelligent Home Energy Metre Unit
RTU Router

connectivity, connected to the WSN via Ethernet. ZigBee-based WSN equipped with ZigBee to GSM/GPRS converter are deployed to form wireless network and base station for collecting data packets transmitted by RS-485 to ZigBee converter. Internet Modem acts as a gateway to host PC via Ethernet connection. The GUI is prepared with professional touch, easy to understand and interesting.

10.8 System Operation

The real-time data regarding energy parameters of generation, transmission, distribution and consumption that are acquired from all the energy metres installed at distributed geographical locations of the installation site in Gandhinagar city are displayed on remote workstation in Ahmedabad city running LabVIEW GUIs. For this purpose, different technologies are employed at different levels. By selecting a specific building, the corresponding data of generation and consumption can be viewed. Switching various appliances ON/OFF can also be done remotely via the GUIs.

Figures 17, 18 and 19 present the GUIs prepared using LabVIEW 8.6 realize benefits such as efficient operation, sufficiently fast response, user-friendly functioning, simplicity and self-explanatory. These LabVIEW GUIs are run on remote workstation with Internet connectivity through which the user can monitor real-time updated values of energy parameters and status of monitored electrical appliances. Figure 17(a) shows the snapshot of GUI that displays the consumption of all buildings with digitized parameter readout values along with the consumption

(a)

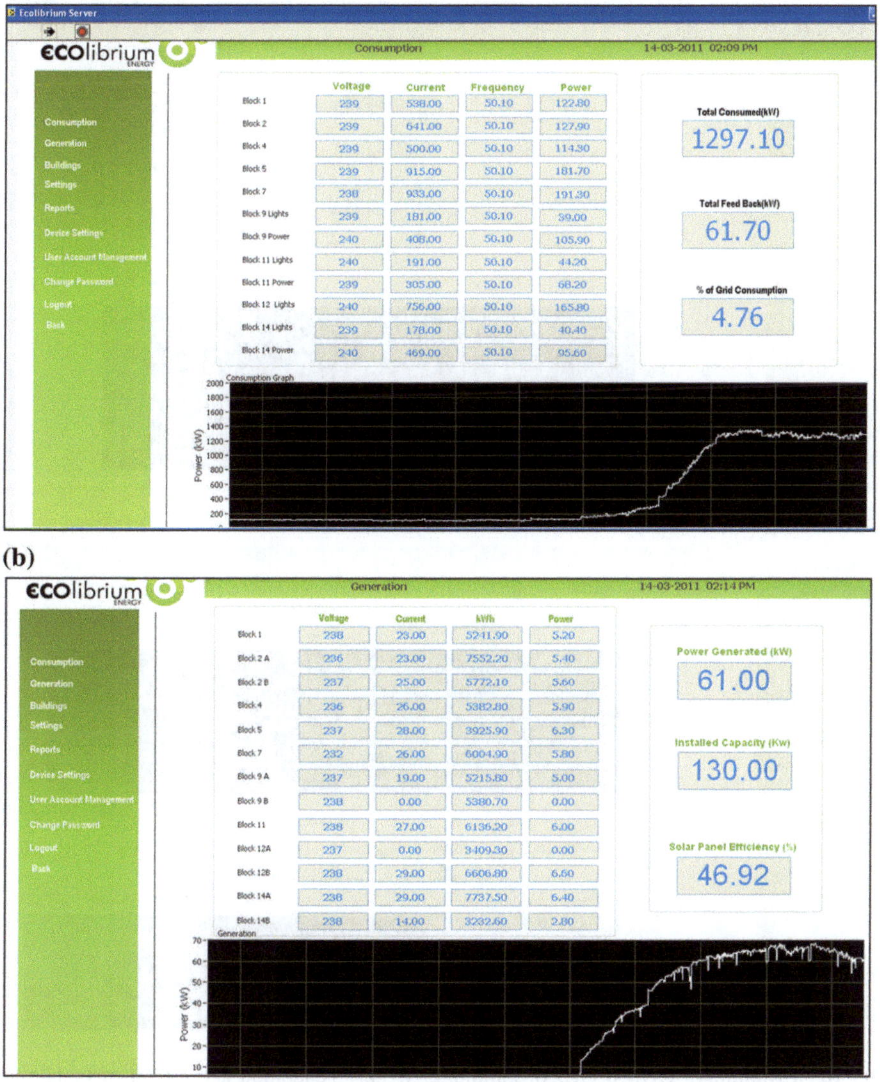

Fig. 17 **a** Consumption of all buildings GUI and **b** generation of all buildings GUI [7]

pattern in form of chart. Figure 17(b) shows the snapshot of GUI that displays the generation of all buildings with digitized parameter readout values along with the generation pattern in form of chart.

Figure 18(a) shows the snapshot of GUI that displays the generation, consumption and tariff data for Building No. 5 with digitized parameter readout values along with the consumption pattern in form of chart. Figure 18(b) shows the

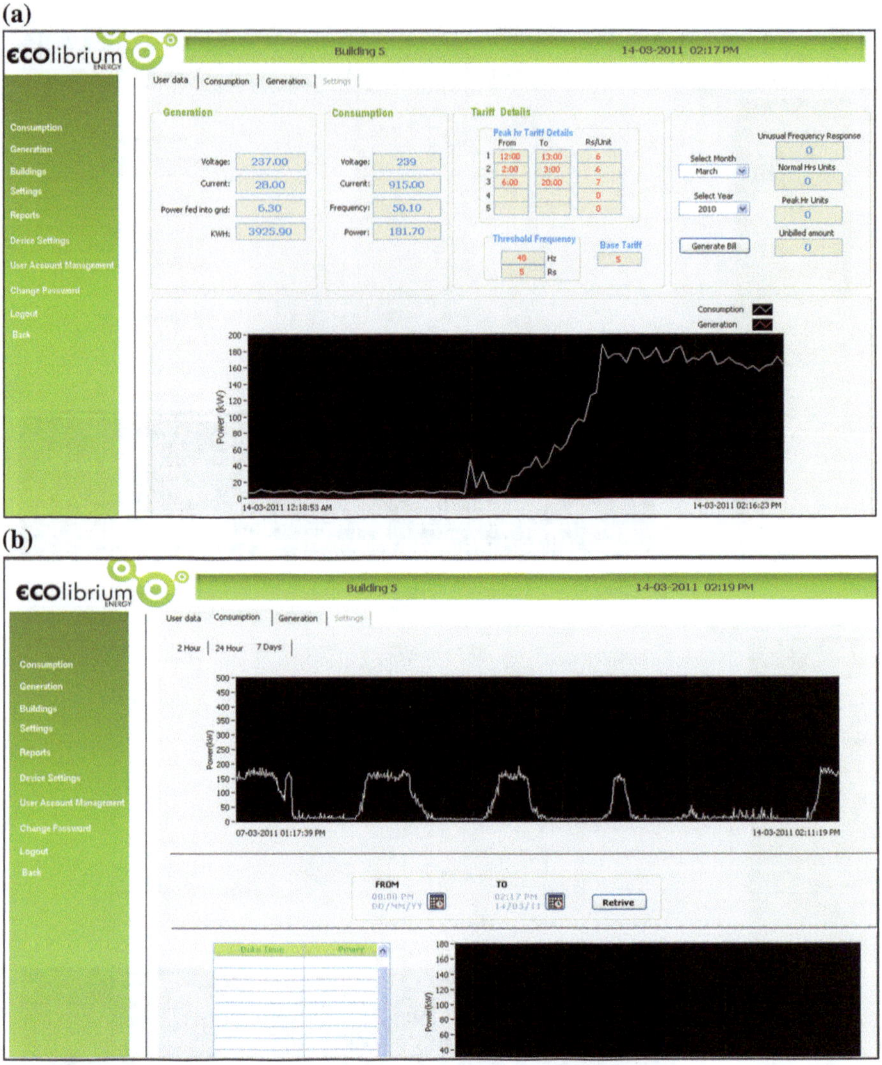

Fig. 18 **a** Building-5 GUI, **b** GUI of building-5 weekly generation and **c** GUI of building-5 weekly consumption [7]

snapshot of GUI that displays the generation of selected period for Building No. 5 with digitized parameter readout values along with the generation pattern in form of chart. Figure 17(c) shows the snapshot of GUI that displays the consumption of selected period for Building No. 5 with digitized parameter readout values along with the consumption pattern in form of chart.

Figure 19(a) shows the snapshot of GUI that displays the Demand settings and Tariffs for selected period in digitized parameter readout values. Figure 19(b)

(c)

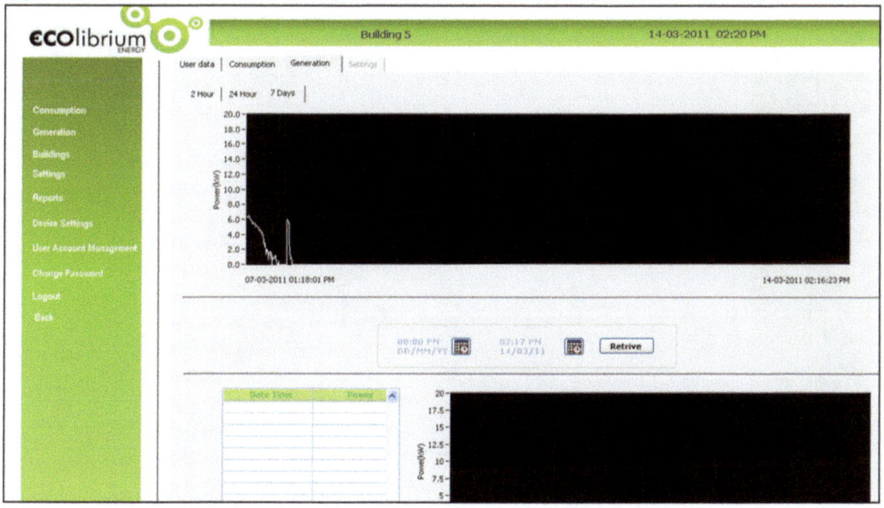

Fig. 18 (continued)

shows the snapshot of GUI that displays the status of various electrical appliances installed in the building. The appliances can be switched ON/OFF locally as well remotely using GUI.

10.9 Advantages

1. Simple and user-friendly design
2. Cost-effective solution
3. Ease of future expansion
4. Less power consumption due to sleep and sniff abilities of ZigBee networking
5. Modular system design
6. Less probability of failures
7. Ease of maintenance
8. Secured data transfer due to ZigBee networking.

10.10 Limitations

1. Frequently signal strength of GPRS found weak, hence data transfer rate is low.
2. The system gets paralyzed if installed in rural areas where connectivity of cellular network is poor.

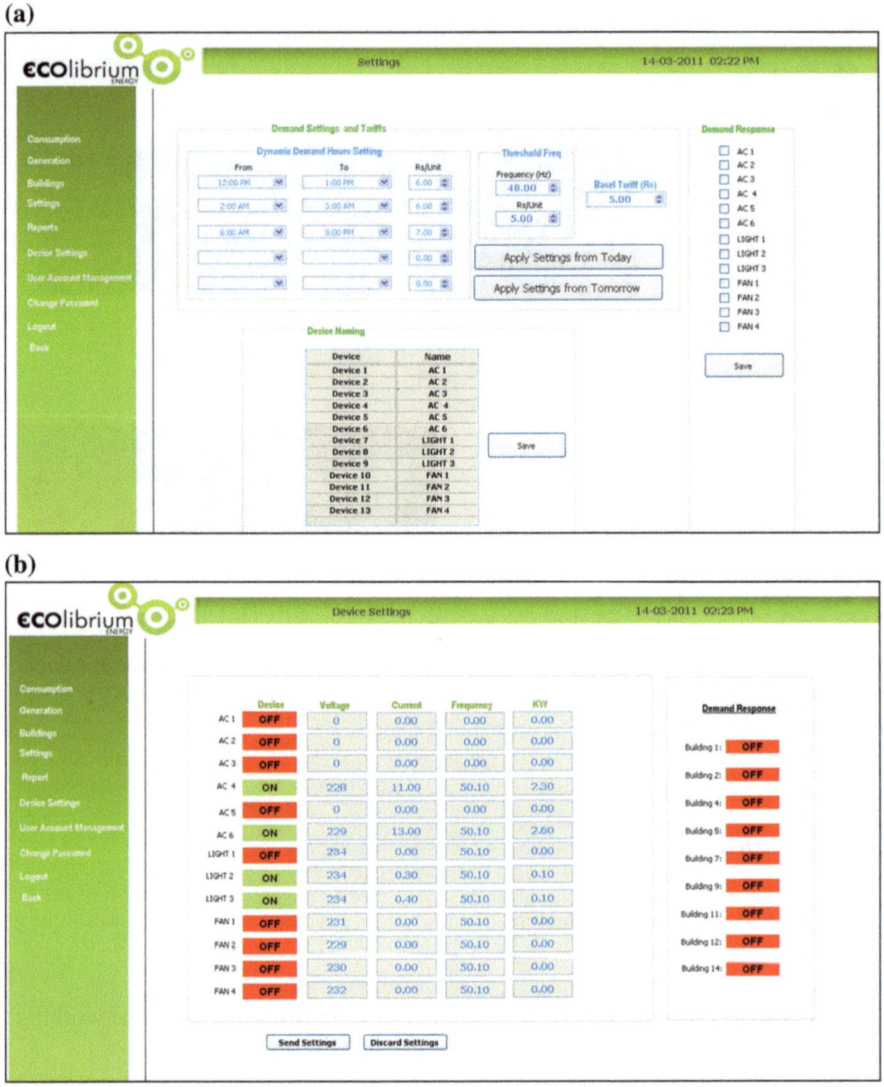

Fig. 19 **a** GUI of demand settings and tariffs and **b** GUI of device settings [7]

3. Supply power requirements are high.
4. Clear Line-of-Sight (LoS) is required, else ZigBee gets interrupted.
5. Spread of ZigBee is relatively less.

11 Transformation: Solar City to Smart City

In context of energy requirements, The Gandhinagar Rooftop Photovoltaic Programme executed under the Solar City—Smart Grid Project is inclusive of well-planned strategic framework for transformation of Gandhinagar from Solar City to Smart City.

Gandhinagar city—the capital of Gujarat, has been located in the centre of the state and enjoys availability of sunlight in large amounts throughout the year. Earlier most of the energy requirements of Gandhinagar were satisfied by thermal power stations of Uttar Gujarat Vij Company Limited (UGVCL) and Torrent Power AEC Limited. The programme aims to transform each *Consumer* into *Prosumer* (Producer+Consumer) by installation of solar panels on rooftops of building, which may be either residential or commercial or industrial or institutional type. By this unique model of Public–Private Partnership (PPP), the programme not only enhances public interest and empowerment, it quite effectively reduces the burden of electricity generation and supply from government and private utility companies. More importantly, the programme shifts the entire focus from expensive Thermal (Non-renewable) resource to Solar (Renewable) resource of power production, resulting in promoting clean and green energy. The other component under the project—Smart Grid—also takes care of technical integration and smooth bumpless transfers between conventional and renewable energy generators based on consumer's choice. The overall requirements and dependency of the city has now been observed having paradigm shift and the city is emerging as self-reliant (not fully, but quite largely) in terms of power needs. This provides a strong platform for transformation of Gandhinagar from Solar City to Smart City.

Acknowledgments The authors thankfully acknowledges the support received with appreciations to all the Faculty Members, Management and Office Bearers of Gujarat Energy Research and Management Institute (GERMI), Gandhinagar–382 007, Gujarat, India, Pandit Deendayal Petroleum University (PDPU), Gandhinagar–382 007, Gujarat, India and Dharmsinh Desai University (DDU), Nadiad–387 001, Gujarat, India, for their kind cooperation, active support and timely help in development of the work presented in the paper. Useful contributions and cooperation from UGVCL, GEDA, Torrent Power, Ms. Avadhi Aghera, M/s Ecolibrium Energy, Ahmedabad–380 015, and all the cited sources of references are also acknowledged with thanks.

References

1. http://powermin.nic.in/indian_electricity_scenario/introduction.htm
2. http://cea.nic.in/reports/monthly/executive_rep/may14.pdf
3. http://www.vibrantgujarat.com/about-event-sesg.htm
4. http://www.energynext.in/rooftop-solar/
5. http://www.vibrantgujarat.com/sed-gujarat.htm
6. http://geda.gujarat.gov.in/
7. Aghera AR (2010–2011) Wireless monitoring of energy generation, transmission and distribution parameters for solar city. M.Tech thesis, Department of Instrumentation and Control Engineering, Faculty of Technology, Dharmsinh Desai University, Nadiad, India
8. http://en.wikipedia.org/wiki/Solar_cell

9. Markvart T (ed) (2000) Solar power generation using photovoltaic cell. In: Solar electricity (chapter 4), vol 6. Wiley, New York
10. http://mnre.gov.in/schemes/decentralized-systems/solar-cities/
11. http://www.mnre.gov.in/
12. http://cwet.res.in/web/html/departments_srra.html
13. http://www.egujarat.net/gg/gandhinagar_solar_rooftop.html
14. http://www.egujarat.net/gprp1/indexmore.html
15. http://www.germi.org/reports/Annual%20Report-2011.pdf
16. Bhatt JG (2013–2014) Literature survey report for doctoral research, PhD thesis, Deptartment of Electrical Engineering, School of Technology, Pandit Deendayal Petroleum University, Gandhinagar, India
17. http://www.energytrendsinsider.com/wp-content/uploads/2010/04/smartgrid.jpg?00cfb7
18. Bhatt JG, Verma HK (2010) RS-485/MODBUS based intelligent building automation system using LabVIEW. In: 4th international conference on computer applications in electrical engineering-recent advances (CERA-09), IIT Roorkee, Roorkee, India, (Section-A1: instrumentation, Paper no. 04 (abstract)], 19–21 Feb 2010, p 02
19. http://en.wikipedia.org/wiki/Photovoltaics
20. http://mnre.gov.in/file-manager/UserFiles/presentations-pwc-workshop-06092012/GEDA.pdf
21. http://mnre.gov.in/file-manager/UserFiles/rtspv_wrtieup.pdf
22. http://seci.gov.in/upload/uploadfiles/files/Sh%20Rajendra%20Nimje%20-Feb%202014.pdf
23. http://www.mnre.gov.in/solar-mission/jnnsm/introduction-2/
24. http://www.mnre.gov.in/file-manager/UserFiles/mission_document_JNNSM.pdf
25. http://www.energynext.in/wp-content/uploads/2013/09/Sept_EN_2013.pdf
26. http://carbonn.org/uploads/tx_carbonndata/Gujarat%20Soalr%20Rooftop%20Program%20_%20IFC_01.pdf
27. http://efficientcarbon.com/wp-content/uploads/2013/07/Net-Metering-and-Solar-Rooftop_Whitepaper_EfficientCarbon.pdf
28. http://www.gpclindia.com/showpage.aspx?contentid=110
29. http://www.egujarat.net/gg/technical_details.html
30. http://www.egujarat.net/gg/general_terms.html

Part III
E-Smart City Governance-Tools and Issues

Can Smart City Be an Inclusive City?—Spatial Targeting (ST) and Spatial Data Infrastructure (SDI)

N. Sridharan

Abstract Smart City is often considered as an all inclusive city integrating various aspects of Quality of Life of the people. However, unless it is planned and managed through Spatial Data Infrastructure (SDI) some of its socio-economic and spatial issues can be left behind in the development process. Through a case study of Hubli-Dharwad City in Karnataka, this paper shows how the poverty hot spots are left behind though many innovative programmes are carried out in the city. It points out the need to introduce SDI in Indian Cities as an essential part of Smart City building process.

1 Introduction

Roberto Boisson de Marca, CEO of IEEE quotes in his article '*Smarter Cities... and Wiser Ones?' (June: 2014, p 13), Steve Jobs statement that 'Technology is nothing. What's important is that you have faith in people, that they're basically good and smart, and if you give them tools, they will do wonderful things with them.*' 'Technology is a tool that governments and societies will employ to transform and improve economies, energy generation and distribution, governance, the environment, and, above all our daily lives'. We are glad that the Indian Government has come out with the concept of 'Smart Cities' concept to develop around 100 cities in India as model cities. Ministry of Urban Development [1] views 'smart city' as the one that attract investments, good infrastructure, simple and transparent online processes that makes it easy to establish an enterprise. It is viewed as an investment destination, a pro-business approach that considers smartness in terms of physical, socio-economic, and institutional infrastructure leaving

N. Sridharan (✉)
School of Planning and Architecture, Vijayawada, Andhra Pradesh, India
e-mail: dr.nsridharan@gmail.com

T.M. Vinod Kumar (ed.), *E-Governance for Smart Cities*, Advances in 21st Century Human Settlements, DOI 10.1007/978-981-287-287-6_8

behind SDI. Often it has been argued 'smart cities' are inclusive as they emanate from two different but related perspectives, namely, 'spatially enabled society' and 'smart city' [2]. While the spatially enabled society comes from the SDI community, the Smart City forms the urban planning base. Both are technology driven and tries to empower urban population in terms of better urban planning and governance.

In the Indian context, smart city is perceived as 'those cities, which are aiming to improve the urban life comfortable and improve the living standards, good governance, efficient health and education, 24/7 power and water supply, efficient transport and quality sanitation system'. 'It also means employment generation to the masses, robust cyber connectivity and good entertainment, smart governance and leadership and above all smart people' [3]. It is also important to note the statement of the Government of India on the development of Smart cities. During the Budget Speech of 2014–2015, the Finance Minister mentioned the development of 'smart cities', as satellite towns of larger cities and by modernizing the existing mid size cities'. These two aspects bring together the idea of what Mitchell [4] calls 'E-Topia', an embedded space or 'digitally inclusive regeneration platforms'. We take this idea of e-topia, and apply this concept to the Indian Smart Cities programme. We use the SDI model through a case study of HDMC area to show how 'exclusion and inclusion' occurs in a city and how it can be addressed for spatial targeting of multiple deprivation. We used three aspects of (a) spatial inclusion, (b) water supply and (c) governance variable to identify areas of deprivation in the city and how are they excluded in the development process. Using sub-variables under each of these parameters identified, we analysed the budget allocation and whether the e-governance platform is used or not, for the budgetary decisions of the Municipal Corporation.

SDI is about facilitation and coordination of the exchange and sharing of spatial data between various stakeholders from different territorial levels [5]. Use of SDI in city governance reduces 10 hours per hour per citizen in terms of time lost in dealing with the public services [6]. However, little attention is paid to how this knowledge could be used in strategic urban governance [7].

Unlike GIS, which is a tool, SDI is an infrastructure platform, which is used by all the stakeholders and is considered as an infrastructure like any other infrastructure such as Water Supply or Road. It is an essential part of urban planning in a futuristic city. In India, Delhi has already initiated SDI through an Act in the Parliament and set up a private special purpose vehicle (SPV) for implementing SDI in Delhi for planning and implementation purpose. The platform for 'smart city' has already been made through this SDI Act in Delhi to be scaled up and replicated in other cities in India.

It is doubly challenging to implement the 'smart city' concept in a spatially distributed way, because of low Internet penetration in India. However, Indian cities such as Hubli-Dharwad, Mysore, Chennai have started using mobile technologies to manage their city's water supply, solid waste management, etc. The paper concludes that the mobile technology combined with SDI will bring efficient governance of our cities.

2 Case Study of Hubli-Dharward

Hubli-Dharwad, a city with a population of 943,788 in 2011, is considered as one of the pioneers in e-governance system in the state of Karnataka (See Map 1 for location). It has an area of 202.3 km^2 [8] spread out in three distinct areas of Hubli, Dharwad and in between area of Navnagar or the new town. Navnagar is located, in an agriculturally rich area and getting converted into urban activity in recent years. Administratively, Navnagar comes under Dharwad Division of the Municipal Corporation. In terms of governance system, Hubli-Dharwar Municipal Corporation area (HDMC) has 67 wards (smallest administrative unit in the local administration system in India) with elected representatives from each ward to the city level council.

Hubli-Dharwad was selected as a case study because of its historical emergence as a twin city and its experimentation in neoliberal urban policy of privatization of water, e-governance system and democratic decentralization. Located on the Mumbai-Bangalore Corridor, Hubli-Dharwad is the third largest urban centre after Bangalore and Mysore in Karnataka state. Hubli and Dharwad, which are separated by a distance of 18 km consists of two different identities (cities) under

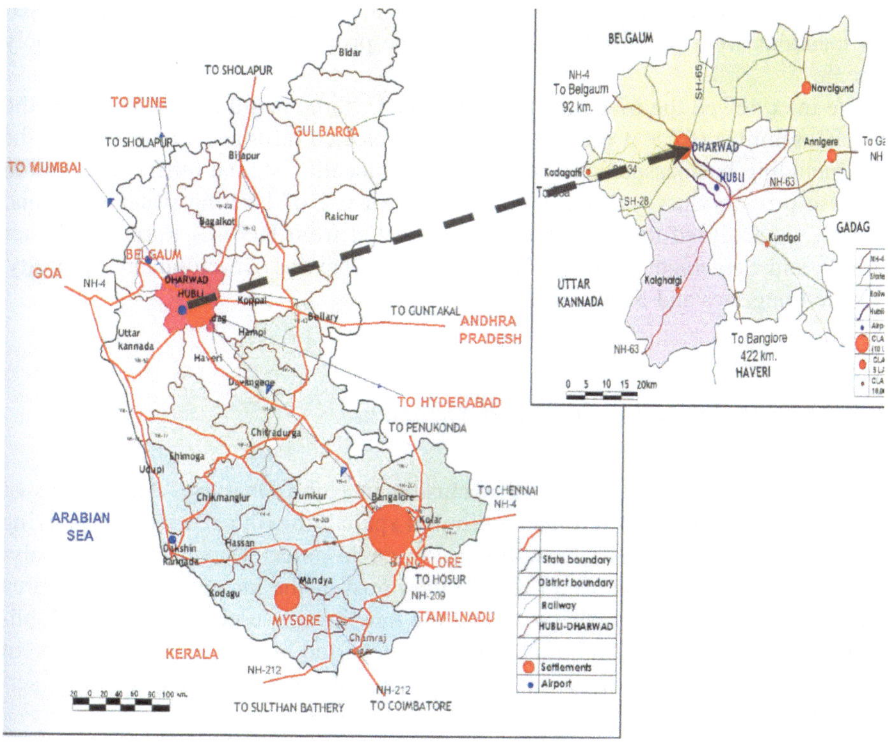

Map 1 Location of Hubli-Dharwad

one administrative unit, which were brought together under the Hubli-Dharwad Municipal Corporation (HDMC) in 1962. It is the first city in Karnataka to privatize water supply through its famous 24 × 7 water supply system in selected wards, to set up ward committees in all the 67 wards, and to set up e-governance system for public grievances.

Hubli accommodates 70 % of HDMC population while Dharwad accommodates 30 % (in 2001). The Planning Area of Hubli-Dharwad area is much bigger than the HDMC area. Using the 'E-topia concept', we analysed three types of inclusion and exclusions, namely, Spatial, Infrastructure and Governance in HDMC area. Spatial inequality was measured in terms of built-up area changes between 1967 and 2009, growth constraints, location of engines of growth (employment generating industries, commercial activity and connectivity) and spatial exclusion process [measured in terms of location of slums, building and lay out sanctions (authorized and unauthorized), land values and property tax collection]. Infrastructure exclusion was analysed in terms of macro- and micro-level issues in water supply (measured in terms of sources of water supply, demand and supply gap at macro-level to micro-level analysis in terms of access to water by slum and Below Poverty Line population, per capita availability of water, budget allocation by Municipal Corporation, and public grievances on water). We used the concept of 'Exclusion by Design'—in which a certain population to be served under potable water was excluded because of faulty design and how certain spatial locations were preferred over others in the provision of 24 × 7 water supply scheme.

We interviewed the 67 councillors of HDMC, three members from each of the ward Committee (67 × 3 = 201 sample), and selected informed citizens across the city including businesspersons, industrialists, educationist, and others. In addition to these primary surveys (that includes GPS survey of locating water tanks and slums), we analysed the secondary data collected from HDMC, Karnataka Water Supply and Sewerage Board (KWSSB), Hubli-Dharwad Development Authority, Google Imageries and Survey of India.

2.1 Spatial Exclusion in Hubli-Dharwad

Map 2 shows the extent of spatial exclusion in Hubli-Dharwad. An analysis of the spatial expansion showed that maximum spatial growth occurred in low-lying areas of Hubli as well as Dharwad causing environmental concerns. The analysis of the built-up area showed (Map 2) that most of the increase in built-up area occurred in Hubli than in Dharwad, and the new layouts especially by the Hubli-Dharwad Development Authority (HDDA) occurred in the low-lying areas of Navnagar (50 %), where land values are lower and access to water and other services are less.

Private layout in terms of housing occurred in Hubli's outskirts (where land values are higher compared to Navnagar), while Navnagar witnessed

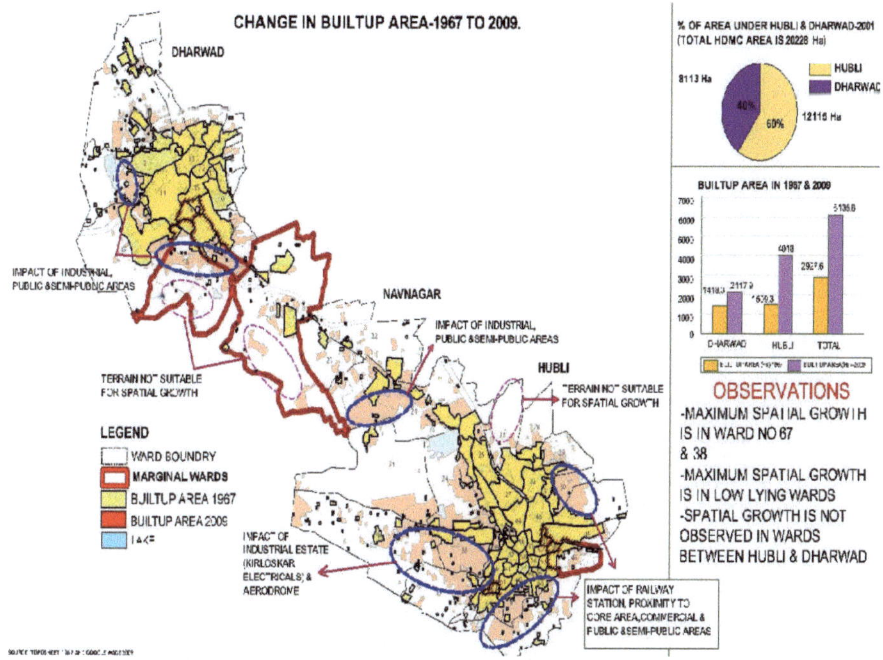

Map 2 Spatial expansion and exclusion. *Source* Author's analysis from Hubli-Dharwad development authority, 2010

developments by HDDA. In fact, HDDA is the only agency operating in the wards where majority of the households are below poverty line. On the other hand, it was observed that the periphery of both Hubli as well as Dharwad showed maximum unauthorized or illegal (100 %) lay out developments with minimum infrastructure access. Though considered as an inevitable part of development process, these unauthorized or illegal layouts provide adequate housing supply to the increasing number of houseless in HDMC area. This is similar to what Sassen [9] writes about 'regulatory fractures' and the 'rent-seeking' efforts by various rent-seekers in the city development process.

Through spatial overlays, we analysed (Map 3) the ward level changes in built-up space between 1967 when the first Master Plan was made to the spatial pattern of 2009 (the latest available spatial data). It shows that Hubli experienced maximum change in terms of built-up space as compared to Dharwad and Navnagar. Even in the case of Hubli, maximum spatial development occurred in the form of informal settlements and also in low-lying areas. If one see the type of built-up space, it is evident that the industrial area and the surrounding areas have changed fast due to the informal settlement growth in these areas. Through the combination of spatial analysis, we have observed that the wards with maximum number of marginalized population were spatially excluded except ward no. 19 and 21, ward no. 29 and 31 have large area of authorized private lay outs, which shows

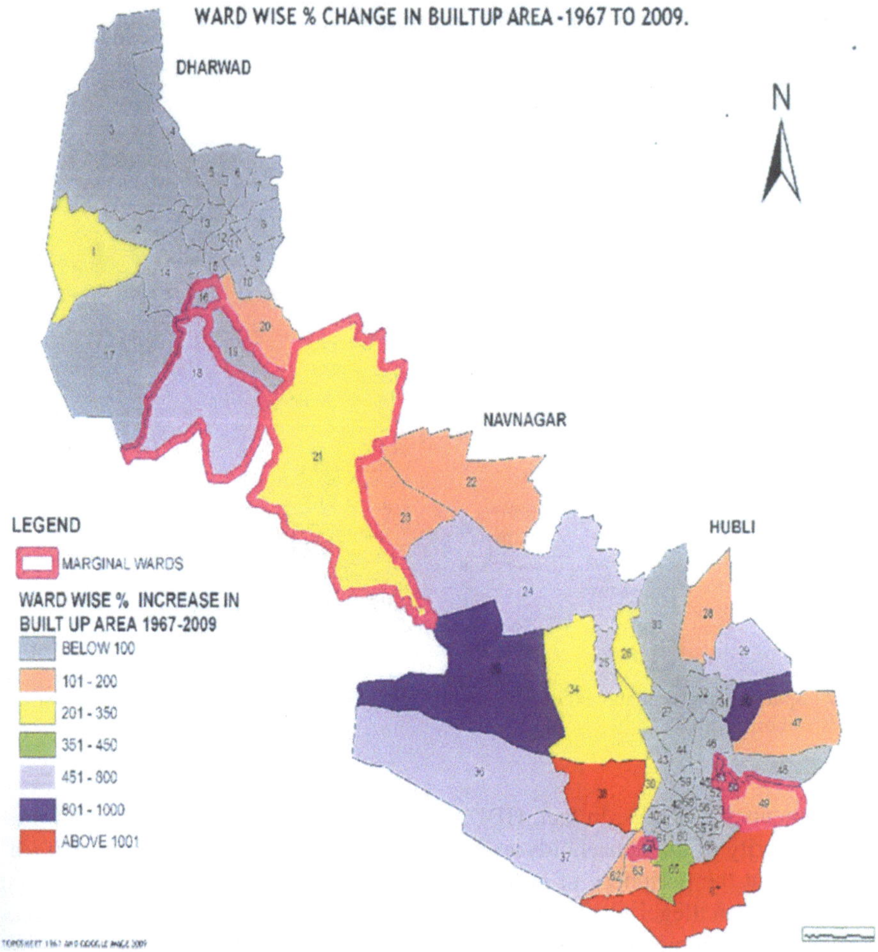

Map 3 Ward-wise % change in built-up area 1967–2009. *Source* Department of Regional Planning [10]: 'Exclusion and inclusion in spatial development—case study of an emerging metropolis: Hubli-Dharwad'. School of Planning and Architecture, New Delhi

that maximum share of increase in built-up area due to private layouts. Ward no. 18 with high poverty line and slum population shows 100 % increase in built-up area due to unauthorized constructions and these occur in low-lying areas as well, there by showing exclusions within spatial exclusions.

When one applies the overlay analysis to various spatial variables such as land values, slum location and expansion, and grievances related to encroachments (see Map 4). Grievances related to encroachments are maximum in the ward no. 34, which has seen little more than 50 % increase in the built-up space. This also highlights the governance issue of linking governance to handling of unauthorized and authorized developments and spatial inclusion and exclusions.

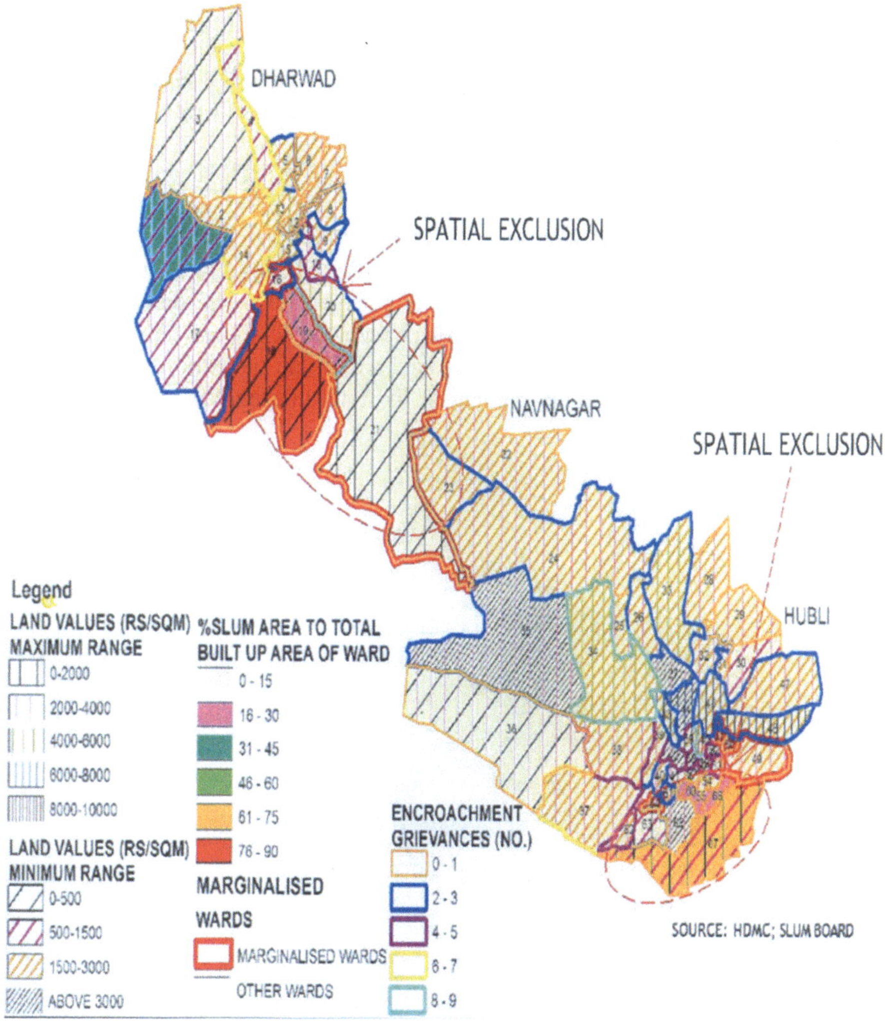

Map 4 Grievances, land values and slum are to total area. *Source* Author's analysis

2.2 *Infrastructure Exclusion*

'Exclusion by Design' [11] kept away almost 380,000 (32 % of total population) of Hubli-Dharwad population in terms of access to drinking water. Though the city claim to be having 24 × 7 water supply, water supply system was designed for 74.5 MLD at the rate of 135 LPCD to cover only 68 % of the total population of around 1.2 million typical 'Exclusion by Design'. Water supply exclusion was more in Hubli than in Dharwad. We analysed the inclusion and exclusion further through the analysis of 'Equity' (Access to tap and hand pump connection by BPL, slum and marginalized group of population), per capita availability of water (in terms of LPCD and budget allocation for water) and 'Efficiency' (in terms Public Grievances for water and low-pressure areas). We also analysed the Water access and related it grievances and budget to arrive at the inclusion and exclusion areas (Map 5).

Even after 24 × 7 water supply scheme carried out in HDMC area (in selected wards), the city as a whole accounts for little more than 50 % unaccounted water. Low water supply access as well as low LPCD provision is in those wards where there is a heavy concentration of slums and BPL households (Ward no. 18 in Dharwad and Ward no. 37 in Hubli are examples of this). The unauthorized lay out areas both in Hubli as well as Dharwad received less connectivity in terms of water. These areas depend on under ground water and have more than 150,000 persons residing in them. Eighty percent of the total households and 30 % of the

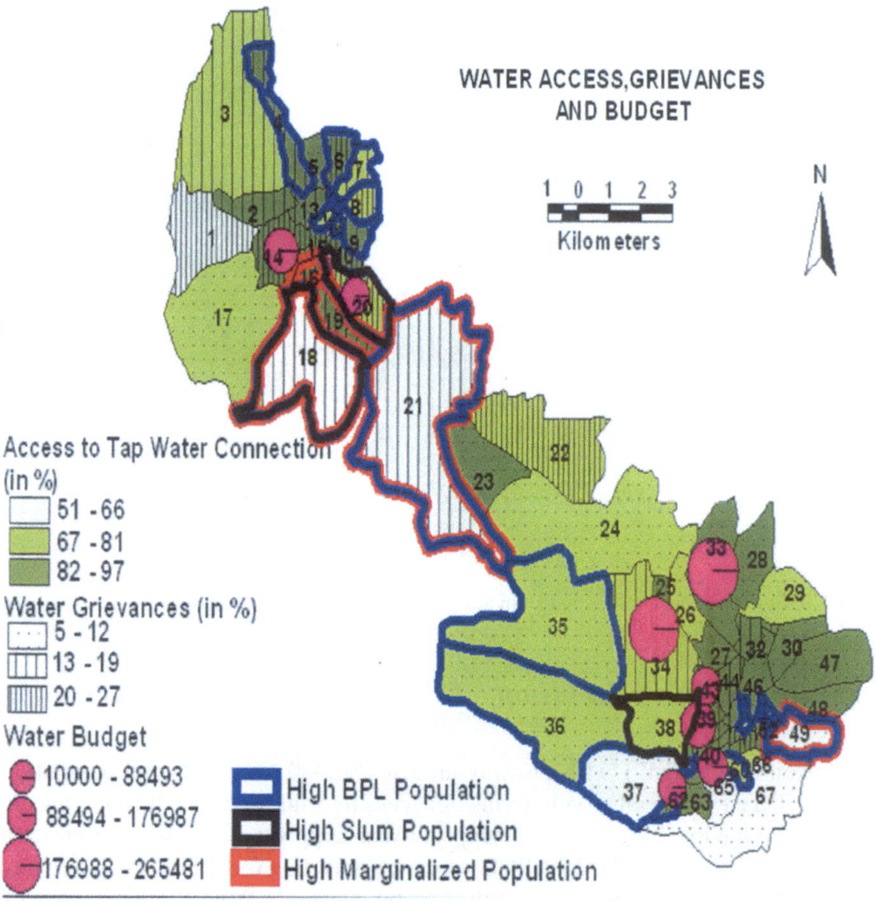

Map 5 Water access, grievances and budget allocation. *Source* Author's analysis based on KSWSSB data

BPL households are getting less than 60 LPCD of water, highlighting the extent of 'design by exclusion'. Dharwad experiences severe low-pressure problems as revealed through the analysis of public grievances data (14 % of the total water complaint). Except Ward no. 20 that is in Navnagar, almost all the slum wards and wards having maximum BPL families are excluded from water budget (political economy equations in budget allocation). These facts show the spatial inequality that emerges through lack of infrastructure access.

2.3 Governance Exclusion

Urban governance was studied through four aspects such as (a) participation (analysed through the frequency of citizen committee's meetings held at ward level), (b) accountability (performance of ward councillor in his/her ward as perceived by the voters), (c) equity (budget allocation) and (d) efficiency (number of grievances in relation to a particular aspect).

Hubli-Dharwad is considered as one of the pioneering cities in introducing citizen's participation model much before the 74th Constitutional Amendment Act, mainly to bridge the gap between the citizens and the Corporation. The citizen's committee is formed at the polling booth level among the registered voters at that polling booth. Nine members are elected by the local groups to meet at one place to discuss about the area level grievances. The meetings are presided over by the Councillor of the ward and officials from Hubli-Dharwad Municipal Corporation. Around 670 ordinary citizens are members of the Citizen's committee. Above this citizen committee is the Zonal Committee formed at every zone consisting of the representatives of the various Citizen's Committee. The representative with the highest score will head the Zonal Committee. About 72 people represent this Zonal Committee. At the highest level is the citizen advisory committee, which consists of 15 prominent members coming from various walks of life such as academicians, professionals, social workers, NGO members and government officials. The main aim of this committee is to guide the citizen's committee as well as zonal committee. Each member of the Citizen Committee represents almost 1,791 people of the city. As one move up in the ladder, one member represents approximately 16,000 people. At the advisory committee level, each member represents 80,000 people.

We analysed through stakeholder's meeting at each ward level regarding participation and accountability in terms of councillor's work in the ward. Total grievances per thousand population have been represented along with party representing at the ward level and marginalized population. We also overlayed the frequency of meetings in each ward over the grievances and level of satisfaction of the end users/voters to arrive at the areas excluded from the political process. The results are given in Map 6.

Map 6 shows the level of satisfaction of the people (accountability/efficiency) is high only in eight wards out of 67 wards. If one associates the number of

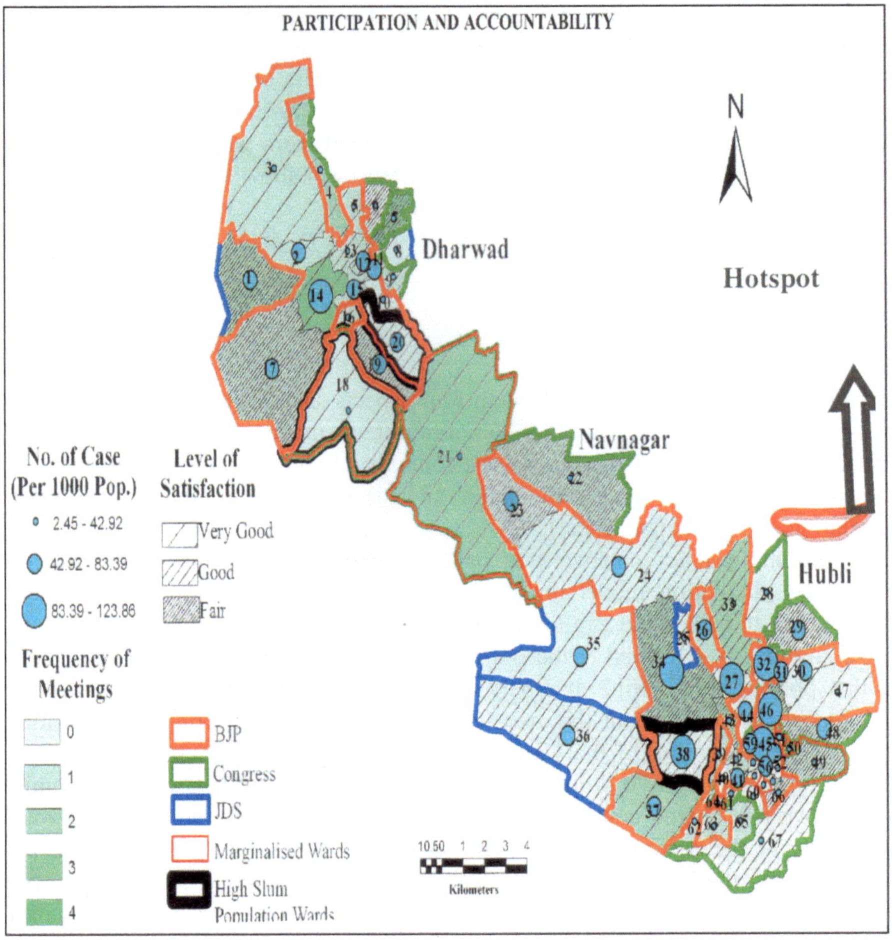

Map 6 Participation and accountability. *Source* Prepared by the author from primary survey and HDMC data

grievances to the extent of satisfaction, one can see that the wards, which reported that the satisfaction as good and very good also reported high number of cases. In these wards, the frequency of meetings is also good. The Map also shows that marginalized wards have lower levels of participation. This shows participation at the lowest administrative level ensures efficiency of service. This is an important lesson for 'smart city' builders. There is a wide spread belief that technology brings in centralization, especially 'smart city technologies' bring in more centralization. However, this can be over come through participatory GIS and participatory budgetary process as an in-built component of 'smart city' programme.

In terms of efficiency, maximum complaints came against public works (60 %) followed by Water Supply (15 %), Sewerage system (12 %), Solid waste (6 %) and health (2 %). In terms of sectoral equity, maximum amount of per capita budget expenditure went to public works followed by health, solid waste, water supply and poverty alleviation. Other sectors followed. Maximum budget allocation to public works and maximum grievances for that sector shows high inefficiency factor in the provision of infrastructure in HDMC area. In fact, irrespective of party affiliation and representation, across party lines, all the parties spent more on Public Works. Based on these analyses, we identified the 'poverty hotspots' in the city.

Over all Ward no. 18 in Dharward was identified as a 'hot spot' (See Map 4) in terms of spatial exclusion for planning intervention. This ward due to its dominance by a particular minority community, representation by opposition and poor budget allocation has been excluded in the over all spatial development process in Hubli-Dharward. This shows clearly 'exclusion by design' even in the budgetary process. This is in contra to what the councillors felt about the issues.

The opinion survey done with the elected representative in all the 67 wards revealed that majority of councillors (98 %) opined that Road improvement is the important issue in their ward. Water and Solid waste Management came second and third respectively in terms of important issue at the Ward level. State bureaucracy (Second Capability according to Sassen [9]) and politicians colluded to expand the development process, in most cases illegally (regulatory fracture) and sought budget allocation to those areas where the high and middle-income group was living in HDMC area. Higher funds were sought for road improvement in wards dominated by illegal colonies/layouts. High Below the Poverty Line households (BPL) dominated areas, undulating terrain and unplanned expansion areas were excluded in the development process in terms of urban governance, and budget allocation (Ward no. 18 in this case). This case study showed how inequalities occur even with in a metropolis, due to the rent-seeking efforts in developing certain areas through fractured regulatory policy, and inadequate water supply due to 'exclusion by design' and faulty budget allocation that were not based on public demand expressed here in terms of public grievances system.

3 Conclusions

Lack of recognition of Spatial Data as an Infrastructure, and including SDI as one of the infrastructure in the planning process has resulted in spatial, infrastructure and governance exclusion in the city development. Though at the national level, there is National Spatial Data Infrastructure (NSDI) as a pioneering organization, its role and relevance has been subdued and its backing through State Level Acts and City level implementation will solve the problems associated with spatial exclusions. Smart city functions on the basis of data, and its backbone is Spatial Data. 'Truly smart cities are long way off' [12].

We have seen SDI creates 'digitally inclusive regeneration platforms' (E-Topia) where various stakeholders such as users of different infrastructure, decision analysts (technocrats) and decision makers (Bureaucrats and politicians) can converge.

However, in the absence of this convergence the decisions on city governance and budget allocation become exclusionary. Added to this, 'regulatory fractures' that are widespread in Indian cities, which further alienates the marginalized groups in terms of infrastructure, spatial and governance access. One hope, the 100 Smart Cities proposed by the present Government, will bring in SDI as an important component of city governance system.

References

1. Ministry of Urban Development (2014) Smart city, ministry of urban development. Government of India, New Delhi
2. Roche S, Nabian N, Kloeckl K, Ratti C (2012) Are 'smart cities' smart enough? MIT SENSEable city lab, GSDI world conference (GSDI 13), Quebec, Canada, 14–17 May 2012
3. The Hindu (2014) State should take the lead in smart cities project: Venkaiah, The Hindu Daily news paper edition: 13/09/2014. Accessed on 13 sept 2014
4. Mitchell JW (2000) e-topia: urban life Jin-But not as we know it. MIT Press, Cambridge, MA
5. Rajabifard A, Williams IP (2001) Spatial data infrastructures: concept, SDI hierarchy and future directions. In: Proceedings of GEOMATICS '80 conference. Tehran, Iran
6. Bernard L, Craglia M (2005) SDI—from spatial data infrastructure to service driven infrastructure. Paper at the international workshop on cross-learning between SDI and information infrastructure, Enschede, 31 Mar and 1 Apr 2005
7. Pfeffer K, Martinez J, Baud I, Sridharan N (2011) Knowledge production in urban governance systems through qualitative geographical information system (GIS). Environ Urbanisation ASIA 2(2):235–250 (Sage, New Delhi)
8. Census of India (2001) Rural-urban distribution of population, registrar general of census, ministry of home affairs. Government of India, New Delhi
9. Sassen S (2006) Territory, authority, rights: from medieval to global assemblages. Princeton University Press, New Jersy
10. Department of Regional Planning (2010) Spatial data infrastructure for urban governance: case study of hubli-dharwad in karnataka, an unpublished report. Department of Regional Planning, School of Planning and Architecture, New Delhi
11. Sridharan N (2011) Spatial inequality and the politics of urban expansion. Environ Urbanisation ASIA 2(2):187–204 (Sage, New Delhi)
12. Roy S (2014) How to make unsmart cities smart? Business Standard, 9 sept 2014

M-Governance: Smartphone Applications for Smarter Cities—Tapping GPS and NFC Technologies

Ummer Sahib

1 Introduction

Come with me for a journey from Dubai to London. Flight departure is 3 hours from now and need to report in 1 hour. I tap the app iGo [1] on my smartphone. The app determines my Home's GPS coordinate, finds my home address and asks for confirmation. iGo then brings a route form in which the Start point is already auto filled as my home and shows three most used destinations; Airport, Office, Gym to choose. Upon choosing the Airport, the app asks to enter the flight number. iGo checks airport departure times and ensures that there is no delay. iGo now checks for the traffic condition and shows three different routes (1 without toll) with the travel time, distance, driving speed and estimated arrival time. I choose Route 2 based on iGo's recommendation.

Get into the car and mount the smartphone with iGo app showing a dashboard with just five direction arrows; Straight, Left, Right, Round About and a U-turn. To make a turn, iGo just pre-empts to turn the indicator to left or right, thus giving a pre-emptive user experience. While traversing a complex intersection, a junction map is displayed with augmented reality displaying landmarks in 3D and direction to take. Notice, neither a map is displayed nor voice guidance. iGo pre-empts on the required speed and upon exceeding legally posted speed limit on that road, it asks to slow down. Additionally, iGo monitors driving behaviour based on the number of harsh breaks and sudden acceleration.

Suddenly, iGo flashes an alert on an accident ahead and displaying a thinking spree icon computes the alternative route and revised arrival time, considering

U. Sahib (✉)
Informap Technology Center, Office HC-2, Al Taawun Tower 1, Al Taawun Street,
P.O. Box 38098, Sharjah, United Arab Emirates
e-mail: us@informap.ae

T.M. Vinod Kumar (ed.), *E-Governance for Smart Cities*, Advances in 21st Century Human Settlements, DOI 10.1007/978-981-287-287-6_9

current traffic. I thank my investment on the smartphone and iGo app and reach the airport parking with just 5 min remaining.

I Enter the airport parking, tap my smartphone on an NFC tag for the access barrier to open. Park the car and tap on an NFC-enabled payment machine to pay the parking fee for 7 days. Just arrived at the airport on time, checked in with an E-ticket and tap the smartphone to proof my presence and board the aircraft.

This is neither a fiction, nor the dystopian ideas of a 1984 novel [2], but an example of how a smartphone with GPS and NFC technology can be used to travel smart, prove one's identity and pay smartly.

Welcome to the chapter on M-governance, the government of the future, to witness the era of 24 × 7 government services using GPS and NFC technology anytime from anywhere at your convenience!

2 Mobile Phone Penetration

The mobile phone, once a luxury, is now in the hands of almost of everyone as a personal accessory. From 1 billion mobile subscribers in 2002, it is nearing the 7 billion mark in 2014 [3]. The percentage of world population with cell signal has dramatically increased from 63 % in 2003 to 90 % in 2010, and is well on its way to the 100 % coverage as illustrated in Fig. 1.

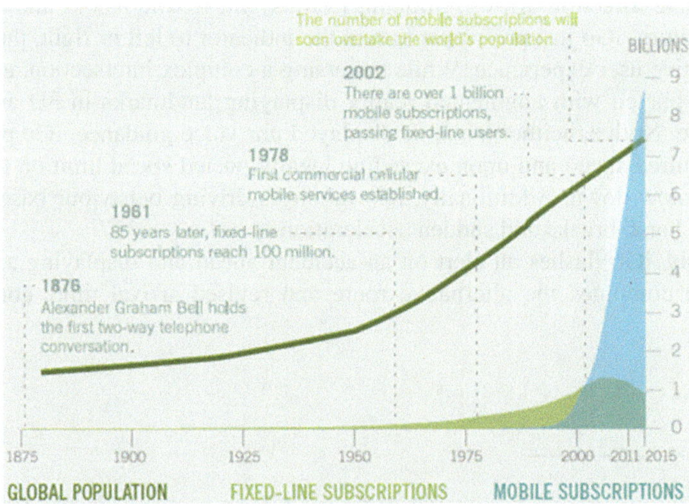

Fig. 1 Global population growth and mobile phone penetration. *Source* World Bank IC4D 2012 Report on information and communication for development: maximizing mobile

Key global telecom indicators for the world telecommunication service sector in 2014 (all figures are estimates)

	Global	Developed nations	Developing nations	Africa	Arab states	Asia and Pacific	CIS	Europe	The Americas
Mobile cellular subscriptions (millions)	6,915	1,515	5,400	629	410	3,604	397	780	1,059
Per 100 people (%)	95.5	120.8	90.2	69.3	109.9	89.2	140.6	124.7	108.5
Fixed telephone lines (millions)	1,147	511	636	12	33	512	70	245	256
Per 100 people (%)	15.8	40.8	10.6	1.3	8.7	12.7	24.9	39.2	26.3
Active mobile broadband subscriptions (millions)	2,315	1,050	1,265	172	92	920	138	399	577
Per 100 people (%)	32.0	83.7	21.1	19.0	24.6	22.8	48.9	63.8	59.1
Mobile broadband growth 2013–2014 (%)	N/A	11.5	26	43	19	21	15	12	16
Fixed broadband subscriptions (millions)	711	345	366	3	12	313	40	173	163
per 100 people (%)	9.8	27.5	6.1	0.4	3.1	7.7	14.3	27.7	16.7

mobiThinking note Mobile subscriptions outnumber fixed lines 7:1 (more so in developing nations); Mobile broadband.
Source ©International Telecommunication Union (May 2014).
via © mobiThinking

International Telecommunication Union (ITU) estimates that there will be almost as many mobile cellular subscriptions globally by the end of 2013 as there are people on Earth. However, this does not mean every person on earth will have their own mobile phone. Some may have two or three and many will be sharing one phone and in extreme cases, a whole village may be sharing a single phone.

Some Facts about Mobile phone penetration [4]:

(a) The developing world is more mobile than the developed world.
(b) Most phones are owned by people living in low income regions.
(c) While 71 % of the mobile phones owned in 2000 were from high Income countries, their share dropped to 23 % in 2010.
(d) Mobile revolution is right at its start of the growth curve.
(e) Mobile devices are becoming cheaper and more powerful, while networks are doubling their bandwidth roughly every 18 months and expanding.
(f) Access to a range of mobile application has increased dramatically throughout the last decade.
(g) Mobile applications not only empower individual users, they enrich their lifestyles and livelihoods and boost the economy as a whole.
(h) Near ubiquity opens new opportunity and the government's ability to connect with people.
(i) Mobile data traffic is exploding with billions of connected devices, demanding more and more bandwidth.

3 Smartphone Penetration

Although mobile phone penetration is far and wide, the smartphone adoption is still in its infancy. This can be attributed directly to the cost of the smartphone and bandwidth. The global smartphone audience surpassed 1 billion mark in 2012 and smartphone users worldwide will total 1.75 billion in 2014 [5]. Mobile users will pick up smartphones as they become more affordable. It is expected that smartphone adoption will continue on a fast-paced trajectory through 2017. Nearly two-fifths of all mobile phone users—close to one-quarter of the worldwide population—will use a smartphone by 2014. By the end of 2017, smartphone penetration among mobile phone users globally will be nearly 50 %.

The global smartphone penetration is shown in Fig. 2.

4 Mobile Phone Eco System

Engaging mobile applications for development requires an enabling ecosystem. The smartphone ecosystem has a number of components as given in Fig. 3.

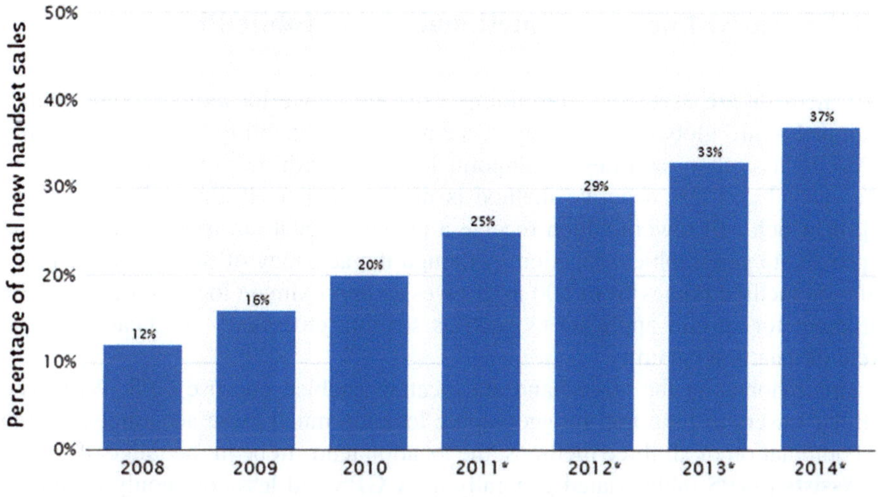

Fig. 2 Global smartphone penetration from 2008 to 2014 (in percentage of new handset sales). *Source* http://www.statista.com/statistics/218532/global-smartphone-penetration-since-2008/

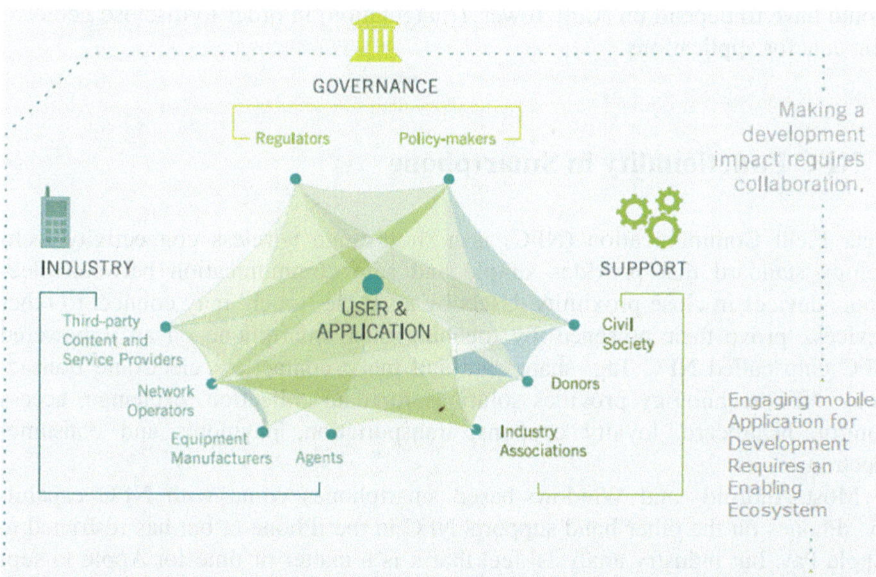

Fig. 3 Ecosystem for mobile application development. *Source* World Bank IC4D 2012 Report on information and communication for development: maximizing mobile

5 GPS-Based Location Functionality on Mobile Phone

Two methods are used for determining a mobile phone location. Mobile phones equipped with global positioning system (GPS) capability, use signals from GPS/Glonass/other satellites to pinpoint locations with accuracy nearly to 5 m. The second and less accurate method is often called 'Cell Tower Triangulation' [6], in which cell towers which receive a phone's signal are used to calculate its approximate geographical location. Although the accuracy of the cell tower triangulation method ranges from 100 m, it gives an approximate location of the phone sufficient for certain applications such as sending emergency notification in the event of a natural calamity.

Smartphones on the other hand are location enabled and use GPS, Wi-Fi and cellular towers to pinpoint the geographic location much more accurately. Due to its combination of all three methods, the location tends to be in the range of 5 m.

Assisted GPS (abbreviated generally as **A-GPS** and less commonly as **aGPS**) is a system that can, in many cases, greatly improve the start-up performance or time-to-first-fix (TTFF) of a GPS satellite-based positioning system. It is used extensively with GPS-capable cellular phones, as its development was accelerated by the U.S. FCC's 911 requirement to make the location of a cell phone available to emergency call dispatchers [7].

According to some industry research estimates, only about 11 % of phones manufactured have GPS capability, so the remaining 89 % of phones without GPS would have to depend on 'Cell Tower Triangulation' in order to disclose geolocation data for applications.

6 NFC Functionality in Smartphone

Near Field Communication (NFC) is a short-range wireless connectivity technology standard that provides simple and safe communication between electronic devices in close proximity. Users by a simple 'touch' may connect to other devices, 'prove their presence' by touching their smartphone on an unpowered NFC chip called NFC Tag, share data and make contactless electronic transactions. NFC technology provides solutions for data collection, exchange, access control, healthcare, loyalty coupons, transportation, payments and consumer electronics.

Most Android- and Windows-based smartphones come with NFC capability. iPhones on the other hand supports NFC in the iPhone 6, but has restricted to Apple Pay, but industry analysts' feel that it is a matter of time for Apple to support NFC for other applications.

From the NFC application context in this chapter, it is essential to understand the three operational modes [8] of NFC smartphone as described below:

(a) **Card Emulation Mode**: This mode allows NFC-enabled smartphones to act like smart cards, allowing users to perform transactions such as purchases, ticketing and transit access control with just a touch. A major Bank in the UAE recently launched The *Tap n Go Sticker* to make low-value purchases 'exceedingly simple' by converting any mobile phone of any generation into a credit card that can then be used to make contactless payments at thousands of retail outlets in the UAE and abroad. In Card Emulation mode, the NFC-enabled device communicates with an external reader much like a traditional contactless smart card. This enables contactless payments and ticketing by NFC-enabled devices without changing existing infrastructure.

(b) **Peer-to-Peer Mode**: This model allows two NFC-enabled smartphones to communicate with each other to exchange information and share files, so that users of NFC-enabled devices can quickly share contact information and other files with a touch. For example, users can exchange data such as virtual business cards or digital photos.

(c) **Reader/Writer Mode**: This mode allows NFC-enabled smartphones to read information stored on inexpensive NFC tags stuck on any surface, equipment, identity card, poster and displays or even on a wrist band. The NFC phone can be used to write information on an NFC tag and at the same time read from it. NFC tags are passive, meaning they do not have any power source. Instead, they literally draw power from the device that reads them by means of magnetic induction. When a reader gets close enough to a tag, it energizes it and transfers data from that tag.

NFC tag contains a small amount of non-volatile memory which is used to store user data ranging from 41 characters to 2,000 characters [9]. NFC tags comes on a wide variety of material starting from a wrist band ($0.5) used in conferences to all weather proof material tag costing $10. The data on the tag can be permanently locked or encrypted so that no one can manipulate the data or can be made to write once only.

This chapter focuses on GPS and NFC technologies primarily to limit the scope and consider applications on two fundamental aspects of smart cities, namely 'proof of location' and 'proof of presence'.

7 Defining M-Governance

M-governance or *mGovernance* literally means making the government mobile. In other words, the government going to its customer (Citizen, Resident, Visitor or Tourist) instead of the customer coming to the government for any service or transaction. There are a number of definitions from varying sources. Three flavours are chosen below to encapsulate the broadband and evolving scope that it covers:

(a) In the 2007 Book on Mobile Government—An emerging direction in E-government; Kuscu, Kushchu and Yu defines thus:

> Mobile government (M-government) may be defined as a strategy and its implementation involving the utilization of wireless and mobile technology, services, applications

and devices for improving benefits to the parties involved in e-government including citizens, businesses and all government units.

(b) In the 2011 OECD/ITU Report on M-Government—Mobile Government for connected societies is defined as:

> M-government—the adoption of mobile technologies to support and enhance government performance and foster a more connected society—can help improve government performance and strengthen public good governance provided that the emphasis is not placed on the 'm'. Focus should be indeed on the needs of the public sector and the end-users, be they citizens or businesses, to ensure that technology is exploited to reorganize the way civil servants work and to meet the needs of citizens through improved service delivery.

(c) The 2012 International Bank for Reconstruction and Development/The World Bank report defines M-government as:

> Mobile government involves using mobile tools to change either the interactions between users and government or the processes of government. In 2012, tools in use included mobile Networks (such as broadband, Wi-Fi and voice-centric), mobile devices (tablets, smart phones, feature phones), their associated technologies (voice calling, SMS text messaging, location detection, Internet access), and software in the form of network services and applications.

From the above definitions, it is evident that M-Governance signifies the usage of mobile phone technologies to serve its users; inter-government departments, employees, businesses, citizens, residents, tourists and visitors.

M-Governance does not replace other means of services that the government delivers through online or offline but offers a radically new method by which government leverages the mobile technologies to radically improve civil engagement and fosters a more participatory and democratic decision-making.

The World Bank's 2012 [4] ICT strategy is built around three themes, namely Innovate, Connect and Transform which summarizes what Mobile Technologies can bring about for the government of the future.

8 Need for M-Governance

Mobile phones have changed the way we live, work, choose and decide during the last decade. Governance has also been transformed as shown in Table 1.

The main factors for M-governance are:

(a) Wider acceptance in the public sector
(b) Smartphone penetration
(c) Ease of use for citizens
(d) Easier interoperability
(e) Bring citizens closer to government
(f) M-government services are cheaper to deliver

M-governance can play a very important role in policy formulation and participation by making it citizen centric and adopting a cooperative model of governance,

Table 1 Overview of conventional, E-Government and M-Government concepts

Concept	C-Government	E-Government	M-Government
Principles	Bureaucratic process paper forms, phone, fax	Process engineering using IT (PC, Internet)	Seamless integration and linkage through wireless mobile phones
Service time	8 h a day 5 days a week	24 h a day 7 days a week	24 h a day, 365 days non-stop
Service space	In-person visit, phone and fax	Customers home and office using the internet	Customers location and physical place can be from anywhere
Service form	Several visits to office	Multi-clicks to web portals	One-time access to needed service

Source Oui-Suk, Uhm (2010) Introduction of mGovernment and IT Convergence Technology, KAIST Institute for science and congress

whereby all stakeholders are given the opportunity to participate and thereby governance is made transparent enabling citizens to measure the outcomes and their impact on their quality of life.

9 Transformation of E-Governance to M-Governance

United Nations Department of Economic and Social Affairs (DESA) [10] have established the idea of connected governance as a means to achieve maximum cost savings and improve service delivery. The idea of connected governance enables public service transformation, innovation, productive and inclusive growth. DESA identifies the following five stages of connected Government .

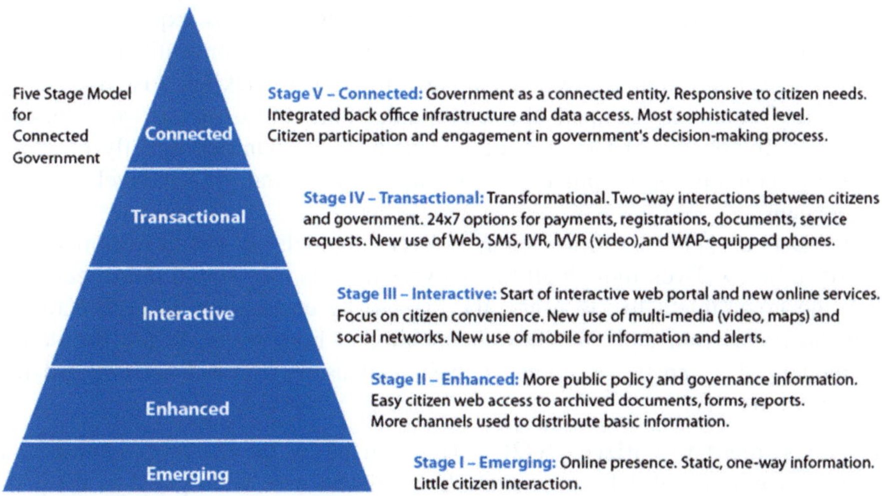

Fig. 4 Stages of connected government *Sources:* DESA (2010), E-Government Survey 2010, United nations, New York

As the government passes through the above five stages, the users may experience each stage as follows:

Stage 1 is characterized by governments having a basic web presence to inform the users of its functions and answers to frequently asked questions. This is equivalent to a reception desk in a government office where the receptionist hands out information or answers telephone calls seeking information.

In stage 2, users are able to download application forms, print it, fill and submit in hand or send by post.

In stage 3, websites become much more interactive in which users are provided an online form to enter the details and submit applications electronically.

In stage 4, the interaction goes to the next stage whereby the government becomes fully bi-directional giving the visibility of the application or transaction status that was previously visible only to the respective officers. At this stage, greater transparency in transaction is achieved and the first level of administrative layer is removed.

In stage 5, the connectivity gains momentum and the government becomes mobile and is now able to go to the users, exactly opposite of what happened in Stage 1. The government becomes proactive and is now able to transmit instant alerts about the expiry of licences or services and encourages users to carry out the transaction instantly or pursue it till it gets done. This signifies the maturity level of an e-government and can be earmarked as having transformed to M-governance status.

10 Delivery Models of M-Governance

The delivery of M-governance for its users or customers is expected to play a significant role in developing countries where the conventional government infrastructure and E-government infrastructure is lacking. Just like many developing countries skipped the communications infrastructure and leaped to Mobile Telephony, M-Governance will help these countries directly to adapt them for economic development, social improvement and greater level of civic engagement.

OECD/ITU [11] identifies four delivery models of M-Government as described below. Even though all the delivery models may sound uni-directional from the government to the users, they are to be understood as bi-directional with the transaction and data flow happening in both ways. While the delivery models of a and b are aimed at individual users, c and d are targeted for Organizations.

(a) **Government to Citizens (G2C)**: This will be by far the most intense and far-reaching of all delivery models. This will become widespread as the penetration of smartphones widens and the smartphone and Internet bandwidth becomes cheaper as we near 2020.

The Mobile apps of G2C will enable the government to push real-time information and alerts and provide interactive services to complete transactions. The users (Citizens, Residents, Visitors and Tourists) on the other hand will be able to access a full range of categorized information, ask questions, get answers, subscribe for a service and report issues and seek assistance, thus a higher level of civic engagement.

(b) **Government to Employee (G2E)**: The government being the biggest employer, its employees are its asset and their well being, cooperation, rigorous engagement and continuous communication are essential for practising good and responsible governance. In addition, M-Governance provides an active channel to encourage retired employees to advice on critical policy decisions and reflect back their valuable experiences. M-Governance will also help the government employees to directly interact with users and provide services at user locations.

The Mobile apps of G2E will facilitate the mobile workers carry a Smartphone or a tablet to carry out their tasks much more efficiency in the field. The current method of first filling all the details on a standard form in the field and then going back to the office to key them in on a computer or seek data entry assistance to complete the tasks can be totally eliminated. Similarly, the mobile workers do not need to come to office and then receive tasks from their managers before going to the field. Instead, they can receive tasks on their smartphones or tablets and go directly from home to work accessing the secured database and pulling the required data and pushing data directly from field to the database in real-time enabling the back office to act in near real time. This will greatly facilitate staffs who are working in remote locations and in difficult circumstances.

(c) **Government to Business (G2B)**: Private sector is the revenue earner and execution arm of government for carrying out almost all the projects. The participation of the Private sector in Public Private Partnership (PPP) is becoming increasingly sought after for governments to engage the private sector in every stage of development. M-Governance will help in monitoring the projects and to promote a business friendly environment, which would bring more investment and help in the overall development.

(d) **Government to Government (G2G)**: This will promote inter-governmental communication and participation which is lacking in many arms of the government. It will help eliminate data duplication and collaboration at inter-departmental level. M-Government will give a common operative picture (COP) during emergencies that require inter-government cooperation.

M-Governance connectivity can be between departments horizontally or vertically between hierarchical administrative divisions and further on to Federal governments. The biggest benefit is of course in security, law enforcement, coordinated inspection and emergency where real-time data is essential from staff working from different government agencies.

11 Benefits of M-Governance

M-Government offers a common platform for good, responsible and transparent governance that is mobile and serves at all time at the convenience of its customers. The benefits of M-Governance on Citizens/Residents/Visitors and Tourists as well as to business and employees are listed below:

(a) **A Mobile Government**: Instead of the people coming to the government for any service, the government is now able to go to the people, an opportunity to virtually connect 24 $\times$ 7.
(b) **Affordability**: Mobile phone comes in a wide variety and economy for all sections of the society to afford a mobile phone rather than a laptop. However, smartphones are at least affordable only to middle and high income groups. But, the smartphone prices are dropping down and we may soon see smartphones becoming an affordable commodity. Same downward trend is witnessed in mobile Internet connectivity charges.
(c) **Ease of use**: Comparing to a laptop computer and Internet access, mobile applications are cool, user friendly and can be self-taught by the user.
(d) **Accessibility**: M-government provides an additional channel for users to access government services. This can attract more users to access government services that are more convenient, especially for people who are located in geographically remote areas or who are physically disadvantaged.
(e) **Availability**: M-government service can be automated to provide 24/7 availability. Users can access these services with virtually no waiting time, whereas completing the same transaction using conventional approaches (such as telephone calls or in-person visits) can take longer.
(f) **Service quality and efficiency**: Success in building interest, enthusiasm and capacity of socially marginalised communities to interact and communicate via online technologies contributes to M-government's success in achieving efficiency gains and improving services.
(g) **Service scalability**: The advantage of scalability is that the provision of M-government services has a far lower cost in comparison to traditional service delivery (e.g., printing materials, especially in regions that have higher population density). Efficiency and effectiveness are improved. Flexibility and scalability can be maintained because functional components can use a set of common interfaces to communicate with each other.
(h) **Stakeholder participation**: M-Government services, optimised by smartphones, allow citizens and businesses to take advantage of the Internet to access government services, resulting in better perception and higher participation.
(i) **Open Government**: This makes the government more open bringing people and government closer for collaboration, coordination and engagement. This will help increase overall 'trust' in the government.
(j) **Integration**, **communication and interaction**: Using information technology allows better integration of functional departments in government processes, and increases customer satisfaction with service delivery across both traditional and electronic channels. The additional electronic communication

channel gives governments the opportunity to interact with specific groups of users who otherwise may not be reached through conventional communication approaches.

(k) **Reduced costs (fixed and operational)**: One of the major benefits to government agencies is the flexibility M-government allows in enabling information storage and presentation. This may lead to far lower operating and maintenance costs, compared to printing all materials. Altering, correcting or updating content can be completed online without incurring costs for re-printing, waste disposal and re-delivery.

(l) **Image and perception**: Research suggests that using online or mobile channels to interact with citizens and engaging them in decision making has a positive impact on trust, as well as public perceptions of government responsiveness. In addition, the use of mobile channels can lead to increased citizen participation, which can in turn make it easier to design and implement policies that lead to better outcomes. Therefore, M-government services may result in an improved image of government operations, so political decisions may push forward the adoption of M-government services to showcase these factors and to create a more positive international image.

(m) **Mobilize Resources**: In the time of emergencies such as earthquakes, it becomes beneficial for government to seek financial assistance and re-distribute to the affected parties.

(n) **Incident Communication**: During terrorist or other natural hazards it is possible to transmit important information to the affected areas and to the specified target groups thus avoiding false information being circulated. This also allows transmitting escape routes and safe passage.

(o) **Mobile Voting**: This is experimented in Estonia and has proven successful. This can allow a larger participation of voters and also to reach the democratic rights to people who are otherwise not included such as voters who are non-residents.

(p) **Reduce corruption**: As greater transparency is brought into the system, mobile reporting acts as a deterrent as instant investigation can be launched and greater evidence can be made to catch red-handed. Automation of processes will give lesser chance to become corrupt. On the other hand, greater transparency can reveal inconvenient truth and result in political fallout.

(q) **Staff Mobility** can break down silos and broaden the skills, experience and mindset of employees and foster greater cooperation in collaborative projects. Horizontality in government operations and service delivery are “musts” in today’s world. They are facilitated by the rapidly expanding potential of technology and critical changes in society and the policy environment.

(r) **Creation of Mobile Platform**: Governments can expand services to previously under-served citizens. This can contribute to a country’s social, territorial and cultural cohesion. Mobile platform can be used for crowd sourcing data, opinion and even funding to raise investment to finance public works improvement projects such as playgrounds, parks and sport facilities.

12 Types of M-Governance

M-Government can be classified into three types or forms [12] depending on the way and degree to which mobile technologies are used for the service delivery and civic engagement. They provide broad criteria to classify the services and to understand the stage of M-Governance.

(a) **Supplementary**: Mobile technology is used to supplement existing E-Government applications. This is equivalent to using mobile Phone as another Electronic channel to provide governments services. This can be considered as the first stage of M-Governance maturity in which access to information available in an existing web site is made available from a mobile phone at all times from anywhere.
(b) **Expand**: Mobile technology is used to cover new users who previously did not have access to E-Government or benefit from existing E-Government services. This could be considered as an inclusive stage and can be realized with simple and easy-to-use apps. The inclusion can be from the perspective of users who are not computer literate or have the resources to own a computer or to have Internet access that derive benefit from simple SMS alerts or through access to information by simple keyword search in a Mobile portal of the government. The senior citizens, farmers, computer illiterates, women, under-privileged, differentially challenged and even children are able to connect from their mobile phone to E-Government services at all times 24×7 from anywhere. This can be regarded as the second stage of M-Governance maturity. Many of the Government Mobile portals today fall in this category.
(c) **Innovative**: Mobile technology is used to provide a new set of services that utilizes the mobile phone functionalities to its fullest extent. This is the stage of creative and value innovation in which true M-Governance is achieved and all the constituents are fully served. Government business processes makes a breakthrough by identifying radically new ways of connected governance through the use of smartphone capabilities such as GPS, Camera, NFC and other sensors that come built-in or can be made available through wireless connectivity. The users in this stage are able to carry out many of the government transactions and to engage more intensively in voicing their opinions, M-Voting, giving feedback, participating in government policy decisions and acting as a facilitator or as a government's extended arm for monitoring, control and development.

This chapter will largely explore the applications that fall in this stage of M-Governance.

13 M-Governance Applications

M-Governance applications use three methods, Push, Pull and Session to connect to the subscriber, which is evident from the evolution of mobile applications. Most M-Government initiatives use 'push' services through SMS to transmit information to the subscriber phone. This inclusive method allowed governments to reach to almost every mobile phone user and has the advantage of pushing in the localized language. SMS has been by far the most used mobile application in M-Governance. SMS also has become the de facto method to respond to the Government by using a dedicated short code. Mobile application moved to the second stage with WAP allowing subscribers to request certain information from the government portal resulting in the 'pull' of required information. With the smartphone and data bandwidth becoming affordable, mobile applications allowed user interaction to constitute both 'push' and 'pull' on a theme-based application that opens a live thread between the mobile phone and the government's mobile platform thus opening an interactive channel for the government either through a mobile-enabled website or through a dedicated mobile app.

14 Review of M-Governance Applications

According to UN 2012 E-government survey [13], 25 countries have developed separate M-government websites and 24 countries have made the option of making payment by mobile [UN e-gov survey 2012]. Asia leads in the utilization of mobile-based channels as is evident from Fig. 5.

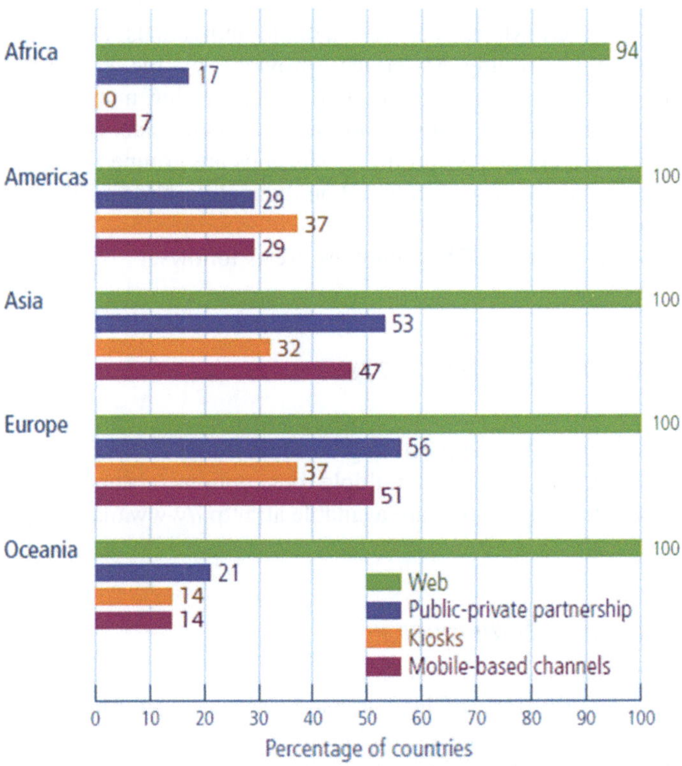

Fig. 5 Breakdown of mobile-based channels by region. *Source:* United Nations E-Government Survey Report 2012: E-Government for the People

Governments during the last 3 years have aggregated all mobile applications launched by different government agencies within their country under one single portal. With different types of M-services available on multiple access platforms, the portal serves as a one-stop convenience to facilitate individuals and businesses to identify government M-services which are relevant to their mobile devices. Most portals have classified their mobile applications into categories, supported operating system, organisations that provide them and a quick search for finding them by keyword.

Since GPS and NFC applications are the focus of this chapter, only mobile applications that make use of it are reviewed below in the context of smart cities.

15 Review of M-Governance Applications: Singapore

Singapore launched its M-government portal [14] in 2011 under its eGov2015 > mGov@SG initiative. 150 mobile applications are listed in its portal. Users can access the smartphone version [15] as well.

The mGov@SG brings together mobile browser-based, native application-based and SMS-based M-services to facilitate individuals and businesses in searching for and identifying M-services provided by the government. Users can access information such as traffic conditions, housing matters and more on the move. The mobile site recognises the mobile device accessing the site, and enhances the user experience accordingly. The apps are grouped under Categories and organizations and can be sorted based on the mobile, SMS or the OS or found through a keyword search.

The applications that use GPS technology are as follows:

15.1 MapSYNQ, by QI PTE LTD

This is a full-featured traffic app offering en-route traffic cameras, incidents, ERP (Electronic Road Pricing) rates and multiple route options (cheapest, fastest, traffic aware). Details of this application are available at: http://www.mapsynq.com.

15.2 My Transport Singapore

A one-stop master directory of land transport services offering essential travel information and alternative choices to plan the journey. Gives real-time traffic and personalised transport information [16].

15.3 *Police@SG*

Locates the nearest police station from your current location, view crime statistics around neighbourhood, read the latest crime news, police appeals for information and missing persons.

15.4 *OneMap*

Search for addresses or find driving directions on the go.

15.5 *OmySCDF*

This mobile application, launched by Singapore Civil Defence Force (SCDF), is for civic engagement in order to report fire safety violations or fire hazards. The app allows capturing up to five pictures and to drop a pin on the location on a map. The app also gives a pictorial guide on how to operate a fire extinguisher.

15.6 *IGov2010 > OneMap*

OneMap is the first major application of the Singapore Geospatial Collaborative Environment (SG-SPACE) initiative. SG-SPACE is a whole-of-government initiative to create an environment in which the public and private sectors and the community can collaborate and create a wide range of innovative applications and services using geospatial or map-based information. It has 60 themes of spatial information, with 34 services available for the public. Built on Web 2.0 technologies, OneMap is very user-friendly and makes it convenient for users to leverage on its base maps to create new applications with their own data through the use of an Application Programming Interface (API). This allows sharing of spatial content using Internet services, accessible on desktop and mobile platforms.

16 Review of M-Governance Applications: UAE

UAE government has launched the M-government initiative in 2013 to make more public services available on smartphones. These applications are aimed to make life easier as the city moves from e-Govt to M-Govt. The apps are put under one single portal by the Telecommunications Regulatory Authority (TRA). The iTunes [17] and Android stores [18] have a dedicated section called UAE M-Government apps, dedicated to all applications for UAE M-Government.

The applications under the apps-gallery [19] are aggregated under the smartphones that they support such as Blackberry, iPhone, Android and other similar devices. The applications are developed by federal and local government entities and there is a dedicated section in iTunes in which apps are classified based on the categories. The smartphone apps that use GPS technology are listed below.

16.1 iDubai, Dubai Municipality

Locate points of interests around and find nearest pharmacies on duty with the help of GPS. Evaluate the entity's services and send remarks or complaints regarding its services along with a picture and location. iDubai also provides information on the weather and high seas and nearest educational institution. There is also another app called mDubai (Dubai E-Government Department) that gives point of interest, their location on the map and their contact details.

16.2 Makani, Dubai Municipality

Smart geographic addressing system app for Dubai. Makani stands for a 10 digit number that can be identified on the interactive map to locate any places. This is based on UAE National Grid (UAE NG) for location and navigation purpose. The app provides auto-conversion mechanism to any other coordinate system to use it with GPS/GNSS-based devices. Makani number of a building also can be obtained by entering the land number or through community-street-building number. Makani also provides routing and find nearest service centers such as health, education and landmarks. Makani provides voice navigation and can view live traffic on main roads.

16.3 Wojhati Dubai Journey Planner, RTA Dubai

Help public transport riders to plan their trips around Dubai. Get detailed instruction to commute from one point to another point with expected timings of public transport modes. User is provided with the list of public transport modes and their timings all over Dubai.

16.4 Smart Taxi, RTA Dubai

Integrated with the Booking and Dispatch Centre, Smart Taxi allows users to book a cab using their smartphone. The App identifies location of the customer and nearest taxi. Customer is intimated when the cab arrives at the location selected by the user. At the end of the trip, the user is provided with a simple 5-point template to assess the performance of the driver.

16.5 Abu Dhabi Taxi, Transad

User-friendly application to book a taxi in Abu Dhabi or Al-Ain region. The LocateME option allows the application to identify the requested location determined by the GPS on the phone. Option is provided to bookmark favourite address and get accelerated access for future taxi booking. SMS notification is sent on the booking with a tracking number.

16.6 CityGuard, ADSIC (Abu Dhabi Systems and Information Centre)

Report incidents and submit complaints related to the Abu Dhabi Emirate directly to the government. CityGuard will automatically create a case for the incident with the Abu Dhabi Government Contact Centre, which will promptly assign the case to the appropriate Government Entity for resolution.

16.7 Darb, DoT, Abu Dhabi

Traveller Information targeted for residents and Tourists. The app provides intelligent transport system using the GPS on the smartphone that includes live data on traffic condition, traffic incidents and road works in Abu Dhabi. Real-time bus arrival information at the nearest bus stop and access to current airport information such as departure, arrival and parking. The user can select a route from current location to a desired destination using on-screen map and choose to receive either the shortest distance or the fastest route. The app will then acquire current traffic flow information in order to provide the estimated travel time to the destination. Users are able to save the favourite routes and receive notification on current traffic incidents along those routes during specific time periods. The app also allows booking a taxi, pay for on-street parking, obtain ferry schedule and to contact DoT through a "Geofeedback" function to report any system malfunctions such as traffic signal bulbs that are burned out, light poles or signs that are broken and other issues.

16.8 SEWA, Sharjah Electricity Authority

Locate nearest payment channel/location, Access to GPS Navigation to locate SEWA Head Office or any of its centers, Locate relevant Customer Service Office through the A/C number, Report faults by attaching Photos and GPS Location. The app also allows viewing bill payment channels, history and receiving important notifications from SEWA.

16.9 Smart Fujairah, from Fujairah Municipality

Allows carrying out land transactions including placing a land request at the required area.

16.10 Education, Ministry of Education

Find nearest educational institution.

16.11 Health, Ministry of Health

Locate a health centre nearest to your current location. SEHA app offers information on ambulatory services and an interactive map service that enables you to find the nearest SEHA facility.

16.12 Dubai Police

Find nearest Police station and Traffic services. The app also provides functions for searching for traffic fines or vehicle registration details and fine payment. Another app SOS allows sending the location of the caller to be used for top emergency situation.

16.13 AvoidJam, UAE

This is a GIS Worx award winning student app that allows registering a traffic jam and to share it with social media.

17 Review of M-Governance Applications: Bahrain

Bahrain's Mobile application portal [20] gives access to mobile applications provided by various government departments and agencies. The smartphone apps that use GPS technology are presented below:

17.1 Fix2Go, Ministry of Works

This application enables citizens and residents of Bahrain to report road faults, traffic signal service reports, sanitary complaints and report road violations.

17.2 eGuide, eGovernment Authority of Bahrain

Provide citizens, residents and tourists with a general overview of the Kingdom of Bahrain covering areas such as business, tourism and health. The map functionality provides the location link to facilities.

17.3 eKiosk and eService Center Locator, eGovernment Authority of Bahrain

Provides a directory listing for eGovernment Kiosk terminals and Electronic Service Centers within in the Kingdom of Bahrain. It allows users to locate eKiosk and eService Centers through the usage of interactive map. The app also facilitates to get directions to the eKiosk or eService Center from the user's current GPS location.

17.4 Health Locator, Ministry of Health

Provides interactive searching facility and location maps for the government and private hospitals, health centers and pharmacies in the Kingdom of Bahrain. The app also allows Drug Search, BMI Calculator and Delivery Date Calculator.

17.5 GEObituary: Developed by Gerome Team of Arabian Gulf University

An app to locate road fatal accident location with their biography and pictures to remember people who died from road accidents. One can help victim's family by sending donations or send flowers through the app. The main purpose is to caution drivers on black spots and to practice safe and responsible driving.

18 Review of M-Governance Applications: Govt of India

The Department of Electronics, Government of India has launched the Mobile application portal [21] under the national mobile government initiative. The applications are categorized into platforms, domains and states. Applications that use GPS are listed below.

18.1 MC CRAMAT, Ludhiana Municipal Corporation

This Android-based Citizens' Reporting app allows reporting civic issues like potholes, garbage, non-functional street lights, etc., with pictures and exact locations on a map. The app captures both picture and location in just one click and allows user to choose category of the problem and to write an optional description. This information is then sent immediately to the concerned official in the relevant department of Municipal Corporation. The citizens get SMS alerts with complaint number. The app also allows users to track the complaint status as it gets assigned to officers and gets resolved. Finally, the complainant can provide feedback about the complaint once it is resolved.

18.2 KVK Khoj, Krishi Vigyan Kendra (Agricultural Science Centre)

This app shows KVK service location to find the nearest the KVK using GPS or Service Provider or internet.

18.3 Rakshak

This one button application is for women and senior citizens' safety. This application can be used in case of emergency where-in on the click of a button, the application sends SMS of the current location to four different relatives/friends. It will also initiate a voice call to an emergency number. The numbers can be specified by the user at the time of installation. This practically converts the mobile phone to a Personal GPS tracker.

18.4 WR Canal Monitoring Service, Rajasthan; Developed by C-DAC Mumbai

Used by WR Staff and concerned farmers of command area to send location and information about Canal breach, leakage, seepage or other illegal activities to WR authorities.

18.5 ERAHAT, Jharkhand

This integrated real-time emergency system is for the whole state of Jharkhand. For any kind of emergency, 1967 number is dialled and eRahat gets activated for immediate intervention on the spot. The real-time web-based eRahat right now takes care of medical emergencies, fire, natural disasters. It is not clear if this has GPS location. Jharkhand Police also has launched another application called JharkhandPOL for reporting that goes into Naxal Information Systems and Criminal Information System for further action.

18.6 Hospital Tracker, CDAC

This application is for the use of frequent travellers who can find the nearest hospital or clinic in the event of emergency. The app determines the current location using GPS and provides the contact details together with the distance from their present location. Calling facility and navigation to the selected hospital/clinic are also provided in the app.

18.7 Accident/Incident Reporting, Bangalore

The Traffic Police Department of Bangalore City, India's third most populous city with 9.6 million people has launched a mobile application for motorists to get real-time traffic updates, report road incidents and check their fine records. Users can also report traffic conditions, including slow moving traffic, jams, bottlenecks, vehicle breakdowns, accidents, potholes and more. Besides entering text details, users can also snap a photo and send it with the report. In the 'PublicEye' feature, motorists can take photos and report road violations directly to the Traffic

Police. This includes complaints of autorickshaws and tracking the status of the complaints. Motorists can subscribe to live traffic alerts via SMS.

18.8 Disease Notification

Mumbai, the Indian city with a population of over 3 million, has launched a mobile application for doctors to notify the central government of new tuberculosis (TB) patients. While it is mandatory for doctors to report any new TB patients since 2012, there are some who are reluctant due to the time taken to fill in elaborate forms. The new app available on Android makes it more convenient for doctors to make a report. It connects directly to Nikshay, a centralized database of all TB patients across the country for monitoring and research purposes with an aim to eradicate TB. This application demonstrates vertical data flow.

19 Review of M-Governance Applications: Other Countries

Following section explores applications from other countries that are of particular interest to smart cities.

19.1 RoadPlus, Highway Traffic Information, Korea [22]

Applications for road navigation and traffic info. The app gives access to real-time CCTV, Traffic MAP, Traffic broadcasting and gas station location.

19.2 Glory, KoRail

Real-Time train reservation with smartphone.

19.3 FillThatHole, UK Dept of Transport [23]

FillThatHole is an easy way to report potholes and road defects. The pothole reports are transmitted to the appropriate council for action. The app sends local authorities up-to-the-minute information about potholes which the council may not otherwise know about, allowing them to identify trouble spots needing fast action.

19.4 Mobile Accident Reporting System, Turkey [24]

Under Bilgi sistemi, the Traffic Police has equipped all traffic patrol cars with Mobile data terminals to locate and dispatch vehicles to accident/incident location. This smart tablet allows making query on vehicle's license validity, insurance, driver's license validity and fines. Vehicle and driver information can be cross-checked for criminal suspicion, cases or the owner can be contacted.

Similar system is also implemented by Abu Dhabi Police through Saeed [25] by which the traffic police can generate accident reports using MDT and transmit them to the back office servers. The Accident analysis experts study these accident locations and do a number of accident analyses and recommend suitable treatment to be implemented on the roads to avoid further accidents in the same spot.

19.5 Monitor Vehicles that Are Doing Insecticide Control, US Florida [24]

Mobile data terminals are used for tracking pesticide spraying vehicles and monitor area covered. In addition, it is used for field reporting, ambulance tracking, emergency professionals, police officers for accident reporting, fire fighters, public works and coordinating complex emergency management and law enforcement.

19.6 www.appmycity.org [26]

The New Cities Foundation is a leading global non-profit organization with a vision to build more inclusive, dynamic and creative cities benefiting people and society. Its mission is to incubate, promote and scale urban innovations through collaborative partnerships between government, business, academia and civil society. The New Cities Foundation team launched AppMyCity (a contest for the world's best smartphone urban app, at the inaugural New Cities Summit in Paris in 2012 after having convinced that mobile apps can make good cities great. The following section reviews the finalist apps and semi finalists of 2014, 2013 and 2012 as all of them are innovative mobile applications and many use cases can be drawn from it.

19.7 Djump (Brussels, Belgium/Paris, France)

A peer-to-peer ride-sharing service that allows drivers to register users to request a ride, and drivers to answer the call. Users can track drivers in real time, and can choose to give a voluntary donation at the end of the ride. Uber is another worldwide private car or rideshare smart phone app that connects you with a driver in minutes.

19.8 Peerby (Amsterdam, The Netherlands)

Enables people to borrow the things they need from their neighbours in less than 30 min. Users post something they want to borrow, and neighbours will get a push notification to which they can respond in a single touch. This is a city program that enables everyone in the world to turn their city into a sharing city.

19.9 Social Cyclist (New York, USA)

Encourages bike riders to map their preferred routes, report hazardous road conditions and vote on their preferred site locations for future cycling infrastructure. This aggregate data can then be used by cities to plan future infrastructure. Social Cyclist is great fun for individual bike riders and also a great tool for cities to organize campaigns to gather data.

19.10 Blue Parking (Valladolid, Spain)

Provides a real-time overview of available parking spots and allows in-app payment. Real-time information is user generated so no additional sensors are required, whilst a printable QR code allows traffic wardens to check payments.

19.11 I Change My City (Bangalore, India)

This app allows users to upload a geotagged picture of an urban problem. This then gets reported to the local government, and users can track all progress as well as vote to draw extra attention to certain issues.

19.12 Mind My Business (New York, USA)

This app uses open data, predictive analytics, and a patent-pending ranking approach to give small shopkeepers updates on everything happening in their area. This includes upcoming roadwork, complaints by nearby residents, or changes in laws which could affect them.

19.13 Parqueate (Bogotá, Colombia)

Provides the biggest database of parking facilities in Bogotá, inviting users to add new parking spots as well as rate existing ones. In the future, users will be able to rent out their unused parking spaces through the app.

19.14 SIGA Chile (Santiago, Chile)

This app provides a platform to map how accessible public places are for people with disabilities. Users can evaluate places and their facilities, upload pictures and discuss how they could be improved.

19.15 Buzzjourney, Thailand

Crowd sourced transportation solution that is sustainable, affordable and provide immediate solution for urban transportation challenge. Using this app one does not need to wait at the station, no more delays in the train and usually drive twice as fast as a bus ride. Payment is made using PayPal.

19.16 Colab, Romania

A social network for citizenship that helps citizens communicates with government. Aimed to promote civic engagement and create better cities with citizens and government collaboration. More than 50,000 citizens and many Brazilian city halls are officially communicating with their citizens using colab. The app allows reporting neighbourhood issues like cars parked in forbidden spots, broken sidewalks or urban sanitation using photographs and post it in order to notify the local government so they can track, manage and solve these requests. User can access Colab, click on 'Suggest' and share their ideas. Find supporters and discuss with friends and other citizens with the same interests. Rating is possible to help improve public services and places by rating the overall quality, cleanness, equipment, service and accessibility. This helps local governments to evaluate their own services and take action to improve overall quality and efficiency in the city.

19.17 B CarbonDiem (London)

This app shows travellers their carbon emissions in real time on their smartphone. The app works in the background 24/7 detecting a user's mode of transport and route. This information is then used to calculate the carbon emissions from the journey. Users can then see their emissions and those of their friends on a bright and simple interface, making it easy to understand the environmental impact of their travel and helping them make smart travel choices.

19.18 City Gardens (Paris)

This app helps urban hikers, tourists, mothers, friends and whoever might be looking for a green space to access pertinent information about the city's parks, gardens and squares. Location, opening and closing hours, closest metro and bus stations, possibility for a picnic, toilets, recommended age for children's games, local history, points of interest, restaurants, etc. All relevant information is combined with a geolocalization feature allowing people to fully enjoy a moment with nature.

19.19 Paris-Ci La Sortie (Paris)

This app alerts the nearest exits at every stop on the Metro or tramway in order to assist the traveller find the exit easily. Waiting times and walking distances are greatly reduced allowing users to save the 2 or 5 min needed to avoid missing their next train or be late for an important appointment. It encourages mobility and helps travellers to optimise their commuting times.

19.20 Emergency Assistance, Nepal

Users can enter two to three emergency contact numbers and two to three emergency email addresses. If ever the user is in a situation they feel threatened or require assistance, the user simply presses the panic button to activate the app. It will send an SMS or email to the designated contacts. Information will also be sent to the police with GPS location of the emergency.

19.21 Checkmyschool.org, Philippines

Development of an interactive mapping platform for project that allows residents to provide formal and informal comments on the proposed projects, plan via an easy to use mapping interface. Public participation in planning can be achieved this way.

Checkmyschool.org in Philippines is able to map 8,000 of 44,000 public schools through civic engagement.

19.22 Application Provided by Private Enterprises—Neatstreet [27]

There are few cases where a private company acts as a facilitator for M-Governance. One such application is Neatstreet. This can be considered as an interface with social media-based crowd sourcing whereby the mobile apps are provided free and users are encouraged to file reports related to day-to-day issues. Neatstreet is an excellent application that allows taking a photo, selecting a category, typing a message, and then pressing the send button. One can file a report and its location within seconds! The report is automatically forward to the appropriate authority based on the GPS location. This is done by the unique location-to-authority mapping service. As for privacy, the reporter remains anonymous. This application offers a totally unified way of reporting issues that concern the community right across the country. Sender can track progress of filed reports.

19.23 Application Provided by Private Enterprises - Seeclickfix [28]

This is another example of a smartphone application to illustrate how a US private company acts as an intermediary to form connection between city residents, media and government to identify and solve non-emergency issues concerning the city. This is a fine model of civic engagement in governance to improve neighbourhoods through a 'self-reinforcing loop'. Through a hosted CRM, city or utility officials are able to manage, engage and analyse issues posted and to communicate with staff and to City residents. First they are able to bulk acknowledge the issues, assign tickets to the concerned staff with work orders, set benchmarks and measure success.

19.24 Smart Bins for Waste Collection [54]

Bee'ah, a Sharjah (UAE)-based waste management company recently made their waste collections bins smart. The smart bins work with the use of small, battery-powered wireless sensors that are firmly attached and hidden under the container lids. The device measures the content fill status from inside the container and the data is sent across the cellular network to Bee'ah's cloud server for analysis and display. From there, a list of the containers that are full and ready to be collected is sent to Bee'ah drivers through the fleet management system. Moreover, the smart bin GPS location, movement, temperature and tilt status are transmitted in real-time using GPRS to the cloud server and can be monitored with real-time alerts.

The smart bin technology [55] uses a UBi sensor that is robust and reliable. The sensors utilize advanced ultrasonic measurement and digital signal processing (DSP) techniques for better accuracy, and superior communication algorithms that guarantee a reliable messaging service. The Ultrasonic Fill Sensor measures glass, general waste and recyclable containers up to 3 m deep, the tilt sensor measures tilt angle, the temperature sensor measures internal temperature and sends alert if it exceeds 85C, the GPS measures bin Location and the GPRS module transmits the data. This is a fine example of how different sensors are combined with GPS for engineering an integrated solution.

19.25 EnviroCar App [56]

This is a traffic data crowd sourcing app for a smart city resident to become conscious of the vehicle usage and to measure the impact of one's driving on environment such as fuel consumption, CO_2, or noise emissions. User can compare their driving statistics with friends and become a better and efficient driver. GPS and telematics data collected by the GPS device are shared with enviroCar citizen science community and Traffic Planners to investigate questions such as What are the spatial characteristics of fuel consumption in the city? How can traffic planning be optimized to effectively reduce emissions? What can we learn from average speed maps?

Based on open platform, the EnviroCar app is able to access car's sensors with an android smartphone and a Bluetooth OBD-II (On-Board Diagnostics) adapter. The app talks to car's on-board computer, records GPS track and automatically stores or transmits data on speed and revolutions per minute, fuel consumption, estimated fuel cost and CO_2 emissions.

19.26 Citybuddy [57]

A mobile app that provides location-based dynamic hotspots ranked by social media posts. Using the app, users can identify areas nearby where popular events are taking place; this determined by likes, reviews, comments and photos posted online. The app uses crowd-sourced technology to analyze social media posts to provide what they call actionable consumer intelligence. The app makes use of users' current GPS location to identify hot events in the vicinity and to view them on a map.

19.27 *Spike [58]*

This is the world's first laser-accurate smartphone measurement solution that enables to safely and accurately snap, measure, model and share any object up to 600 ft (200 m) away. With a simple snap from a Smartphone, one can measure height, width and area from a photo, and capture target location. Measurement and location are saved with the picture, and can be easily shared via email as a PDF. From a single photo, spike can capture accurate data for an object including: height, width, area and location. Spike's 3D compass and laser rangefinder integrates with standard mobile technologies, and easily attaches to the back of the smartphone case for commercial iPhone and Android devices. Connection to the smartphone is made through Bluetooth and utilizes the Smartphone's digital camera and GPS. This is a great app for a Building Evaluator or Inspector.

19.28 *Terrago Edge [59]*

This app is an open collaboration solution for mobile workflows that immediately syncs field data with headquarters and other mobile workers. It enables field crews to easily take notes, photos, assign tasks and share data in real time. The easy-to-use app enables the enterprise to embrace the bring-your-own-device (BYOD) promise of improved productivity, ease of use and seamless networking.

20 Smartphone Applications for G2C

The above section demonstrates the use cases of mobile phone application that uses GPS technology. It is important to note that a vast majority of the applications reviewed above are released by the respective governments through their common portals belong to the category of G2C. They are aimed at their citizens, residents, visitors and tourists and very few G2E, G2B and G2G. This could be due to: (a) Governments restrain from publishing applications that they use within their government both horizontally and vertically, i.e. between departments within a state and across the hierarchy of departments from the village to the state and from the state to the center; (b) Use of Mobile GPS technology applications between intra-government and inter-government has not yet taken roots due to number of reasons. Same is the case with G2E and G2B applications as there are very few applications published. This aspect is dealt in detail on the challenges part later in this chapter.

Following sections present potential innovative smartphone applications that can be used in M-Governance by combining GPS technology for G2E, G2B and G2G delivery models.

21 Smartphone/Tablet GPS Applications for G2E

M-Governance offers tremendous opportunity for automating the tasks of Mobile staff and field crews whose work involves field work in uncomfortable, remote and challenging environment. The workflow in conventional governance required employee reporting to work and receiving work order and paper forms, e.g. Check transformer, Give Utility connection, Check utility meter, etc., and then go the field to carry out the tasks. Field paper forms are filled at work site, brought to office for auditing and filing.

E-governance has taken this step a bit further by having the back office print tasks together with the data entry forms and the check list. The field staff goes to the field, fills them by hand and brings it to the office, now having to enter them in their office desktops or hand them over to Data entry encoders or clerks who have to enter them. This has effectively doubled the work leaving lot of room for input errors, lack of accountability and work ownership.

One of the key eroding virtues in the modern government work culture is lack of 'trust'. Lack of trust in a government employee leads to trust deficit in the entire government as the employee is the interface for the user's experience. CCTV Surveillance, Access control and sensors are installed inside office to counter internal and external challenges of trust in employee and the work done. However, monitoring field staff remains a challenge.

Using smartphone with GPS technology in this context serve as 'Proof of Location' with a time stamp and solves the most essential proof of "Time" and "Space" for unsupervised work carried by a government employee in the field. Equipping field staff with smartphone-based field applications increases the credibility and transparency of the tasks carried out by field staff in the field and inculcates a smart city culture on both the employee and the customer who get to experience Mobile Government.

Following sections illustrates how innovative mobile applications can save time and money, increase efficiency of field staff, reduce errors and achieve a higher quality of work at the same time improve transparency, accountability and ownership. Pseudo *names used for the following applications to name the idea and if there are any applications by the same name, it shall be treated as pure coincidence.*

21.1 Field Task Master

This application allows field worker to login to the application from home and view all the tasks that are scheduled for that day. The field worker can in effect go to work site from home or proceed directly to work site immediately upon reaching the office.

By logging into the application (*transmits work start Time and Location to backend server*), all the scheduled tasks for that day are shown. They are sorted based on the route optimized for visiting them in order or priority and sequence. Customer location, contact details, expected arrival time, duration of stay and expected departure time are shown for each visit point. Tapping on the address, the app computes route from current location and directs the worker to take the optimized route. Upon reaching the field, Field Task master app brings up a Task execution form in which the worker enters the completion of the work and reasons if there is a delay. Task master sends GPS position and time stamp of the worker every 5 min so the back office staff knows work in progress and staff presence. This helps to ensure the safety of the field worker and intervene in case of emergency. The back office automated report server generates daily work report and sends it for confirmation and then aggregates the confirmed report for weekly or monthly salary reimbursement. The system also computes time at each customer point for billing purpose and to take action if an unreasonable delay is detected. The Mobile worker data flow is illustrated in the Fig. 6.

21.2 Field Check Master

90 % of the time, field staff that carry out field inspection or verification require checking various regulations, status and customer up-to-date records from one or many departments in order to take a decision or to complete the task. Since the mobile application gives the worker read access privilege with controlled visibility to horizontal or vertical government records, the worker is able to cross verify or check with multiple records while making a field decision. If the worker is unable

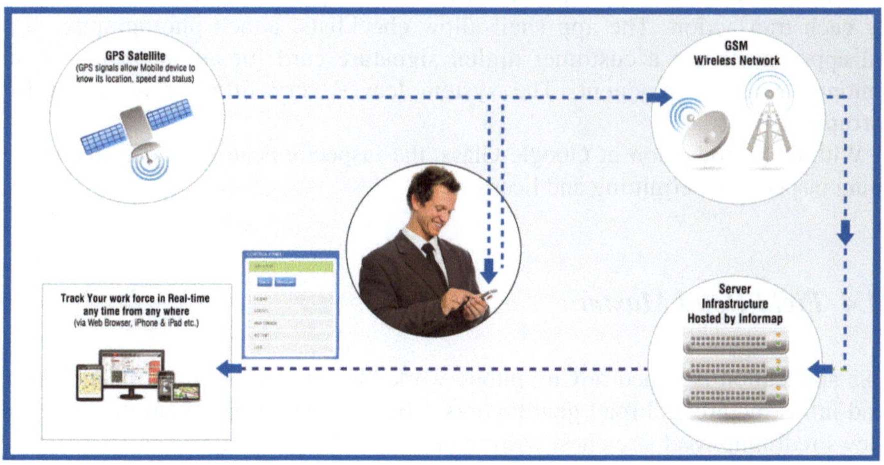

Fig. 6 Mobile workforce management system process flow. *Source* Informap MiTrack Brochure

to make a decision, the app allows using an instant messenger service to the supervisor or domain expert for further checking or for support. The field check master is a read-only app and does not allow any edit operation. It provides direct, real-time access to data in the agency's database from a mobile device and increases internal productivity and helps agencies bring critical government services directly to customers at their convenience.

21.3 Field Inspector

The workers who use this app are those responsible for permitting, licensing, building inspection and code-enforcement. This app brings the respective customer record and allows the worker to add new or edit existing records and complete data entry in the field. The app U/I is made field data entry friendly so that data entry is limited with on-screen options and check boxes. Very few absolutely essential typing have to be done. All other data is read from the ID card of the customer automatically which acts as a proof of identity. App allows viewing history of inspections and remarks or any black points.

The transaction time and GPS coordinates are automatically determined by the app and autofilled in the respective field. The application further allows issuing the permit in the field itself by sending a digital permit through an SMS notification or a digitally signed permit through a secured email link. Thus the task of issuing the certificate in the field is carried out without a single piece of paper. Great user experience and greater trust of people for the government. M-Governance in a Smart city is in action!

Upon completion of the transaction, all data sets are transferred instantly to the government database and updated in the various horizontal database and vertical databases thus creating a harmonized and consistent database record for each transaction. The app shall allow checklists, attach photographs, digital approval or use a customer digital signature card for authorization or an e-money card for payment. The system leaves very little or no room for corruption!

With the introduction of Google Glass, the Inspectors can be hands free while doing inspection, permitting and licensing.

21.4 Field Road Master

This application is aimed at City public works department for road maintenance, road infrastructure and road guard works. The application brings up the map and allows redlining road stretches, writing notes or taking a picture and attaching to transmit the reports to back office.

21.5 Field Accident Master

This app is for motor accident data entry and issuing ticket. Police officers attending accidents have to fill a triplicate accident data entry form at the accident scene under challenging and tensed arena. Separate forms are used for vehicle damage and injury accidents as necessitated due to legal and insurance claims. Several agencies have automated this activity using Mobile data terminals and have streamlined the process by outsourcing the damage reporting to third party service providers and handle only Injury reporting. For e.g., Abu Dhabi Traffic Police and the outsourced agency Saaed [25] uses an MDT for accident referencing and reporting. The accident report is issued as a small printed ticket instead of the old triplicate paper report. The accident location report goes to the police database for further process and analysis. Post processing of data and analysis [29] is carried out to determine accident causes and to identify black spots or stretches that are dangerous or objects such as a garbage skip that caused the accidents. The GPS location of accidents is essential for diagnosis (determine the causation), administer treatment and to monitor treated accident sites in order to prevent further accidents from occurring at the same sites. M-Governance will help achieve Smart city's vision to significantly lower fatal accidents in the city.

21.6 Field Safety Auditor

Safety and security is of utmost concern in a smart city. Seldom a road safety audit is carried out on routine basis and road safety hazards and violations are noticed only when a fatal accident occurs. Although a video mapping or a mobile mapping system can be used for a stream check, detail safety audit would require safety auditor making a field visit and carrying out the safety audit. The Field Safety Auditor Application is aimed at providing a live map on a smartphone/tablet. With GPS determining the current location in the field, the officer is able to identify and mark the safety hazards on street furniture or hazardous objects. The safety form in the app allows entering the hazardous details with the photograph to prove the nature of hazard. The geotagged photo can be used to further analyze and to implement corrective measures.

21.7 Field Utility Master

Utility companies have significant work force in the field. Mobile application for each task will help automate their field tasks and to carry out data entry completion with GPS position and geotagged photographs from the field itself. This application will have similar functionality described in the above applications such as field staff going for regular manhole inspections. This app will eliminate paper forms and long delays due to the elapsed time between filling out paper forms in the field and data entry back at the office.

21.8 Field Survey Master

Field data collection of staffs include collection of statistics data, GIS data, establishment data, etc. This mobile app will automate their field data collection and transmit in real-time with their location and geotagged photos.

22 Mobile Applications Using GPS Technology for G2B

Outsourcing many of the government business processes is becoming or will become a norm in a smart city. The government will play the role of establishing work process rules, guidelines, checklists, and to ensure time and quality. Under this scenario, process re-engineering may be required to identify the tasks to be outsourced and to establish contractor expectations, ensure data security, real-time data transmission, monitor and audit work specifications. Mobile applications will play a vital role in this scenario. Using GPS technology for proof of location and time keeping will become essential for accounting, payment and audit purpose.

Many of the above apps that are designed for G2E would now have to be re-engineered for G2B with an additional data layer for contractor security authentication and data transmission protocols.

23 Mobile Applications Using GPS Technology for G2G

These are mobile applications that will have the functionality to fetch data from multiple government departments using interdepartmental access protocol and security. This would require security layer, e-signature and authentication for inter-operability. Use case of an emergency application is presented below to understand the complexity.

Emergency application requires live communication and collaboration among field workers in challenging circumstances under severe stress. Live photographs, video and audio of the Incident have to be transmitted in real-time. There has to be a common operating platform so that all members involved in the operation have one common operating picture (COP).

One force Raython [30] is an application provided by a private company that aims to provide reliable and secure real-time communications, situational awareness and a suite of robust collaboration capabilities for groups of users on smartphone tablets and Mobile data terminals.

One Force has released a mobile app for first responders that provide reliable and secure real-time communications, situational awareness and a suite of robust collaboration capabilities for groups of users on smartphones, tablets and mobile data computers. It is a complete end-to-end system with a rich set of capabilities, including voice, maps, drawing tools, chat, real-time position tracking with GPS, streaming video and image sharing.

For most first responders, the personal cell phone they carry has significantly more communications capabilities than their traditional police radio. However, a civilian cell phone is not optimized for public safety use. The One Force application brings together the best of both worlds: access to smartphone technology and the customizable features that first responders need to get their jobs done.

By improving communication between groups, through both voice and data collaboration, One Force enables faster decision cycles and greater efficiency of department assets for public safety professionals. The application supports an evolving incident in the field through the integration of existing and emerging technologies to maximize response effectiveness.

24 Smartphone Applications Using NFC Technology

NFC Technology provides "Proof of Presence", a vital requirement in many government work processes that are carried out in the absence of personal verification. Currently, NFC is largely known for its application for mobile wallets, cashless and other payments, but its true potential in mobile governance is yet to be realized. NFC offers a very wide range of applications from identification, time keeping, attendance, physical access, ticketing, loyalty and a secured login. Although this chapter will focus on NFC applications from the perspective of M-Governance in the context of a smart city, the true potential of NFC applications can be understood from the following Bill and Ted's NFC adventure [31] in which Bill meets his old friend Ted in London for a drink.

Bill is in London on a business trip, staying at a hotel in the centre of town. He has pre-booked his hotel online, receiving his *digital room key* over-the-air to his NFC-enabled smartphone. Upon arriving at the hotel, he taps his phone on the terminal at the front desk *to check in* and heads for his room, tapping his mobile phone on the reader to *unlock the room.*

Ted leaves work and sets off to meet Bill at his hotel for a drink. He *buys a newspaper* on the way to the train station, making a low-value NFC contactless payment in seconds. When he arrives at the train station, Ted taps his NFC smartphone—with his ITSO (Integrated Transport Smartcard Organisation) ticket stored on it—on the turnstile *to gain access to the platform* and travels into town.

The pair meets for drinks in the hotel bar to catch up on old times. At the end of the evening, Bill taps his phone on the point-of-sale to *transfer the bill to his room.* Ted uses an application, such as PayPal Bump, to *transfer money* via *NFC directly to Bill's account* by touching the two handsets together to split the bill. Ted had prepaid for his taxi online and walks outside to find the car waiting. On entering the taxi, he holds his phone close to the contactless terminal *as proof of presence* and informs the driver that the fare has already been paid. In the taxi, a smart-poster advertises a pizza company. Tapping on it *redeems the deal* and a voucher is sent to the handset; Bill *orders Pizza* and makes a secure EMV(Europay, MasterCard and Visa) payment, all via the mobile phone. Five minutes after arriving home, the pizza arrives and one tap on the delivery man's mobile phone (configured as an NFC terminal) acts as *proof of delivery*. Dinner is served.

In the above usecase, NFC technology is used for access control, contactless payment, person-to-person payment, prepaid NFC ticketing, proximity marketing and EMV certified remote mobile payment. In order to make this happen, a number of stakeholders (banks, mobile network operators, handset providers and service providers) have to operate in tandem.

The above usecase also serves to illustrate the three modes of NFC operation. **Card Emulation Mode** is when NFC phone is used to purchase a news paper, **Peer to Peer Mode** is when money is transferred from phone to another phone and **Reader/Writer mode** when tapped on the Pizza Poster. When two NFC phones are tapped, communication happens between two powered active devices and when the NFC phone is tapped on the Poster, one powered active device is communicating with a non self-powered passive NFC tag. While NFC technology will unfold lot more applications in the years to come, one can touch a camera to an inkjet printer to start printing a photo, playing an online game with a friend by simply tapping and pairing, tapping a notebook against a router to create instant secure Wi-Fi connections without a password and affixing an NFC health monitoring tag to a human skin that transmits temperature and sugar level to smartphone and on to the hospital patients record are exciting applications in the making.

25 Difference Between RFID and NFC Technology

There are a number of applications that uses RFID (Radio Frequency Identification) technology and a discussion about its difference with NFC is important due to its widespread use all over the world.

RFID tags contain an antenna and a memory chip that stores data. An RFID reader is required to read that data. These RFID tags and readers are used in a mind-blowing array of applications [32]. RFID tags are embedded into retail products to help stores keep tabs on inventory and on animals and ID cards to track the wearer. Some airlines use RFID tags to efficiently track and control large loads of baggage. The RFID highway toll tag stuck in the car automatically identifies the car to the toll reader from 2 to 12 m far, even at top speed, for billing. RFID appears in so-called smart passports and credit cards, as well as on identification badges that let employees' access secure areas. RFID often works well at distances of many feet and is a one-way communication system, in which data flows from tags to the RFID reader.

The comparison of NFC technologies with various existing technologies is given in Table 2.

NFC is a newer and a finer extension of RFID technology that combines the function of a smartcard and a reader/writer into a smartphone. It operates at a maximum range of about 4 in (10 cm) and can be set up for one-or-two-way communication. NFC tags are much like RFID tags and are simply tuned to work with an NFC reader (smartphone), instead of an RFID reader.

26 Benefits of NFC Technology

NFC provides a range of benefits [33] in the context of a smart city as described below.

26.1 Ease of Use in Having Contactless Payment Systems

No need to carry multiple credit cards or dig through the wallet for the right one. No more lost movie tickets or subway passes and finally, no more individual rewards cards from different stores to track. Instead of carrying a card for every store, it can be loaded on the smartphone and coupons and rewards points go straight to account without ever digging out the card.

Table 2 Comparison of NFC with various existing technologies

Ref	Parameters	NFC	RFID	Bluetooth (IEEE 802.15.1)	WiFi (IEEE 802.11)	Zigbee (IEEE 802.1.5.4)
1	Range	10 cm	3–12 m	10 m	100–150 m	30–100 m
2	Throughput	106, 212, 424 kbps	Varies	721 kbps	6 Mbps	100 V kbps
3	Operating frequency	13.56 MHz	Varies	ISM band 2.4–2.485 Ghz	2.4 Ghz	862, 915, Mhz, 2.4 Ghz
4	Latency	<0.1 s	<1 s	6 s	1.5 ms	20 ms
5	Cost	Low	Low	Moderate	High	Moderate
6	Power consumption	Moderate to low	Low	Low	High	Moderate
7	Security	Fairly secure	Secure	PIN 64 bit, 128 bit (Less secure than WiFi)	More secure than bluetooth	128-bit AES

Source http://en.wikipedia.org/wiki/Near_field_communication

26.2 Security

If a physical wallet is stolen, the thief has instant access to all the credit cards, debit cards, and photo IDs. Storing this information all in one place on the smartphone may sound dangerous at first, but actually provides a safer environment than physical wallet. While wallets cannot be password protected, smartphone can be password protected. No thief can use ones' card if they can't unlock the phone to get at them. NFC uses a secure channel for communication with data encryption when sending sensitive information between two phones or a card reader. The chances for a GPS and NFC enabled smartphone to locate and recover is better than a physical wallet.

26.3 Versatility

NFC technology in this respect can be compared to a Digital Versatile Disc (DVD). The range of usage from checking out at a store, purchasing and loading a movie or concert tickets to smartphones, board the subway, read information from a smart poster, and many such tasks from a smartphone just with a tap makes it a really versatile technology for the smart city.

26.4 Inter-operability

NFC facilitates fast and simple set up of wireless technologies and the smartphone acts as a pass through to transmit data to the cloud allowing NFC's use in low resource areas under challenging situations. For example, the 2009 NFC Forum's First place winner used NFC for a pneumonia surveillance study in young children in Karachi, Pakistan, for "Interactive Alerts for Childhood Pneumonia," which is a real-time patient tracking and referral system used in low-resource settings [60].

26.5 Cost-Effective

Since there is no separate reader required and the smartphone can be used as a reader and writer, the user can be given one single device to carry out multiple functions. NFC tags comes in varying formats starting from a wrist band costing less than 50 cents to an all-weather proof tags costing less than $5. NFC tags can be easily fixed on street furniture in the city or equipments in remote locations.

26.6 *Proof of Presence and Social Media Integration*

Smartphones with NFC can easily be integrated with social media allowing people who meet in a physical space to exchange profile data through their phones. The idea can be extended in an M-governance situation where the government staffs meets the owner of a restaurant and exchanged their respective profiles. If NFC tags are installed on event locations, just tapping the phone enables to let others know about one's check in.

26.7 *Paperless Transaction*

NFC technology help reduce the use of paper to a very large extent. It enables government to move closer to the idea of paperless office.

27 Potential Smart City Applications Using NFC Technology

NFC use cases for NFC mobile services takes advantage of mobile phone functions that complement and enhance the current services enabled by contactless cards. The following table gives a snapshot of the typical day in the life of an NFC Mobile Phone user and how a mobile device will be integrated into everyday life of a smart city resident (Fig. 7).

The NFC applications presented below are in the context of a smart city. Since there are very few documented NFC case studies, the applications listed are potential innovative applications offered from commercial off-the shelf application developers. The applications provide a very good insight to comprehend the degree to which NFC can be applied across different government departments.

28 Field Inspection Workforce Management (G2E)

NFC enabled smartphones plays a vital role in automating field inspection workforce [34] who are assigned a certain number of visits to carry out various inspection tasks. In this case, the field inspection staff is equipped with an NFC enabled smartphone or a tablet. The field staff does not require going to the office. Instead, logs into the smart city mobile work force account and receives the day's field visits that are allocated with their location, task details and expected time to reach the site and duration of stay in each site. Following scenario's illustrating the application.

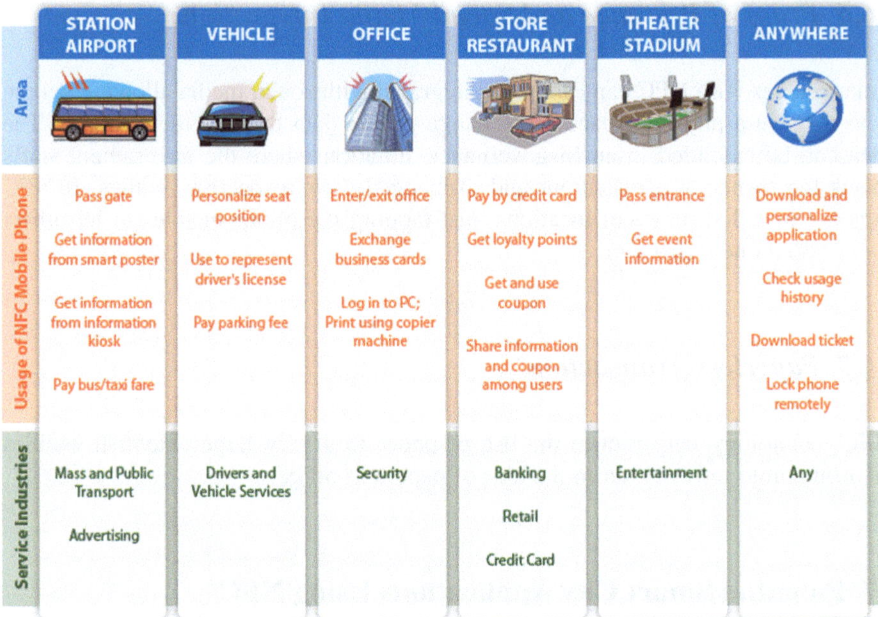

Fig. 7 A day in the life of a smart city resident. *Source* Essentials for successful NFC mobile ecosystems: NFC forum white paper

28.1 Building Inspection of a Building Under Construction

Inspector visits the construction site to ensure building rules are not violated. The inspector opens the smartphone app that has internet data service connection and taps on the NFC tag stuck on the building upon reaching the construction site. The application transmits the proof of presence of the Field Inspector in the construction site to the server and it is shown on the control room map. The Inspector proceeds with the compliance check list listed on the app ticking one by one. In case of any violation, the inspector takes a geotagged photograph along with his comments on the nature of violation. Upon completing the inspection check list, the inspector returns back to the NFC tag and taps on it to signal the completion time of inspection. The app gives a brief summary of the amount of time spend by the Inspector in the site and shows the next site to visit. The app transmits the inspection report in real-time to the back office and forward to the respective authority to initiate action if any violation is found. The inspector's detailed daily and monthly report shows if he/she have met the expectations of the department. Incentive system rewards the inspector for meeting the expectation and penalized upon failure to meet.

28.2 *Food Control Inspection in Restaurants*

The use case in this also is much similar to the above case, but uses a different app and the inspection check list that is applicable for food inspection. In this case as well, the inspector receives the inspection orders as soon he/she logs into the system. The NFC tag can be affixed by the inspector during the first visit and the NFC phone is used to write the Restaurant details on to the NFC tag as directed by the NFC app. Inspection is carried out using the dedicated smart phone app and photograph is attached in case of any violation.

28.3 *Maintenance of Street Furniture and Utility Infrastructure*

This include carrying maintenance works on traffic lights, light poles, transformations, sub-stations and other equipment installations. The workflow involves equipping the mobile workforce with an NFC enabled smartphone or tablet with the app. The Maintenance technician receives work orders with the visit location on the map and an optimized route that takes care of the traffic and the prevailing weather conditions being outdoor work. If the weather does not permit working outdoors, the technician is instructed to report to the office with the pending back office tasks that are scheduled to complete.

The technician taps on the NFC tag to register the time of arrival and carries out the maintenance as per the check list. Now, the key difference between above inspector apps and the maintenance app is the ability to seek higher order technical support from the back office. For e.g. if the technician is unable to trouble shoot, technician is able to call for the back office higher order support engineer to help identify the problem. The technician in this case will make a picture, video or a sound recording and transmit back or even transit a live video of the functioning of the equipment. The support engineer in the back office is able to assist the field technician to trouble shoot and rectify the defect. The advantage here is to have one higher order support engineer supporting a multiple field technicians thereby having the freedom to have a lower skilled field technician to carry out the grunt work and the higher order qualified resource is distributed to step in when required remotely.

Another benefit of this application is the proof of defect as the technician has to make a picture of the faulty equipment and transmits back to the office for replacement approval. The back office system checks if this part is ever replaced before and the service history is transmitted back to the technician with note if the part is covered under warranty from the supplier. The proof of defected part and the inventory management thus happens in the field. The replaced part fitted on the equipment picture is also send back with the replacement report. When the technician taps the NFC tag before leaving the site, the system at the back office is able to compute the time taken for maintenance and added to service history and the inventory is adjusted accordingly, all happening in real-time.

28.4 Conservation of Art Items

The use of NFC in this application is to safeguard cultural Heritage and to manage artefact collections. NFC is used in three points; Art objects are fixed with NFC enabled artefact identification tags, Museums or excavation locations are fixed with NFC tags installed on registered locations; Workers are provided with NFC tag IDs for registered personnel thus providing a comprehensive monitoring of all assets. The application allows Over-the-air, real-time Electronic Registration and documentation of artefacts at historic sites and conservation of cultural and heritage assets. The assets can be cross checked during transit and final delivery. Routine stock taking and management is also made easier. The system also improves security through a unique real-time scanning and patrolling feature used either in museums or remote sites. As the data and photo is transmitted to the back office, museum and site managers can have a clear remote overview of activities in the field for assets in storage, in transit, undergoing conservation procedures or in exhibition. The system provides real-time monitoring, statistics and automated reports related to registered artefacts as well as different controlled locations or sites.

28.5 Campus Management

NFC application in campus such as schools, universities and other government institutions is now replacing the current RFID-based applications. Smartphone-based campus management utilizes NFC technology for attendance from the time of boarding the bus and drop off, tracking student and staff activities, facilitating in-campus service delivery and enhanced security and proves the presence of student in the campus during school hours and during activities.

Students and staff receive personal 'SofTouch' NFC badges (cards or other formats). By touching their badge to a mobile NFC phone, students will record attendance to classes, record their presence in the school-bus, enrol to courses, access in-campus services or areas and present their badges to campus' security staff as proof of their identity during random security checks. The staff will record their presence and activities in a similar manner, as well as gain access to restricted areas.

The advantage is the analytical capability by combining GPS and NFC technology on to a smartphone. Without any special devices, the conductor or attendant in the bus is able to receive the student record with a photo and thereby do a supervised check-in. In the case of RFID, the attendant is not able to view student details online and also to take care of the boarding from different point or boarding without the ID card. The NFC smartphone app is able to alert parents from the smartphone application itself in real-time when there is any mismatch of scheduled picks ups and drops off. This ensures instant alerts as the event unfolds without any delay.

In NFC-based system, teachers are able to take attendance and grade the student using a smartphone and transfer the marks to the database. The teacher is also able to see the previous exam marks in the class in order to see if the student has been continuously having poor grades and have better visibility or alert the parents in real-time on the students' performance.

Students can optionally participate in an in-campus e-purse scheme, through which NFC ID cards can be topped-up with a limited amount of money to be used for cashless transactions in points-of-sale accepting this secure form of payment. The e-purse system of 'NFC Campus' enhances security and ensures that funds will be spent on designated goods offered within the controlled campus environment.

29 Outsource Contractor Management (G2B)

More and more work from the government is being outsourced to the private sector. The government in this scenario has to manage the contractor's work and to ensure the required quality is achieved. The challenge here is to ensure that the contractor follows the approved process and procedures and to have the proof of work being carried out.

Using smartphone/tablet with NFC technology, both the objectives can be achieved. Following use case scenarios illustrate the application.

29.1 Periodic Building Maintenance/Cleaning Services

This is one of the most challenging service contracts to monitor, audit and account particularly if the service contract is awarded in remote locations such as schools or primary health centers owned by the government where there are low resources. NFC offers a near perfect solution to monitor contractor's work in real-time and to produce automated audit and account reports where reimbursement is based on the service duration or per visit.

In this context, the government requires the contractor to use government's dedicated app on smartphone/tablets as part of the service contract. While the supervisor or manager of the service who goes to the site is required to carry the smartphone and tap the NFC tag at the work site, the workers are provided with smart NFC identification cards that have to be tapped at the work site on the NFC smartphone before starting work and after completing the work. The smartphone is able to transmit the worker deployment at site with the GPS coordinates and the proof of presence. The supervisor uses the app check list to tick mark the maintenance tasks and is supposed to attach photographic evidence of the repair work carried out and submit the task report from the field.

The system at the back end is able to quantify the duration of the work and payment dues to the contractor as a daily record and transmit to the finance department for payment processing. The NFC smartphone in this case becomes a mobile attendance and workforce management system.

29.2 Outsourced Maintenance of Security Equipments

Take the case of speed radars that are outsourced to multiple contractors for maintenance in different areas of the city. These equipments require unfailing visits to ensure that the equipment is operating in order; their time is fine-tuned, etc. The contractor is provided with the smartphone app developed by the government agency and NFC tags are affixed in all the radars. The NFC tags in this case is encoded with the RADAR unique ID, location GPS coordinates and the contractor code who is commissioned to carry out the maintenance work. The contractor is required in this case to use the radar maintenance application that gives the check list for maintenance. The contractor staffs visits the radar and taps using the smartphone app on the NFC tag to register the proof of presence. All the maintenance tasks are carried out as per the checklist and the maintenance report is sent from the field instantly to the government servers. If there is a defect in the equipment, photographic evidence is made on the faulty equipment and replaced accordingly. The contractor is now able to invoice the government for the replaced part upon returning the faulty part. Upon completion of the work, the contractor technician once again taps on the NFC tag to end the work and send the maintenance report. The duration and the report are transmitted in real-time to the government servers.

The most important point to note here is the current practise of raising a faulty part report and then obtaining approval before it can be changed. This Process takes a very long time and the administrative hurdles followed by paper work to prove that the part is replaced. In the case of smartphone application, both the contractor and the government save the time for paper work, approval and the administrative overheads to process the transaction. Another advantage of the smartphone usage is to consult the experts at the back office in the event the fault requires intervention of experts.

The police patrol vehicles that carry a smartphone with maintenance app also is able to register a fault report by simply tapping on the NFC tag with their smartphone. The system transmits the fault report to the contractor instantly for corrective action. The fault report ticket is error free and there is no field data entry.

29.3 Security Services

Proof of Presence of the security personnel in the designated points during designated time is a challenge that government departments have on the outsourced security staff. The security service providers allocate the security staff to the respective building, but at what time the staff reached the designated point and his/her presence at different times require government to have internal resource for monitoring and at night, it is largely left to anyone's guess.

NFC technology can be used to monitor the security personnel by equipping the security personnel with NFC enabled phones and affixing heavy duty 'Securi-Touch' NFC tags at cardinal points surrounding the building.

Upon reporting to work, the security personnel will log into the NFC security application and then tap on the NFC tag at the main point and then walk around other points and tag on them thus completing one round. This has to be repeated every hour or decided timing and at completion of duty. This ensures that the security person is carrying out the duty as required and as per the timing. If the security personnel notice anything unusual during their rounds, security staff opens a form from the application and files a report and attaches a photo, video or audio. The app sends an alert to the security company supervisor and at the same time the office in charge of the government department. Depending on the severity of the observation form, the alert goes to the police station, civil defence or fire station with the location, full address and NFC tag point. The app also sends an SOS if the security personnel's life itself is in danger. NFC enabled application is extremely effective in securing remote and low resource locations under challenging circumstances.

All the data gets automatically transmitted in real-time to the government servers with Multi-tier encryption and data security module. The NFC security system provides visibility in the control room that alerts the staff of any security personnel's absence or not adhering to the set rules. Automated reports are generated at the desired interval.

30 Civic Engagement Using NFC (G2C)

Following section illustrates NFC applications that can be used for other applications that involve government services provided to residents in a smart city.

30.1 Event Management

NFC smart tags combined with NFC smartphones can be used for innovative event management providing value added services to attendees and organizers, in exhibitions, congresses, sports and other events. The NFC enabled smartphone in this case assumes multiple roles and provides an innovative user experience and ease of use. An event organizer can define event parameters in the app and collects attendee information on-line through SofTouch NFC enabled badges upon registration in the event. All access can be controlled including the meal booths. The organizer is able to the record the presence at the event venue(s) and in specific areas such as exhibitor stands and meeting rooms. NFC also allows transmitting promotional coupons to the visitor and to dispense brochures and product documents saving significant amount of paper for both the exhibitor and the visitor. The SofTouch NFC badge is a personal identity allowing every attendee to get personalized service from various NFC-enabled spots (e.g. info-kiosks, vending machines or points-of-sale) deployed in the event venue. The information collected over-the-air from the field, in real-time, is stored and aggregated in an event database providing value added statistics and reporting to all parties involved; the organizer, the visitor, the exhibitor and the merchants.

30.2 Building with NFC Tags

If NFC tags can be handed over when a home owner/tenant or an establishment comes for a government transaction such as renewing the tenancy contract or payment of utility bills, it can be affixed on their house or apartment unit or even in business establishments and warehouses. In such a scenario, there can be G2C application that allows the owner to tap on the NFC tag to request government services including in the event of emergency so that giving ones location becomes extremely easy. The government agency who is handling the request also gets error free data and is able to provide the services with greater accuracy and ease. In the event of an emergency, articulating ones' address becomes very difficult in the case of children/old age people/differentially challenged and visitors. The NFC tag in this case makes a big difference.

The building NFC tags affixed in homes and private establishments becomes useful for the government employee who visits the unit for providing the service related to utility (Electricity or water issues) to prove their presence and to send real-time repair reports. In the case of business establishment, the establishment inspection carried out by the Economic Department, statistics Department or Municipality for permitting certain functional use becomes easier. The visit becomes paperless, easy to prove their presence, easy to perform and real-time reporting allows the officer at back office to process the transaction quicker. Thus the smart city benefit is directly felt by residents and businesses.

30.3 Resident Reporting and Civic Engagement

This gets a bit smarter with NFC technologies. Now, the user is able to tap the smartphone on the NFC tag affixed directly on the object for e.g. broken street light, ATM kiosk or transformer and choose the fault shown on the app. Since there is no data entry and very little effort required to submit a report, people tend to do it with ease. At the same time, the government is able to get error free report so that the time required at the back office to sort out the problem and take action gets shortened. In many instances, civil engagement becomes difficult to attend due to the error and inaccuracy in the reports filed by the citizens.

31 NFC Applications Use Cases in M-Government

NFC applications are in its infancy. In the UAE, Du Telecom has launched NFC-based SIM to allow mobile payment and soon Etisalat is expected to follow suit. RTA, the Dubai Roads and Transport Authority has announced plans to introduce NFC-based ticketing system on its metro, bus and water services. Dubai Police uses NFC in its Al Ameen E-tag services. Conferences and exhibitions in the US and Europe have started using NFC bands quite successfully and NFC tags are used in posters to interact with people. Google, Samsung, Microsoft, Master card, Visa and leading banks are pushing NFC into the mainstream. As per ABI Research, the number of NFC devices will reach 2 Billion by 2017.

Introducing any new technology in the government poses a number of challenges due to the outdated process and procedures and the long winded implementation process. While government make mission statement in their lobbies wanting to become the leader and one of the top five smartest cities in the world, they frequently ask the question; where is it used now? Which country or agency is currently using it? If they are told that it is used nowhere, the government agency is reluctant to embrace the idea and have the courage to implement it. So, the question of where is it used? in fact contradicts their mission statement whether they want to become a leader or a follower.

All the above applications in the context of a smart city are innovative application use cases and new ideas that are doable and practical. But, a leader must take the risk to apply it and demonstrate leadership for other countries to follow. Inaugurating Dubai Internet City HH Sheikh Mohammed Bin Rashid Al Makhtoum, the VP of UAE and ruler of Dubai said: "We all take risks in life, but do you know what is the biggest risk? Of not taking risks at all" [35].

32 Risks and Challenges of M-Governance

Smartphone Transformatory technologies come with risks and challenges and it is further more when combined with GPS and NFC technologies and on payment processing. While risks are seen as effective blockers, challenges can be treated as something that can be overcome. The potential risks and challenges in the context of using GPS and NFC applications may vary depending on the delivery models; however, they can be summarized as follows.

33 Risks in Using Smartphone Applications in M-Governance

Using smartphone application in the various delivery models (G2E, GB, G2G and G2C) described in this chapter poses a number of risks as listed below:

(a) Privacy concerns in ‘proof of location’ and ‘proof of presence’. Users may not want to reveal their identity or their location. Lot of government data on citizens is sensitive (e.g. tax-related or health related) and therefore, there is a ‘fear’ in letting this out.
(b) Exposure to hacking and security threat particularly in using smartphones for payment. Hacking can happen at the handset level or at the server level where they are stored.
(c) Impersonalization and manipulation of transaction is a serious risk, although several biometric and security procedures are being developed to counter this threat.
(d) Data loss due to network outage or power or equipment damage caused due to dropping or mishandling while carrying out a transaction.
(e) Prevailing government laws: Certain government laws may prohibit the usage of GPS or may be very ambiguous. Some countries may restrict importing a smartphone with a GPS and WIFI. Until recently there are countries that did not permit phone with camera.
(f) Abuse of hardware or personal data due to the handing over of government property to staff for taking it home, as it may lead to its loss or damage due to carelessness or deliberate vandalism. As such, managing and inventory control becomes a very serious matter. This can get out of proportion and even Trade Union interference can stall its entire use. NFC tags damaged due to vandalism also stand as a threat to affixing hundreds and thousands of NFC tags on street furniture and equipments without sufficient protection.
(g) Lack of penetration of smartphones and affordability. Provision of public services is citizen centric and governments must meet the digital divide challenge. So, smartphone-based services may cause severe discontent among residents not having access to a smartphone.
(h) Change of technology, hardware models, and operating system poses a serious threat to the government investment on something that is not certain can lead to total abandonment of Mobile technology. A number of phones are available in grey market and risks in their security.
(i) Service to be delivered ubiquitously beyond any temporal and geographic constraints. The basic government requirement to ensure equitable access to technology stands may act as a stopper that has happened in the case of early computerisation and e-governance. Innovative applications native to smartphone may be seriously questioned if they cannot be offered in other alternative channels such as desktop.
(j) GPS accuracy: Using smartphones in cities and indoors may not give satisfactory accuracies due to urban canyons or structural hindrances for certain

applications. Assisted GPS and a new technology named TIMU—Timing and Inertial Measurement Unit [36] may solve this as the latter is miniaturized to go in the GPS module of a smartphone.

34 Challenges in Using Smartphone Apps for M-Governance

Using smartphones in the government business process as well as enabling customers to carry out the transaction anytime from anywhere has significant challenges. Changing the point of customer contact from a 'counter' being helped with a personal touch to a 'self-service' hand-held terminal, requires a thorough re-engineering of business processes. Major challenges are given below:

(a) **Real-time data exposure**: This is one of the biggest challenges that governments have to face. Normally the smartphone end users are "unforgivable" customers due to their ignorance on the complexity of the Mobile ecosystem. Providing up-to-date, accurate and reliable data that can stand public scrutiny with direct or contingent liabilities is indeed a big challenge. Similarly, there may be several errors or inconsistencies in data and systematic data cleansing is prohibitively expensive.

(b) **Dependency on multiple Technologies**: GPS technology for instance relies on GPS satellites and several factors influence the data accuracy. The users out of ignorance believe that the location shown on the phone is accurate and therefore carries out certain actions. So, the applications have to address this and rigorous computation may be required to warn the user before certain action is performed. In many cases, cross verification and cross referencing may be essential.

(c) **Inter-operability between multiple agencies and their compatibility**: Many government transactions require multi-agency data access. For instance, land transaction would require to access revenue records or bank records for determining the encumbrance. This may pose significant challenge in rolling out mobile applications that requires multi-agency data access. The agencies may be in different stage of E-governance and may not grant access privileges.

(d) **Application user interface**: This is a maker or breaker of any mobile application. Several challenges need to be overcome such as very little skills requirement, small screen, operated in open space and in sunlight, highly intuitive, touch interface and little or no data entry.

(e) **Data Bandwidth**: Due to the speed between several networks, the app speed is very critical. Acquiring GPS data and transmitting NFC data for verification requires certain time and speed which has to be factored.

(f) **Data handling Policies**: Data duplication, data sharing, data standard, data custodian, data harmonisation, Replication, data reversal and change management need to be addressed.

(g) **24 × 7 service availability**: This poses a significant challenge for staff that is responsible for server and application maintenance. High availability with failover and redundancy with firewall, physical Security, power back up requires dedicated server farms with cloud architecture. Emergency procedures and round the clock surveillance also poses a big challenge.

(h) **Applications guidelines, quality assurance, certification**: With several government departments developing their own applications, a central agency has to ensure that the applications adhere to the standards and pass the tests before they are rolled out. This may take time and resources and the current approval mechanism has to be re-examined. Upgrading applications. Developing application for multiple devices, supporting multiple OS and the decision to make a dedicated App or use platform independent technology such as html5 becomes highly debatable and resource intensive.

(i) **Training of staff** on following M-Government procedures and in using applications require significant investment and resistance to change may require change management strategies, particularly among staff who are not well versed with smartphones whose mindset is that a phone have just one function i.e. to make a call.

(j) **Application development**: Recruiting, training and retaining high talent staff in smartphone applications is very challenging and government has to create separate departments and salary structure that would require several challenges to be overcome.

(k) **Responding to the speed and huge inflow of information**: In G2C applications such as users sending in notification and complaints may need to be addressed with dedicated staff and process and procedures for rapid action. Following the current style of handling a complaint or suggestion will lead to the users simply getting disillusioned and abandon the whole idea of crowd sourcing or civic engagement and thereby total withdrawal.

(l) **Scalability, adaptability and expandability**: Mobile applications can become viral and the speed at which it happens is quite unpredictable. Therefore, mobile application supporting server infrastructure needs to be highly Scalable at short notice warranting the use of cloud architecture.

(a) **Open data policy and 24 × 7 customer support**: Several government data sets may have to be opened to the public. Smartphone application will give access to government data from the government staff to the hands of customer terminal in seconds. Support for this data and application will have to be made round the clock and significant challenge need to be addressed. The role of the Government lies in identifying the right balance between data transparency, privacy and security.

(m) **Cultural resistance**: Lack of self-confidence in new technologies and risk avoidance can seriously prevent users from using smartphone applications. Failure in providing support or the application failure may lead to lack of trust in the M-Government services and perceived quality of Mobile services.

35 Strategy for Implementing M-Governance

It is evident from the above section that the risks and challenges posed for M-Governance requires a well crafted and well articulated national and state level policy and commitment. Establishing Institutional framework for development and implementation of guidelines, application testing, training, certification and quality assurance is vital for M-Governance to succeed. Countries such as South Korea, Singapore and UAE who have advanced in M-Governance implementation have established strategies with a clear vision and commitment.

36 Articulating Vision, Commitment and Road Map

Success of M-Governance depends on the government's recognition and the joint efforts of all the stake holders; Government Departments, Business Establishments, Customers and the media who have to support such an initiative. UAE government targets to make 1,000 government services to go on smartphone in the next 3 years.

Unveiling the Strategic Plan to transform Dubai into world's smartest city, HH. Sheikh Mohammed Bin Rashid Al Makhtoum, Ruler of Dubai and Vice President of UAE articulated Dubai's ambition as follows [37]:

> "Our ambitions is that this project touches the life of every individual in our country, or every mother in her house, or employee in his work or investor in his project, or a child in his school or a doctor in his clinic; our goal is to achieve a happier life for all and we ask God to help us achieve this".

The above statement sets the scope, commitment and expectation. The UAE government has been in the forefront of demonstrating this along with the media who have helped educating the users on the benefits and addressing their concerns.

In the UK, the technology strategy board provided financial resources for Glasgow to become UK's first smart city providing real time traffic, apps to check when buses and trains are arriving, and pothole reporting service.

37 Establishing Institutional Framework

The institutional framework strategy is described below, reviewing the case study of UAE.

(a) **Developing Institutional Framework**

The Government has to establish the institution in advance with an expandable and adaptable framework, as the technology is itself evolving at a rapid pace.

Appropriate measures have to be established to address privacy, risks and challenges and how to mitigate them through institutional framework.

In the UAE, The TRA has established the Arab region's first laboratory to test the security and usability of new E-government and M-Government applications to start operation in October 2014 [38].

The M-Government Lab is set up to verify the quality and security of M-government applications, and ensure they comply with global standards and best practices. The lab is equipped to provide full security screening, load testing for applications and accounts for government entities that want to test applications remotely. This is an important step towards 'protecting both the users and government entities from security threats that might arise from unauthorised applications'. As per the UAE's Computer Emergency Response team (CERT) at the TRA, the lab performs automated tests on applications and then manual tests on those that pass the trial to assess its strength taking into account the ease of use, security and many other important functions.

(b) **Developing Training Center**

As a part of an integrated strategy aimed towards successfully achieving the smart government transformation process, the UAE M-Government Initiative in Feb 2014, launched a special e-portal to register for M-Government training programmes [39]. These programmes are aimed at preparing and training employees from various government entities at both federal and local levels. The training courses includes various important topics including the management of smart services in the UAE, the M-Government IT infrastructure, security of smart applications and smart application developers.

A master's degree [40] in Smart City has been offered by the RIT University in Dubai for the first time in the UAE with a tie up from Polytechnic University, Spain.

Another initiative is the starting of Smart City Forum at the Mohammed Bin Rashid School of Government [41] (MBRSG) to bring industry experts and specialists to share expertise, discuss challenges and review best practices as part of the Dubai Smart city initiative. MBRSG aims to make a prototype of a global smart city by leveraging expert advice from around the world.

(c) **Establishing Government Cloud Infrastructure**: M-Government required a dedicated cloud infrastructure with the necessary physical security that can be scaled as per the needs.
(d) **Common Application Portal**: It is important to have all the applications that are offered by various government departments under a common M-Application portal. This is evident from the review of applications from around the world. Dubai government has made an application portal [42] in iTunes and in android for easy access.
(e) **Hardware and OS Alignment**: The applications developed for G2C will have to support multiple hardware clusters. For G2E and G2G applications, the government has to test and approve models that are aligned with the

government's requirements and security policy. This has to be reviewed often so that they can be updated and new models are supported as they become available.

(f) A **common authentication and verification policy** need to be defined. In Thailand through apps.go.th portal, a single sign-on feature for apps will come in place as soon as EGA consolidates the various registration requirements of different apps. Once completed, users will need to register their username and password once and will be able to access all apps through the same username.

(g) **Holding regular annual smartphone application competitions** to instil competitive spirit among the stake holders have proved very helpful in developing innovative applications. Annual competitions to choose best apps have been followed in the UAE and various other countries.

(h) Technology trends suggest that for governments and public-access applications in general, a strategy based upon HTML 5, CSS 3 and JavaScript will deliver the best balance between features and cross-platform compatibility.

38 Establishing M-Government Guidelines

A very clear and well articulated guideline is essential to help government entities prepare themselves for M-Government. The guideline should address the following areas:

(a) Identify and prioritize current services and select services that should be offered as part of the M-Government services and the respective delivery models. Slice the services into dedicated apps for better and simple user experience.
(b) Illustration of types of mobile services, associated issues and the target audience.
(c) Mobile application development factors such as platforms, APIs, user and usability issues, mobile content, user interface and user adoption.
(d) Clear guideline on security issues considering enterprise wide mobility.
(e) Data update, data protection and privacy issues.

The Telecom Regulatory Authority (TRA) in the UAE has released a guideline document for Mobile Government to help government entities in their Mobile transformation.

39 Star Rating for Government Establishments

Dubai has initiated the idea of rating government departments [44] with the implementation of a new set of standards and rating system to improve government services and enhance people's satisfaction level. The system will measure each department's performance on the scale of 'two to seven' star ratings. The standards for measurement will be expanded to Middle East and North Africa and thereby the entire region is benefited by this initiative. Arab Accreditation, a standardisation arm of Arab League, will oversee implementation of guidelines at regional level. However, it won't be binding on the member countries. At the UAE level, the standards are being implemented and the performances are being measured by the Prime Minister's office in coordination with Emirates Standardisation and Metrology Authority (ESMA).

There will be a set of common guidelines for all services. The ratings will be an annual process, which will involve measurement of the performance, giving a timeline for the departments to adjust to the requirements standards in case of unsatisfactory results. This ambitious project is based on a set of criteria which includes dealings with customers, speed of service delivery, reduction of waiting time, accessibility of the services, office ambience, and use of technology and location of service delivery outlet. The rating will be a continuous process. There will be dedicated staff to continuously monitor the performance and ensure the standards are maintained and if a department fails, its ratings would be dropped.

The idea of star rating illustrates the institutional motivation and an avenue for recognition and thereby increasing the overall performance of the government. This will motivate the departments to embrace M-Governance.

Dubai has recently set up Happiness Meters in government departments with 3 buttons (Smile, OK, Sad) for citizens to rate services instantly.

40 Strategy on In-house Development or Outsourced

Due to the challenges posed by the requirement of applications and its nature, it is essential to take a strategic decision on whether the infrastructure and applications need to be developed in-house or to be outsourced.

Governments that have already established E-Government infrastructure will find it easy to add M-Government applications by starting a separate section. But, governments that have low resources might have to outsource it to a private company with legal binding. Another model would have the government providing its premises and asking the outsource agency to operate from its premises and hand over to government in certain number of years during which the capacity building of the government can occur.

Another strategic decision is for the government to opt for Commercial off the shelf (COTS) applications that are emerging from various private enterprises or develop In-House. Using COTS may help roll out the applications quickly and to achieve results rather quickly during the early stages.

The use of a private company under the supervision of a government entity is tried in the Indian state of Kerala. The government uses a private company [45] for providing its M-Government Service Delivery Platform (SDP). MOBME is successful model in providing M-Governance Solutions in India and their implementation in Kerala was recently recognized by World Bank as one of the Best Practices globally. All government departments access the SDP to enable cost-effective design, development, and deployment of various M-Government applications. This arrangement avoids duplication of effort and cuts capital spending on stand-alone systems. By integrating with all telecommunication companies, the SDP eliminates the need for individual coordination by government agencies.

MOBME's SDP also provides technical assistance to public departments to design and launch mobile applications. Since its launch in December 2010, the program has involved more than 60 government agencies, facilitated more than 3 million interactions between the government and citizens and has deployed at least 20 M-Government applications. One notable application is the Mobile Crime and Accident Reporting Platform that is used by Kerala police to enhance public safety. This multimedia messaging service-based accident and crime reporting has captured some 200,000 photos for crime and accident reports.

41 Establishing Urban Operating System (UOS)

A large number of wired and wireless sensors such as GPS, NFC, M2M (Machine 2 Machine) devices and combined with video surveillance, UAV (Unmanned Aerial Vehicle) delivering packets to homes and audio surveillance to detect some equipments throwing an alarm are set to dominate future smart cities. It then becomes absolutely essential that all these connected objects are controlled from a central system. UOS is an emerging smart city control tower (Fig. 8).

The proliferation of sensors in the urban environment has given birth to the idea of internet of things in which the sensors [46] typically send their data to a sink node, which in turn sends it to some sort of processing centre. These two components (sensing and processing) are not explicitly unified, from a Wireless Sensor Networks (WSN) perspective. Some unified solutions such as Thingspeak [47], acts a platform that enables users to upload information from their sensor nodes directly to the Cloud. Users create an account on the cloud platform, register their devices, and get access to their sensor data as desired. The system allows querying

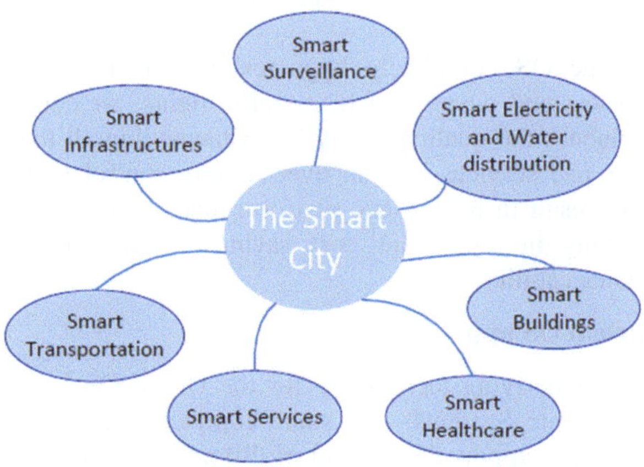

Fig. 8 Advanced sensors in smart cities. *Source* [46]

by location, enabling the user to have access to data from various locations in the world. iOBridge [48] in contrast to Thingspeak where libraries for different sensing platforms are available, develops its own hardware modules that are connected to the Cloud, which in turn can be accessed through web interfaces, for remote monitoring and API's allow integration of the collected information into other web services or third-party applications. For example, mobile applications for controlling these ioBridge devices remotely from smartphones such as Android and iPhone or even from a smart watch [43] or through a Google Glass.

In the WEF Report on future of government smart tool box [49], the Strategic Foresight team has identified three visions of government in 2050.

In scenario 1, named **e1984** is a world in which the promise of big data is realized; economic, geopolitical and cyber threats are omnipresent; and collective solidarity is a core societal value.

In scenario 2, named **Gated Community** is a world in which Big Government is broke; political power rests with individuals and private sector organizations.

In scenario 3, named **CityState** is a world in which authority is decentralized to city level and pragmatism trumps idealism in addressing collective issues. Urbanization leads to growth in the number of cities and their size and consolidation. The scale of cities means they can innovate and get things done more easily, which raises their relevance and power vis-à-vis national governments. Globally connected city states of this magnitude will in turn give birth to the idea of a Global City.

In the e1984 scenario [50], the smart cities with devices chatting to each other may dot the planet in the near future requiring the need for an Urban OS that works just like a PC operating system but keeps buildings, traffic and services running smoothly. The software takes in data from sensors dotted around the city to keep an eye on what is happening. In the event of a fire, the Urban OS might manage traffic lights so fire engines can reach the blaze swiftly. The idea is for the Urban OS to gather data from sensors buried in buildings and many other places to keep an eye on what is happening in an urban area. The sensors monitor everything from large scale events such as traffic flows across the entire city down to more local phenomena such as temperature sensors inside individual rooms.

The OS completely bypasses humans to manage communication between sensors and devices such as traffic lights, air conditioning or water pumps that influence the quality of city life. Channelling all the data coming from these sensors and services into an over-arching control system had lots of benefits. This will result in having one platform managing the entire urban landscape of a city helping in significant cost savings, implementation consistency, quality and manageability.

Urban OS apps

The underlying technology for the Urban OS has been developed by McLaren Electronic Systems—the same company that creates sensors for Formula One cars. To support the myriad of different devices in a smart city, an extensive set of application services will have to be developed to run Urban OS, the urban environment equivalent of apps on a smartphone. Independent developers will also be able to build their own apps to get data and provide certain services around a city.

Eventually applications on smartphones could hook into the Urban OS to remotely control household appliances and energy systems, or safety equipment to monitor the wellbeing of elderly people. It could also prove useful in the event of a fire in a building. Sensors would spot the fire and then the building would use its intelligence to direct people inside to a safe stairwell, perhaps by making lights flicker or alarms get louder in the direction of the exit.

A test bed for the Urban OS is currently being built in Portugal. For its work in developing smart cities, Living PlanIT [51] was selected as one of the World Economic Forum's Technology Pioneers of 2012. There are other proprietary systems available from HP called HP Central Nervous System for the Earth [52] (CeNSE) and smarter planet [53] from IBM.

42 Conclusions

Smartphone penetration has only touched the outer ring. As smartphones and data bandwidth become affordable, people shifting from basic handsets to smartphone is just around the corner. Just like the switch to a flat panel TV, smartphone will become ubiquitous and will act as a self-service terminal for every citizen, resident, visitor or a tourist to carry out government transaction at their convenience, anytime from anywhere with comfort. While GPS in the smartphone will determine the location, NFC will prove the presence, both so transparent and hidden that no one notices it.

The young generation goes beyond the application defaults to customizing the look and feel of their own self-service terminal. M-Government applications will have to face this reality and find innovative ways to enrich the user experience and make it simple to use for a wide range of users. Personalised digital App assistants would have to appear from the corner of the terminal to help the user in the chosen language. Should a mistake be made, suggestions mined from user's social cloud must pre-empt. Applications have to be served on a silver platter; complexity must be wrapped in beauty!

M-Government forces city administrators, planners and managers to slice/modularize their services and make them tightly coupled to be served as a self-serving app. New processes and procedures have to be innovated in view of the self-service terminal as the service counter. Every single form that required the user to enter their personal data will have to be auto-filled and authenticated. User's geo code will be auto-filled too and the proof of presence is done by a smartphone hi-fi tapped on an NFC tag. The one-stop common M-Government application portal will become the reception desk in a smart city and the Olympia for annual competition to determine who is smarter?

Mobile applications are going to influence every single aspect of our lives and the future may appear to be a bit scary for some, but exciting for others. Notwithstanding, mobile phone has empowered mankind from all corners of the world regardless of their colour, race or status. It will further act as the connection point for the government to realize what its real purpose is; to serve the people and to make them partners in change and development thereby engaging them in true governance.

References

1. IGo does not refer to any app, a fictitious name used to denote a navigation app, although there is a navigation app called iGO Prime
2. Nineteen Eighty-Four, sometimes published as 1984, is a dystopian novel by George Orwell published in 1949
3. https://itunews.itu.int/En/3741-Mobile-subscriptions-near-the-78209billion-markbrDoes-almost-everyone-have-a-phone.note.aspx
4. www.worldbank.org/ict/IC4D2012
5. http://www.emarketer.com/Article/Smartphone-Users-Worldwide-Will-Total-175-Billion-2014/1010536
6. http://electronics.howstuffworks.com/gps-phone.htm
7. http://en.wikipedia.org/wiki/Assisted_GPS
8. http://nfc-forum.org/what-is-nfc/what-it-does/
9. http://rapidnfc.com/which_nfc_chip
10. DESA (2010) E-Government survey 2010, United Nations, New York
11. M-government: Mobile technologies for responsive governments and connected societies—© ITU, OECD 2011
12. Making Government Mobile; Siddhartha Raja and Samia Melhem with Matthew Cruse, Joshua Goldstein, Katherine Maher, Michael Minges, and Priya Surya, Maximizing Mobile, OECD/ITU
13. United Nations E-Government Survey 2012, E-Government for the people
14. http://www.egov.gov.sg/egov-programmes/programmes-by-citizens/mgov-sg
15. http://mgov.activate.sg/aboutus.aspx
16. http://m.mytransport.sg/redirect/index.html
17. https://search.itunes.apple.com/WebObjects/MZContentLink.woa/wa/link?mt=8&path=apps%2fUAEmGovernment
18. https://play.google.com/store/apps/details?id=info.mgov.app
19. http://mgov.government.ae/web/guest/apps-gallery or apps.gov.ae
20. https://apps.bahrain.bh/CMSWebApplication/action/AppStoreAction
21. https://apps.mgov.gov.in/index.jsp
22. http://english.msip.go.kr/index.do
23. https://www.gov.uk/government/news/government-backs-smartphone-app-to-pinpoint-pot-holes
24. OECD/International Telecommunication Union (2011), M-Government: Mobile Technologies for Responsive Governments and Connected Societies, OECD Publishing
25. http://www.saaed.ae/en-us/home.aspx
26. http://www.appmycity.org/
27. http://www.neatstreets.com.au/NeatStreets
28. http://en.seeclickfix.com/
29. Dr. Gharib and Sahib, Accident Referencing and Analysis System for Abu Dhabi

30. http://gcn.com/articles/2013/02/14/one-force-app-first-responders-collaborate.aspx?admgarea=TC_Mobile
31. http://www.slideshare.net/bellidcom/bill-teds-nfc-adventure?qid=d93efcd3-b3f6-414e-b5b6-aa8b196e5db6&v=qf1&b=&from_search=1
32. http://electronics.howstuffworks.com/difference-between-rfid-and-nfc.htm
33. http://www.nearfieldcommunicationnfc.net/benefits.html
34. www.hotech.com
35. www.sheikhmohammed.co.ae/
36. Geospatial world, March 2013 Issue page 10
37. http://www.emirates247.com/news/government/mohammed-launches-dubai-smart-city-smartphone-to-be-key-pivot-2014-03-05-1.540576
38. http://www.fiercewireless.com/europe/story/uae-begins-testing-m-government-smart-apps/2014-01-26#ixzz34OJTTJcf
39. http://www.smeadvisor.com/2014/02/mgovernments-training-portal/
40. http://www.khaleejtimes.com/kt-article-display-1.asp?section=educationnation&xfile=data/educationnation/2014/june/educationnation_june7.xml
41. http://www.mbrsg.ae/
42. https://itunes.apple.com/app/id619712783
43. http://www.khaleejtimes.com/kt-article-display-1.asp?xfile=data/government/2014/June/government_June57.xml&section=government
44. http://gulfnews.com/news/gulf/uae/general/uae-to-introduce-star-ratings-for-government-departments-1.1289766
45. http://www.mobme.in/
46. Erhard P. Hancke, Bruno de Carvalho e Silva 1 and Gerhard P. Hancke Jr. 1,2 The Role of Advanced Sensing in Smart Cities
47. https://thingspeak.com/
48. https://www.iobridge.com/
49. WEF_GAC_FutureGovernment_SmartToolbox_Report_2014
50. http://www.bbc.com/news/technology-15109403 report on Living PlanIT—the company behind Urban OS
51. http://www.living-planit.com/
52. http://postscapes.com/hp-cense-platform
53. www.ibm.com/press/smartercities
54. http://gulftoday.ae/portal/7af56920-a3c7-4068-b26b-ab0dc5fce08e.aspx
55. http://smartbin.com/images/Pdfs/SB130r00%20Datasheet%20-%20UBi%20Sensor.pdf
56. https://www.envirocar.org/about.php
57. https://www.pathgeo.com/citybuddy.html
58. http://www.ikegps.com/spike/
59. http://lbszone.com/2014/07/23/terrago-brings-mobile-note-sharing-collaboration-edge-enterprise/
60. http://nfc-forum.org/newsroom/nfc-forum-global-competition-2009-announces-winners/ (http://www.wima-nfc.com)

Bibliography

1. Al-khamayseh S, Lawrence E, Towards citizen centric mobile government services: a roadmap. University of Technology, Sydney
2. Bhatnagar S, Exploring conditions for delivery of successful m-government
3. Services to the Bottom of the Pyramid (BOP) in India. International Development Research Centre, Canada and the Department for International Development, UK

4. Georgescu M (2010) Mobile government: an emerging direction. Alexandru Ioan Cuza University of Iasi, Faculty of Economics and Business Administration, Iasi, Romania
5. Government of India, Ministry of Communications & Information Technology, Department of Information Technology. 2009. Framework for Mobile Governance. Notified in The Gazette of India, Weekly No. 8, February 25–March 2, 2012, Part I—Section 1, 471 GI/2011
6. Harbor Research, 2011. Machine-To-Machine (M2M) & Smart Systems Market Opportunity 2010–2014. Harbor's market analysis on how networks of companies, customers, people and intelligent devices are driving the convergence of the internet of things and the internet of people. Harbor Research, Inc.
7. Hussain H (2013) Dialing down risks: mobile privacy and information security in global Development Projects. Open Technology Institute
8. infoDev (2013) Mobile usage at the base of the pyramid: research findings from Kenya and South Africa. World Bank, Washington, DC
9. infoDev (2013) Guiding ideas from mind to market: learning from infoDev's Mobile. Microwork innovation competition. The World Bank, Washington, DC
10. infoDev (2012-2013) Growing innovation, Annual Report. 2014. infoDev. World Bank, Washington, DC
11. infoDev (2014) The business models of mLabs and mHubs—an evaluation of infoDev's Mobile Innovation Support Pilots. World Bank, Washington, DC
12. INIT, Services for the esociety. eGov2Go; Seven Propositions for Mobile Government
13. Jotischky N, Nye S (2011) Mobilizing public services in Africa: The m-government challengeYour, Informa Telecoms & Media—Head Office, Mortimer House, 37-41 Mortimer Street London W1T 3JH, UK
14. Kerala Mobile Governance experience and roadmap for a Comprehensive mGovernance Strategy, M-Governance Initiative by Department of Information Technology, Keral State IT Mission, GoK
15. Kushchu I (2007) Mobile government: an emerging direction in e-government. Mobile Government Consortium International, UK. IGI Publishing
16. Nordfors L, Ericson B, Lindell H, Lapidus J, Gullers Grupp (2009) VINNOVA Report VR 2009:28 eGovernment of Tomorrow: Future Scenarios for 2020
17. Misra DC, Make m-government an integral part of e-government: an agenda for action
18. Moon J (2004) From e-government to m-government? emerging practices in the use of mobile technology by State Governments. IBM Center for The Business of Government
19. OECD/International Telecommunication Union (2011), M-Government: Mobile Technologies for Responsive Governments and Connected Societies, OECD Publishing
20. United Nations Development Programme (2012) Mobile technologies and empowerment: enhancing human development through participation and innovation, UNDP
21. United States Federal Government (2012) Digital government: building a 21st century platform to better serve the American people
22. Vijayakumar S, Sabarish K, Gokul K, Innovation and m-governance: the Kerala mobile governance experience and road-map for a comprehensive m-governance strategy. MobME Wireless Solutions
23. World Bank (2012) Information and Communications for Development. 2012: Maximizing Mobile. World Bank, Washington, DC. doi:10.1596/978-0-8213-8991-1

Participatory E-Budgeting Using GIS-Based Spatial Decision Support System: Kozhikode Municipal Corporation

M.A. Naseer, P. Bimal and T.M. Vinod Kumar

Abstract The 73rd and 74th Amendment of the Indian Constitution has brought the power to prepare development plans to the grass root level – to the *Grama Panchayat* on the rural side and to the *Municipalities* on the urban side. Annual budgeting in these institutions is of utmost importance as the implementation of development schemes is possible only through a logical and balanced allocation of available funds. At the municipal level the budget provides a balanced and coordinated approach to municipal activities so that requirements and responsibilities of all sections are analyzed while preparing a budget. One important innovation, Smart City embarks on is a web-based participatory e-Budgeting which empowers citizen. E- Budgeting will use an interactive and dedicated annual Municipal budgeting web site and will also use web-based Geographic Information System (GIS). A hundred percent E-Literate Smart City with smart people would like to be part of E-Budgeting. Participatory budgeting allows local citizens in a municipal ward to identify, discuss, and prioritize public spending projects, to make decisions about how money is spent every year. Kozhikode was practicing Participatory annual planning and budgeting since the mid-1990s. This chapter explores how the existing participatory budgeting can be converted to e-budgeting befitting a Smart city. It also demonstrates how Spatial Decision Support System (SDSS) that uses Geographic Information System (GIS) can be deployed in Kozhikode using most up-to-date data for budgetary decision making to help the community to arrive at the most rational budget allocation.

M.A. Naseer (✉) · P. Bimal
Department of Architecture, National Institute of Technology Calicut,
NIT Campus Post, Kozhikode 673601, Kerala, India
e-mail: naseer@nitc.ac.in

P. Bimal
e-mail: bimalp@nitc.ac.in

T.M. Vinod Kumar
School of Planning and Architecture, New Delhi (SPA-D), Calicut, Kerala, India

T.M. Vinod Kumar (ed.), *E-Governance for Smart Cities*, Advances in 21st Century Human Settlements, DOI 10.1007/978-981-287-287-6_10

Keywords E-Budgeting · Participatory budgeting · SDSS · GIS

1 Introduction

One important innovation, Smart City embarks on is a web-based participatory E-Budgeting which empower citizen. E-Budgeting will use an interactive and dedicated annual Municipal budgeting website and will also use web-based Geographic Information System (GIS). GIS data provider would be selected members drawn from Ward Committees or Residential Welfare Association. They will also help in producing GIS database using GPS (using GPS enabled Cameras) as well as excel sheet of attribute data base for E-Budgeting GIS. It is recommended to use Linux-based freeware's as Q GIS to reduce the GIS cost. A hundred percent E-Literate Smart City would like to know the rationale, how annual budgets of Municipalities are allocated and also change it if required based on social and spatial justice. Smart people would like to be part of E-Budgeting and would have embraced Participatory Budgeting (PB) much earlier. PB is advancement in the process of democratic deliberation and decision-making. It facilitates participatory democracy. Here ordinary people decide how to allocate part of a municipal budget. This is a share from the national and state budget plus own municipal earnings. PB allows local citizens in a municipal ward to identify, discuss, and prioritize public spending projects, to make decisions about how money is spent every year. PB is based on mutual trust between local governments and citizen for mutual benefit. In some cases PB even raised people's willingness to pay taxes [1].

PB generally involves several basic steps: (1) Community members say Ward committee identify spending priorities in the ward and Municipality identify task force members to help in the scientific aspect of budgeting as related to specific sectors as practiced in Kerala, (2) Task force members develop technically sound specific spending proposals, with help from outside local experts, if required, (3) Community members vote on which proposals to fund, (4) The Municipality and higher level institutions such as District Planning Committee or Metropolitan Planning Committee approves the budget for implementing proposals. They all work after assessing how much money is available from different sources and directives of higher authorities like State Planning Board as regards to general way of distribution of budget in different sectors as per State Government priorities. The above steps roughly conform to Kerala but in different Federal States and countries approach may differ.

Although most of the literature praises PB, it also has its negative aspects based on authors' observations in Kerala. They are

1. The felt needs of the community at ward level are the basis for recommended projects for annual budget provision. These felt needs are often not based on hard and verifiable data well recorded. A GIS-based database with annual update is required to remedy this situation.
2. Often in Ward Convention for annual budgeting, the most articulate speakers impress what is required in the ward leaving the felt needs of poor, lower caste,

less educated and less articulate persons or those who are unable to attend ward convention totally ignored, although they may be most relevant project for the disadvantageous. The ward convention for PB shall be based on most up-to-date information showing the requirement of all classes, sex, religion and castes, which is not available more often.
3. Mostly what is budgeted for the ward is more than the funds available to execute these projects. Because of the financial constraints over the budget, some of the project for funding selected by Ward Conventions, needs to be excluded at the higher level such as Municipality and District Planning Committee/ Metropolitan Planning Committee levels. Mostly, it has been observed that the excluded projects are from Section where opposition party is elected in Municipal Corporation.
4. In the community-based assessment of projects for annual budget, there is no scope to consider the needs of adjoining wards with respect to higher level projects involving many wards. For example in an urban watershed project, upstream wards have to take more environmental measures but the benefit of water shed is mostly accrued by the downstream ward. There is a need for give and take in this situation with respect to downstream and upstream wards in PB. Based on most up-to-data, a GIS Based Spatial Decision Support System (SDSS) for annual budgeting can be a tool for such consideration.

Kozhikode was practicing Participatory annual planning and budgeting since mid-1990s. It is a Smart City in the making with smart people. After, attaining 100 % e-literacy in Kozhikode district, utilization of ICT in annual planning and budgeting is highly feasible and can give out a better PB than that is practiced today. This chapter explores how the existing PB can be converted to E-budgeting befitting a Smart city. It also demonstrates how SDSS that uses GIS can be deployed in Kozhikode using most up-to-date data for budgetary decision-making by people. This chapter presents the art of PB that is practiced internationally and then nationally in the state of Kerala. A study is conducted for Kozhikode Municipality Corporation which includes annual planning and budgeting system based on PB. In order to prepare for E-Budgeting, PB practiced need to be converted into a web-based GIS. The chapter therefore demonstrates how SDSS can be prepared to help community to arrive at most rational budget allocation.

2 State of the Art of Participatory Budgeting (PB)

PB practice is the foundation for E-Budgeting. One of the pioneers in PB is Brazil and it is worthwhile evaluating their experience first. A comprehensive case study of eight municipalities in Brazil analysing the successes and failures of PB budgeting has suggested that it often results in more equitable public spending, greater government transparency and accountability, increased levels of public participation (especially by marginalized or poorer residents), and democratic and citizenship learning [2]. In a brief survey, we start the first city in Brazil which started PB and then other cities world over and conclude with the detail study of India and Kozhikode before discussing GIS-based E-Budgeting and SDSS in annual municipal budgeting.

2.1 Porto Alegre (Brazil), the Pioneer

The first full PB process was developed in the city of Porto Alegre, Brazil, starting in 1989. One-third of the city's residents lived in isolated slums at the city outskirts, lacking access to public amenities (water, sanitation, health care facilities, and schools) [3]. PB in Porte Alegre was a part of various innovative, reformist programmes undertaken by the local administration to reduce socio-economic inequalities in Brazilian society. PB in Porto Alegre occurs annually, starting with a series of neighbourhood, regional, and citywide assemblies, where residents and elected budget delegates identify spending priorities and vote on which priorities to implement [4]. Porto Alegre spends about 200 million dollars per year on construction and services; this money is subject to PB. Annual spending on fixed expenses, such as debt service and pensions, is not subject to public participation. Around fifty thousand residents of Porto Alegre now take part in the PB process (compared to 1.5 million city inhabitants), with the number of participants growing year on year since 1989. Participants are from diverse economic and political backgrounds [4].

The PB cycle starts in January and assemblies across the city facilitate maximum participation and interaction. Each February there is instruction from city specialists in technical and system aspects of city budgeting? In March there are plenary assemblies in each of the city's 16 districts as well as assemblies dealing with such areas as transportation, health, education, sports and economic development. These are large meetings—with participation that can reach over 1,000—elect delegates to represent specific neighbourhoods. The mayor and staff attend to respond to citizen concerns. In the following months, delegates meet weekly or biweekly in each district to review technical project criteria and district needs. City department staff may participate according to their area of expertise. At a second regional plenary, regional delegates prioritize the district's demands and elect 42 councillors representing all districts and thematic areas to serve on the Municipal Council of the Budget. The main function of the Municipal Council of the Budget is to reconcile the demands of each district with available resources, and to propose and approve an overall municipal budget. The resulting budget is binding, though the city council can suggest, but not require changes. Only the Mayor may veto the budget, or remand it back to the Municipal Council of the Budget (this has never happened) [4].

A World Bank paper suggests that PB has led to direct improvements in facilities in Porto Alegre. For example, sewer and water connections increased from 75 % of households in 1988–1998 % in 1997. The number of schools quadrupled since 1986 [3].

The high number of participants, after more than a decade, suggests that PB encourages increasing citizen involvement. Also, Porto Alegre's health and education budget increased from 13 % (1985) to almost 40 % (1996), and the share of the participatory budget in the total budget increased from 17 % (1992) to 21 % (1999) [3].

The study concludes that PB can lead to improved conditions for the poor. Although it cannot overcome wider problems such as unemployment, it leads to "noticeable improvement in the accessibility and quality of various public welfare amenities" [3].

Based on Porto Alegre more than 140 (about 2.5 %) of the 5,571 municipalities in Brazil have adopted PB.

2.2 PB Implementation Around the World

Since its emergence in Porto Alegre, PB has spread to hundreds of Latin American cities, and dozens of cities in Europe, Asia, Africa and North America. More than 1,500 municipalities are estimated to have initiated PB [5]. In some cities, PB has been applied for school, university and public housing budgets. These international approaches differ significantly, and they are shaped as much by their local contexts as by the Porto Alegre model [6].

The United Kingdom and the Dominican Republic have implemented PB in all local governments [7], and a number of towns and cities in France, Italy, Germany and Spain have also initiated PB budgeting processes [8]. In Canada, PB has been implemented with public housing, neighbourhood groups and a public school in the cities of Toronto [9], Guelph, Hamilton [10] and West Vancouver. In India, a village called Hiware Bazar has served as an epitome of the process. The village being highly deplete with water, education and basic needs for life at one point, is now self-sufficient with a high per capita income. Arvind Kejriwal, the national convenor of *Aam Aadmi Party* in India is trying to introduce the concept of PB (*swaraj*) in places where he may be elected [11, 12]. He was 49-day long chief Minister of Delhi state and then he resigned. Similar budget processes have been used in communities in Africa. In France, the Region Poitou-Charentes has launched an experience of PB in its secondary schools [13]. The first recorded PB process in the United States of America is in Chicago, Illinois [14, 15]. Led by the ward's Alderman, Joe Moore, Chicago's 49th Ward is undertaking this process [14] with the Alderman's "Menu Money." Menu Money is a yearly budgeted amount each of Chicago's 50 wards receives for use on capital expenses. This money in other wards is typically allocated at the complete discretion of a ward's Alderman. Since 2011 more examples have been occurring in the US, in the city of New York and now in Vallejo, California.

Other cities covered are briefly reported here.

2.3 North America

2.3.1 New York City

In 2011, four New York City Council Members—Brad Lander, Melissa Mark-Viverito, Eric Ulrich, and Jumaane D. Williams—launched a PB process to let residents allocate part of their capital discretionary funds. In 2012, Council Members David Greenfield, Dan Halloran, Stephen Levin and Mark Weprin joined PBNYC, giving the community real decision-making power over around $10 million in tax-payer money. How New York City Participatory Budget works is diagrammatically shown in Fig. 1.

2.3.2 Chicago's 49th Ward

Starting in 2009, residents of Chicago's diverse 49th Ward have decided how to spend the $1.3 million capital budget of Alderman Joe Moore. Residents identified spending ideas and selected community representatives in neighbourhood assemblies, these representatives developed full project proposals from these ideas, and then residents voted on which projects to fund.

2.3.3 Vallejo, California

In 2012, the Vallejo City Council established the first city-wide PB process in the United States. Through PB, the community is helping decide how to spend over $3 million of revenue from the Measure B Sales Tax. Vallejo residents and stakeholders propose spending ideas and develop project proposals, residents vote on

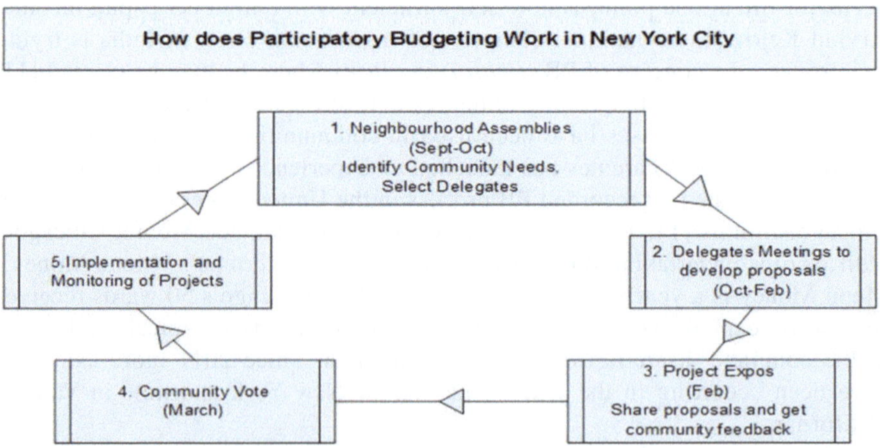

Fig. 1 Participatory budgeting in New York

projects, and the list of projects that receive the most votes is submitted to City Council for approval.

2.3.4 San Francisco

PB started in San Francisco's District 3 in 2013. The process began with community meetings in January and February, followed by voting in March. In this abbreviated pilot programme, residents directly decide how to spend $100,000 of discretionary funding. Programmes and activities are eligible, along with capital projects. While this initial programme is small, the potential is considered big.

2.3.5 Toronto Community Housing

Since 2001, Toronto's social/public housing authority has engaged tenants in allocating $9 million of capital funding per year. Tenants identify local infrastructure priorities in building meetings, and then budget delegates from each building meet to vote for which priorities receive funding.

2.3.6 Montreal's Plateau Borough

The Montreal borough Plateau Mont-Royal implemented a PB process in 2006, 2007, and 2008 for its capital budget. The process evolved each year, starting as one large assembly and later incorporating a series of meetings and the election of neighbourhood delegates. Up to $1.5 million per year was allocated.

2.3.7 Guelph Neighbourhood Support Coalition

A coalition of grassroots neighbourhood groups in Guelph, Ontario, has been allocating a pot of public and foundation funds since 1999. Each year, the groups meet separately and then together, to decide how to spend roughly $250,000 from diverse sources. The funding is generally used for services and programmes, which are delivered by the neighbourhood groups themselves.

2.4 Latin America

2.4.1 Belo Horizonte, Brazil

Belo Horizonte, population 2.5 million, has had a district-level PB since 1993, a Housing PB since 1996, and a digital PB (e-PB) since 2006. Through both local assemblies and online voting, residents allocate over $50 million per year.

2.4.2 State of Rio Grande Do Sul, Brazil

The Southern state of Rio Grande do Sul, home to Porto Alegre and over 10 million people implemented PB between 1999 and 2002. During each annual cycle, people met in assemblies in each of the state's 22 regions, to identify priorities for public works and services. Delegates from each region then worked to harmonize the proposals into a single budget. 1.2 million people participated over the 4 years, deciding on over 12 % of the state's budget.

2.4.3 County of Paysandú

In 2005, the Department (County) of Paysandú began implementing an annual PB process. The county is a mix of rural and urban areas, with 113,000 people, one city, and eight local town boards. The process includes a mix of both local territorial assemblies and thematic assemblies focused on issues of particular local concern.

2.4.4 Rosario, Argentina

Rosario's participatory budget consists of an annual cycle in which over 4,000 city residents decide how to allocate $8 million of the city budget. In this city of 1 million people, residents discuss spending ideas at neighbourhood assemblies, elected delegates develop full budget proposals, and then residents vote on the proposals at another round of voting assemblies. The funds can be spent on both capital projects and services or programmes.

2.4.5 La Plata, Argentina

In 2009, La Plata launched a PB initiative in which citizens gather in neighbourhood assemblies to debate their needs, and to develop projects that propose public works, services, and programmes. This is followed by a larger process of voting, where a secured system allows votes to be cast through either paper or electronic ballots. In 2012, it became the first city in Argentina to enable participants of the public assemblies to decide on rules and regulations for the PB process.

2.5 Europe

2.5.1 Tower Hamlets, London, UK

The Tower Hamlets 'You Decide!' project began in January 2009. In the first 4 months, 815 residents spent almost £2.4 million through eight events. The money was from the central council budget and was spent on services.

2.5.2 Newcastle, UK

In 2008, Newcastle launched a PB process in which 450 young people helped decide how to allocate the city's £2.25 m Children's Fund. After months of preparation, youth age 5–13 attended a PB event at which they voted electronically for services targeted at young people. Their votes were incorporated into the Fund's complex procurement process, weighted to count for 20 % of the final spending decisions.

2.5.3 Seville, Spain

Seville, with 700,000 residents, is the largest European city to implement PB. Since 2004, residents have decided on roughly 50 % of local spending for their city districts, for capital projects and programmes. They can submit project proposals online or in neighbourhood assemblies, and after a series of meetings, locally elected budget delegates deliver the participatory budget to city hall for implementation.

2.6 India

Since the concept of PB first originated, its appeal in India has been limited. Only a few cities in India namely Bangalore, Mysore and Pune have experimented with PB after Kerala. In 2001, Bangalore implemented PB due to efforts by a local NGO, *Janaagraha*. The campaign resulted in citizens' budget priorities being approved in over 20 % of the city's wards but as time passed, the concept lost its ground in the city. In 2006, Pune implemented PB for the first time in the city and attracted a massive response from the citizens as well as the city-based NGOs—*Janwani, NSCC, CEE, and Nagrik Chetna Manch* amongst others.

The 74th Constitutional Amendment directs state governments and urban local bodies to form ward committees, comprising of citizens of the locality and preparing the ward-level budgets in consultation with them. But these initiatives have not been implemented in some states, with the result that there were few channels for citizens to participate in local governance.

There has been a lot of enthusiasm amongst the civil society organizations but the most basic requirement for successful implementation of PB is the political will that is lacking in India barring Kerala with its sustained effort. Pune has taken some positive steps towards this direction and it has gained its due attention within the administrative and executive circles but nationally. PB remains an alien concept to most of the 7,933 census towns in India. Most of them have no Municipality there.

2.6.1 Pune's Experiences with Participatory Budgeting

PB was initiated in Pune in the year 2006 under the leadership of Dr. Nitin Kareer, the then commissioner of Pune Municipal Corporation. Currently under PB in Pune, each *Prabhag* (comprising of two electoral wards) is allocated a budget of Indian Rupees 50 lakhs. Each *prabhag* can implement any number of projects with Rs. 50 lakhs but the total amount cannot exceed Rs. 50 lakhs and cost of each of the project cannot exceed Rs. 5 lakhs. A total of 38 crores were allocated for 76 *prabhags* towards PB. PB process of Pune is diagrammatically shown in Fig. 2.

It is to be noted that New York City's annual budget is approximately $50 billion and the allocation towards PB is $10 million; whereas Pune Municipal Corporation's annual budget is Rs. 4,167.5 crore and the allocation towards PB is 38 crores, a much higher proportion than New York City. Even though New York City allocates a much lesser share of its budget towards PB, the proportion of participation by the citizens is much higher than Pune and knowledge management on the process and results is commendable.

While the initial years saw a lot of enthusiasm among citizens, the participation graph has been declining since the past few years. PB for the year 2012–2013 in Pune witnessed only 600 suggestions from the citizens, for a city with a population of over 3.5 million people. With the purpose of engaging citizens in the PB process for 2013–2014, *Janwani* (*Janwani* meaning Voice of the people), the social wing of Mahratta Chamber of Commerce, Industries and Agriculture, took the initiative to promote and facilitate the process with the support and guidance of Pune Municipal Corporation.

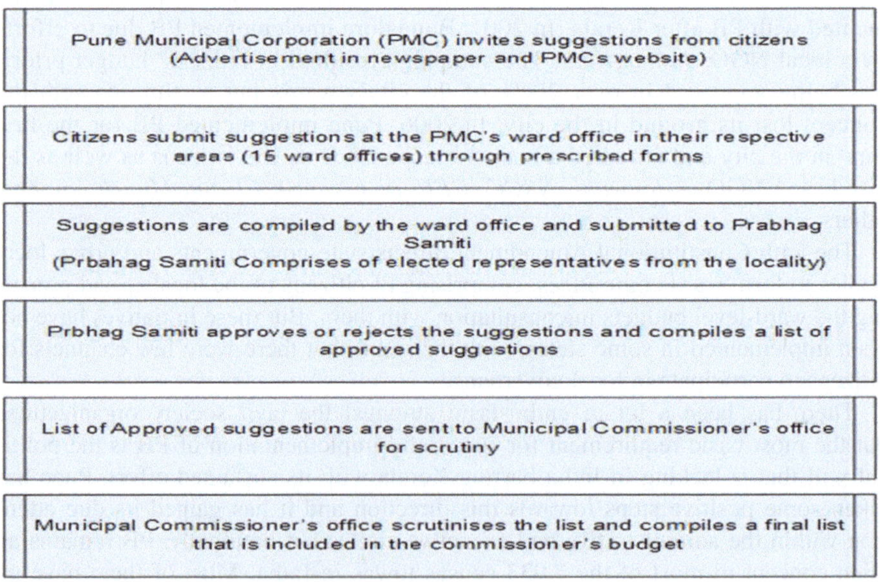

Fig. 2 Participatory budgeting in Pune

3 Participatory E-Planning

Planning is required for budgeting. Participatory E-Planning is relatively new way to engage citizens in urban planning using ICT. E-Participation is defined as "technology-mediated interaction between the civil society sphere and the formal politics sphere [16]". According to Horelli and Wallin (2010), "Participatory E-Planning, similarly to E-Participation, can be an important instrument of E-Democracy and E-Governance [16]". Participatory E-Planning is also related to engaging the general public to use tools traditionally used by urban planning experts, such as GISs and Planning Support Systems. Participatory E-Planning research has focused on incorporating forms of participation with existing governance and urban planning processes. The E-Planning research is also limited to needs of current participatory planning. The original investigations used tools like online questionnaires, surveys, and polls to consultant citizens. The feedback from participating citizens is then used or *not used* by experts and professionals. Before e-planning, citizens could provide only their opinion via direct confrontation, snail mail, phone calls, or e-mails. The E-Planning participation tools allow for more organized and substantive participation from the interested public. The tools are:

1. "Plans-on-the-map", which is a website that allows citizens to get acquainted with existing plans.
2. "Tell-it-on-the-map", which is a questionnaire-based online tool to collect citizens' comments on specific issues presented by planners.
3. The planning competitions website, where citizens can get acquainted with ongoing planning competitions and comment on them.

These tools allow feedback, but still do not allow the public to spotlight issues they find important.

Critiques challenge the E-Planning tool to empower citizens to collaborate on the same level with experts. More and more citizens own devices that allow them to produce media. Participatory E-Planning can not only use customary collaborative urban planning tools but must delve into sharing media content.

4 Participatory Budgeting for Kozhikode Municipal Corporation

Participatory Governance at grass root level has been the focus of discussion for many decades in India. Many concepts are generated in several States of India, but attempt to implement these concepts by State Governments have not been successful because of severe obstacles faced. They are non-availability of planning expertise at local level, lack of experience, weak and highly centralized administrative setup and inadequacy in administrative procedures and insufficient spatial database at grass root level. Because of these limitations, many State Governments in India felt it difficult to pass on the financial and management functions of urban governance to the grass root level. However, Government of Kerala took a bold initiative to embark on people's planning movement. The strategy adopted, is based on well thought out procedures. Action learning from people's planning experiences was utilized to further strengthen procedures. The lessons learned are used in the subsequent annual plans. People's Planning Movement of Kerala, demonstrated the art of decentralization of power to plan, manage, make financial decisions and implement area development by local people within a multi-level planning frame work. This experience is fascinating and unique, both in the developed and developing nations. It is a noteworthy example of a model of translating the 74th and 73rd Constitutional amendment of India without any reservation of holding on to power at the State Capitals. Trust in local populace is the foundation of such a venture of Government of Kerala. This is in contrast to the existing practice, which does not allow local people to govern themselves with public finance, planning and management, and if forced to allow then they over-administers them with checks, counter-check as if they are "outsider native Indians in a British Colony". This has been by and large non-productive for urban and regional development in India. At present, in many States decentralized planning is essentially planning at district level as a technical, academic or professional exercise without massive mobilization and active participatory decision-making at Municipal Ward and Gram Panchayat levels. Kerala's attempt was to empower people at ward, municipality, village, Panchayat and district levels to undertake area development planning and implementation. This also opens up new emerging role for Professional Planners [17].

This chapter is partly based on a research project "Annual Planning and Budgeting System Using GIS" undertaken by T.M. Vinod Kumar with a grant received from Ministry of Human Resource Development, Government of India. Further work was executed in National Institute of Technology, Kozhikode on SDSS using the most recent GIS database for Annual Budgeting of Kozhikode Municipal Corporation. It is designed to further strengthen People's Planning Movement in Kerala and thereby improve Urban Governance Vinod Kumar et al. (2001) Annual Planning and Budgeting System Using GIS for Calicut Corporation Research and Development Project, Ministry of Human Resource Development, Government of India, School of Planning and Architecture, New Delhi (unpublished), [18].

5 Peoples Planning Movement in Kerala, the First Three Years

The people's planning movement of Kerala started in the financial year 1997–1998. This research project was completed in May 2001 after studying in depth annual planning and budgeting of Kozhikode Municipal Corporation for first three years. Kozhikode City in the past had two Master Plans and the third official plan is under preparation [19, 20]. Constitutional amendment has spelt out and confirmed the role and responsibilities of local government such as Panchayat and Municipalities. These institutions are being converted into strong constitutional institutions of Governance in Kerala. Mandatory periodic elections and assured share of State Government grants through Finance Commission [21, 22] recommendations are two pillars of its strength. Local population shares the power to decide developmental activities and implementation. There is considerable scope for improvement in people's planning movement of Kerala, and this research and development project aims on the upgrading by expanding its scope, methodology and utilization of state of the art technology of GIS [23–29].

The most important contribution of the people's movement in planning was the way in which Kerala mobilized the participation of people. Furthermore, State Planning Board in Kerala steered complex activities of people's planning with direction, training programmes, documentation and evaluation of past experiences. Beneficiary Committees, Task Forces and Working groups worked side by side with the elected members of local bodies. They showed that they have an important role to perform in people's planning movement. The initial mistrust between elected and non-elected local people melted away as the planning movement started showing positive results. Many obstacles faced by these Committees and Task Forces had been removed for their better functioning. In the past such committees and officials were subjected to the rigor of administrative controls. These controls have not resulted in better governance. Experience shows that more the controls, more the corruption and administrative inefficiencies. The result is exhibited in the web-sites of Central Vigilance Commission of India that shows a long list of corrupt senior administrators, a by-product of the existing system. It looks like that corruption is a part of Indian administration and culture of paying bribes is accepted even for poverty alleviation programmes. Government of Kerala took a bold stand that these Task Forces will not misuse power given to them since they are closely watched by local people in their transparent functioning. It is difficult to be corrupt when several local persons cutting across political parties are involved in decision-making for a local level project, which benefits many. Furthermore, there is scope for reviewing decisions at different levels. In a situation where all are equal as far as decision-making power is concerned, corruption gets minimized. On the other hand administrative secrecy and model of old British Colonial administration practiced in India breeds corruption among people, politician and administrators. Past style of administration elevated India as one of the most corrupt Country in the world. In this context the very bold attempt of Government of Kerala to undertake People's planning is praiseworthy.

During 1998–1999, Rs. 950 Crores were set aside by the Government of Kerala to implement plans prepared by the Local Governments. This amount was 25 % higher than the sum allocated during the financial year 1997–1998. The budget for 1999–2000 has earmarked a sum of Rs. 1,020 Crores as plan grant (Rs. 40 Crores for Tribal Sub Plan, Rs. 200 Crores for Special Component Plan, and Rs. 780 Crores for the General Sector Plan) for Panchayat Raj/Nagar Palika institutions. Further, local planning bodies may mobilize local resources other than the State Government grant. Earlier, unspent State Government Grant in the financial year was known only to Government officers, but now populace of the ward or municipality knows about it and its reason, as they are part of participatory planning. The first instalment of the second year plan will be available to the local Government only if 90 % of the fund allotted is spent. If funds are not available the populace gets critical of all in power.

6 Project on Annual Planning and Budgeting Using GIS

In this project no attempt was made to prepare an alternate annual planning process or better plan and budget for Kozhikode. This would be totally negating the participatory people's planning movement. On the other hand this project was mostly a system development exercise for user friendly spatial methodologies and technologies that help make people decide on the basis of factual information in participatory planning and budgeting. The main issue in PB is how to allocate funds spatially in geographic areas as per assessment of that area within the constraints of funds allocated for such schemes as per state planning board guidelines.

The planning process adopted by Government of Kerala is given in Fig. 3.

Figure 4 shows how spatial information flow can be managed in the ongoing planning process. Spatial information of parcel data base and indicator data base at ward level related to annual plan budgets can be updated annually and utilized for creating a GIS-based decision support system which gives spatial orientation missing in the annual planning and budgeting practice. Annual area plan and structure plan can also be prepared with this database. This involves mapping and spatial data base generation, annual structure plan, updating of existing land management database, and updating urban indicators and area plan. This keeps the present planning process intact. It strengthen and expand the scope of annual planning to include land management, use of urban indicators and many useful tools in the ambit of annual planning which is not practiced in Kerala.

7 Deriving Annual Budget from GIS

The annual Plan document prepared by Peoples' Planning Movement for Kozhikode Municipal Corporation strictly follows directives of the State Planning Board. The three annual plan documents studied in this project are brief and often telegraphic in nature to convey the plan with economy of words and necessary tabulations. Plan document gives considerable importance to financial aspects and project formulation in a simple and practical way. The annual plan looks more like annual budgeting for Kozhikode Municipal Corporation. The spatial aspect of urban development planning is not visible at all in these documents. There should

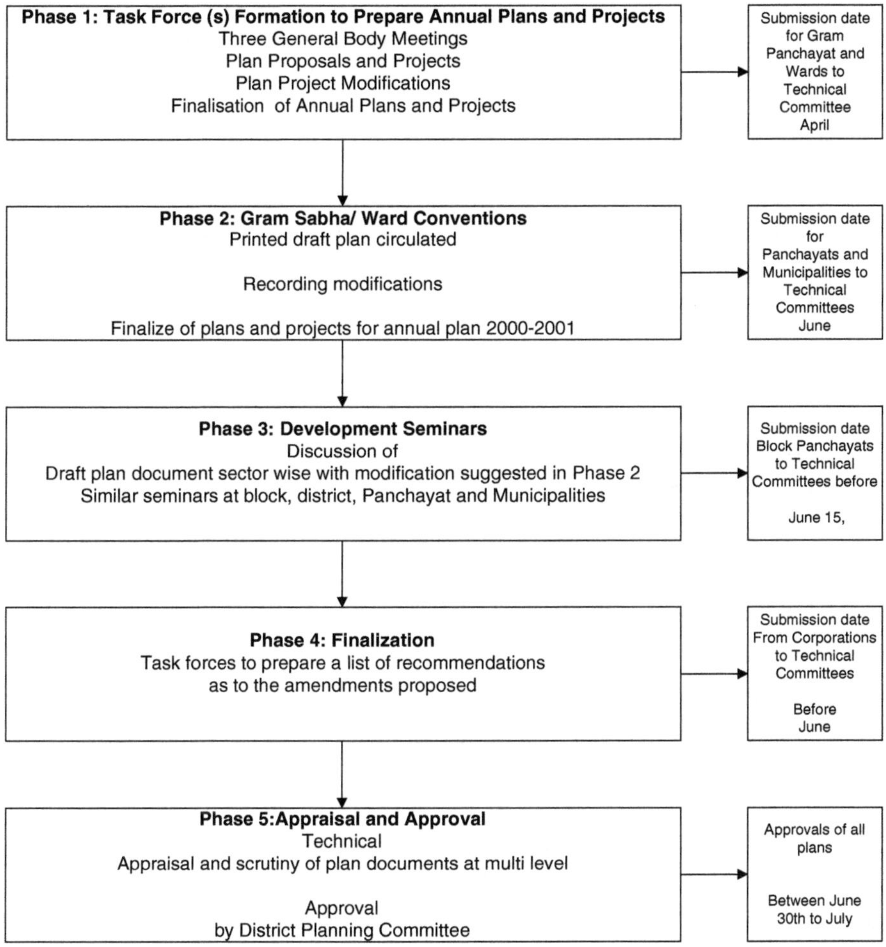

Fig. 3 Plan formulation procedure for the annual plan

be a strong relationship between annual spatial plan and budget. This relationship can be utilized for urban development management effectively.

This chapter reviews the annual planning process in Kozhikode for the first three years ever since People's planning movement came into existence and suggest how adoption of GIS would give much needed spatial planning orientation of the annual planning. People's Planning Movement in Kozhikode started for the first time on 17 August 1996. Ward conventions were over by October 1996. In January 1997 Development Seminar took place. Reviewing the annual plans for Kozhikode, it can be seen that about 90–95 % of the plan document is devoted to budget. At present annual plan can therefore be considered as budget. It also gives an overview of development issues in different sectors of urban economy. It monitors the past pattern of expenditure by sectors. It carefully records sectors,

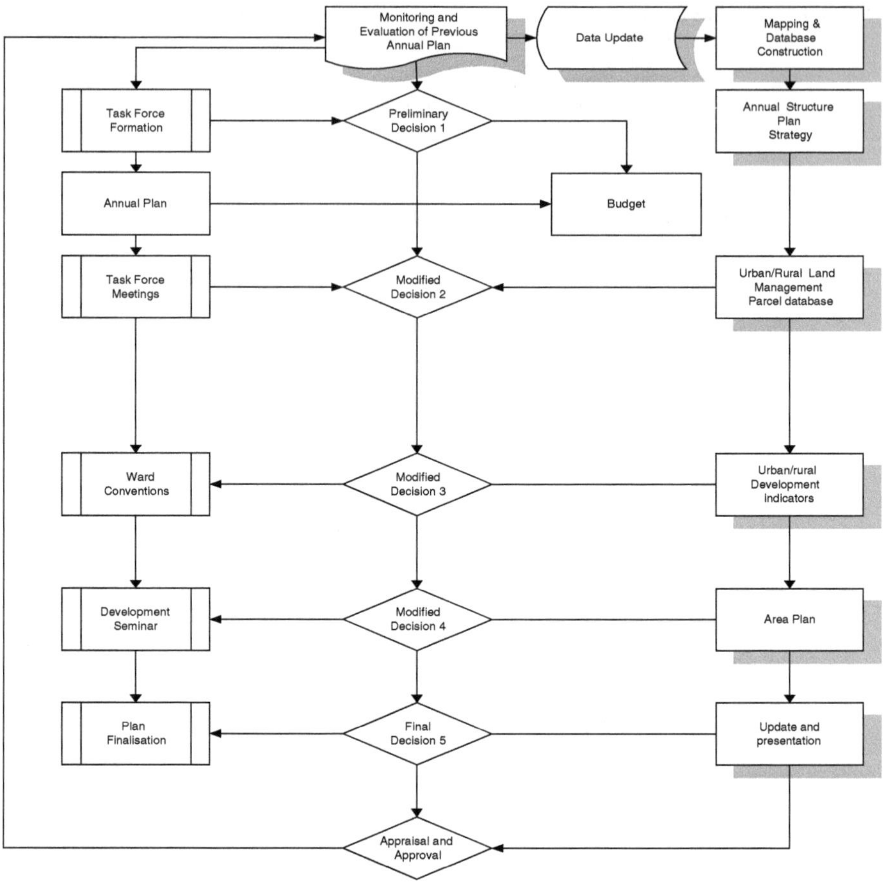

Fig. 4 Integration of annual planning and budgeting with GIS activities

sub sectors and wards where money was overspent and also underutilized. The balance amount from the last plan is then allocated to different sectors. Here, those sectors that overspent get negative money in this year allocation from the balance amount of the last year. Annual planning is more than annual budgeting and project formulation involving people. Annual planning and budgeting system is considered integrated. GIS provides an easy approach for achieving plan integration.

To start with, an attempt is made to describe the budgeting system practiced in Kerala under the Peoples Planning Movement of local bodies. The process of using GIS in this budgeting exercise follows thereafter. This proposal is only possible in an environment of open planning and budgeting system that is highly participatory involving thousands of people. These preconditions exist in Kerala now. If what is proposed is implemented it would be possible to have a paperless GIS budgeting system and will form the foundation of E-Governance.

8 Municipal Budget

The Municipal Budget is a comprehensive and coordinated plan expressed in financial terms for the operations for 1 year. It is a financial document showing the amount of estimated revenues and expenditures in detail. It expresses the plan of the Municipal Body in financial terms. The Budget is also, a managerial tool for Municipalities. A Municipal budget serves different purposes. Five Objectives of Municipal Budget are:

1. To forecast the future and avoid losses due to lack of planning.
2. To bring co-ordination among different functions and departments. While drafting the budget, responsibilities and requirements of each department is determined. This will contribute to a smooth operation of the municipal activities.
3. To clearly state expected performance of the municipal body in financial terms. The budget communicates to all concerned sections/persons, targets, and methods of operation and levels of performance expected from various units and employees in clear and formal terms. It helps all to understand, support and implement the intended activities.
4. To control the financial activities. The budget provides a means of measuring and controlling the activities of each department/section. It helps check whether actions are in tune with the target. The Budget sets a limit of expenditure under each head and shows the amount of revenues to be collected by the municipal body to fulfil its tasks.

The budget stimulates thinking in advance, which helps forward planning that provides an opportunity to avoid problems and to take advantage of any particular situation. The budget also makes it possible for top-level management to delegate authority and responsibility without sacrificing overall control, because the limits of each department/section are clearly laid down in the budget. The Budget thus provides a balanced and coordinated approach to Municipal activities so that requirements and responsibilities of all sections are analysed while preparing a budget and collaborative plan of operation is evolved. The Budget also communicates the financial objectives and policies of the Municipality to Ward Committees, Municipal Council, District Planning Committee, the State Planning Board and State Government.

9 The Main Actors

The main actors in budget making are the Task Force members, planning and budgetary committee of the Municipality, and line agencies. The budget is prepared and discussed at ward conventions, task force meeting and development seminars and finally approved by District Planning Committee.

10 Step-Wise Preparation of the Annual Budget

Step 1: Collation of budget of annual plans from task force members, ward committees and line agencies by a participatory process.

Step 2: Assessment of potential income of the municipality based on State Planning Boards directives for assessment. A detailed estimate of expected revenue next year needs to be computed accurately. The State Planning Board had developed a method of deriving budget estimates from a series of Municipal indicators. There are also different ways of projecting the revenue. The first is based on growth rate and inflation and the second based on the average of last five years. If the population growth rate is 6 % and annual inflation rate is 10 %, then for a revenue head of Rs. 100,000, the projected revenue is 100,000 + 100,000*6 %*1 0 % = 160,000. In the second method if the average increase of revenue is 15 % for the last five years, then the expected revenue of Rs. 100,000 under that head is 100,000(1 + 15 %) = 115,000.

Step 3: Estimation of the administrative expenditure can be easily available at the Municipality and so also debt servicing cost.

Step 4: Deducting administrative expenditure and debt payment from the total revenue the municipality can arrive at the Development Budget. The Municipal development projects should be divided into long, medium and short duration. Projects should be selected based on the recommendation of district planning committee. The expenditure needs to be sub divided by sectors and spatially by Municipal wards. However it may be noted that the habit of dividing the budget equally into the number of wards does not correspond with a sound method of budgeting.

Step 5: Finalization, Approval and Presentation of Budget. This step involves presentation of the budget at the ward conventions, development seminar at town level and municipal council as per the legislation to get approval, before placing it for final approval to District Planning Committee. The approved budget is then published.

11 Budgeting Under People's Planning Movement

The starting point of making annual budget is the study of local conditions. Involvement of Task Forces who represents the subject matter specialist and ward committees generate realistic assessment of local situation spatially. If local area is mapped or create GIS database it helps further. The next stage is to prepare a strategy statement of designated plans under designated sectors and sub sectors as identified by the State Planning Board. These strategy statements adopted for three sample annual plans are tabulated below. It can be seen that strategy adopted in many sectors remain the same for the three annual plans. However, there are changes in strategy for selected sectors.

Table 1 Annual plan strategy adopted by sectors for Kozhikode Municipal Corporation 1997–1998, 1998–1999 and 1999–2000

Sectors	Plan 1997–1998	Plan 1998–1999	Plan 1999–2000
Agriculture	Flood control Spared of vegetable flower cultivation	Flood control Improve productivity of agriculture through group approach	Homestead farming Flood control
Animal Husbandry	Increase availability of milk, egg and meat	Increase availability of milk, egg and meat	Increase availability of milk, egg and meat. Income generation for women below poverty level
Energy	Efficiency in the utilization of electricity Spread smokeless stoves	Extend electric supply where it is not available	Extend electric supply where it is not available
Health	Mini-water supply schemes Extend healthcare activities garbage disposal	Mini-water supply schemes garbage disposal and cleanliness	Mini-water supply schemes garbage disposal and cleanliness
Education	Improve physical facilities Quality improvement	Improve physical facilities Quality improvement	Improve physical facilities Quality improvement
Transportation	Improve transportation services Solve traffic problems	Land for roads Improve transportation services	Land for roads Improve transportation services
Social welfare	Mother and child care	Mother and child care	Mother and child care
Slum improvement	Improve basic services	Improve basic services	Improve basic services
SC/ST programmes	Improve employment opportunities Improve basic services	Improve employment opportunities Improve basic services	Improve employment opportunities Improve basic services
Industries	Mini Industrial estates	Mini industrial estates Software technical park	Mini industrial estates Software technical park

Source Annual plan for Kozhikode Corporation 1997–1998, 1998–1999 and 1999–2000

The next stage of annual plan is formulating projects in each ward or a group of wards based on development strategy by sectors. These include designating the name of the project and specific objective of the project and sources of finances available for implementation of the project. The strategy adopted in three sample years is given in Table 1.

11.1 Budget of Kozhikode Corporation at a Glance

To get an overall idea of Budget of Kozhikode, Budget at a Glance for 1997–1999 is given in Table 2.

Table 2 Annual budget of Kozhikode Corporation 1997–1998 and 1998–1999

Budget head for the year	1997–1998 in Rs	1998–1999 in Rs
Opening balance	28,097,900	14,539,780
Revenue receipt	260,683,800	359,001,000
Capital receipt	246,602,800	639,703,400
Total receipt	535,384,500	1,013,244,180
Revenue expenditure	243,557,100	307,351,300
Capital expenditure	277,287,620	693,126,100
Total expenditure	520,844,720	1,000,477,400
Balance	14,539,780	12,766,780

Source Calicut Corporation budget 1997–1999

It can be seen that there has been almost doubling of total income and also expenditure because more state government funds were transferred to the Municipal Corporation as a part of financial decentralization. All of total receipt does not go in development expenditure. Part of this goes to administrative expenditure and another part goes into debt servicing cost. Various sources of financing annual plan of Kozhikode Corporation are given (Table 3).

Funds for annual plan have been increasing since people's planning movement had started. It can be seen from that there has been an effort to mobilize more own funds ever since State Government stepped up their contribution that has risen to 30 %. During the three financial years studied, there has been an increase in the mobilization of funds from the financial institutions. Account of expenditure and income of Kozhikode Corporation for 1996–1997 is given in Table 4.

It can be seen that 46 % of the income goes to administrative cost and the remaining for development and maintenance of Municipal functions. By outsourcing activities it is possible to reduce the administrative cost. Further a certain amount of positive balance should be set aside as per the Municipal Act of Kerala.

Table 3 Source of financing annual plan for Calicut Corporation 1997–2000

Source of finance for annual plan	1997–1998	1998–1999	1999–2000
State government contribution (untied funds)	90,057,000	113,020,000	122,121,000
Grant for State Government Sector Plan	22,514,250	4,344,000	7,878,100
Grant for Central Government Sector Plan	18,011,400	24,805,022	27,364,371
Own fund of Calicut corporation	44,842,000	37,042,750	55,839,600
Funds from financial institution	7,632,200	19,319,376	41,200,000
Beneficiary share of fund	32,104,655	19,749,620	23,848,993
Previous year balance after expenditure	25,517,776	8,415,248	48,671,566
Voluntary service/donation	1,389,700	6,131,611	4,427,063
Other sources		13,962,000	982,510
Grand total	242,068,981	246,789,627	332,333,203

Source Calicut Corporation Annual Plans 1997–2000

Table 4 Items of expenditure of Kozhikode Corporation 1996–1997

Items of expenditure 1996–1997	Rs.	Percentage of expenditure
Administration expenditure	101,444,356	45.98
Roads and building	33,070,563	14.99
Green clean city	1,016,627	0.46
Grave yard	303,671	0.14
Construction office complex	15,999,363	7.25
Purchase of vehicles	2,953,221	1.34
Special component plans	320,929	0.15
Purchase of land	2,443,900	1.11
Mini water supply scheme	439,102	0.20
Drainage construction	13,413,741	6.08
Street lighting	666,178	0.30
Advance deposits	48,538,248	22.00
Total	220,609,899	100.00
Total income	246,127,675	
Total expenditure	220,609,899	
Balance set aside	25,517,776	
Previous balance	3,099,583	
Total	28,617,359	

Source Kozhikode Corporation, peoples planning document 1996–97

The expenditure pattern from the untied grant from State Government is spent on a pattern suggested by the State Planning Board. This percentage split changes from year to year as per the change in the development policy of the State of Kerala. The total expenditure of the untied grant from State Government is divided into several sectors and sub sectors. This is as given below.

The allocation for Plan preparation is a lump sum of Rs. 75,000. Special sub plans and SC/ST sub plans are based on the allocation by the State Government.

It is likely that some time the budgeted amount cannot be spent as well as there are instances of excess expenditure over and above the budgeted amount. The unspent amount under each sub sector is then added to the next year's budget and subtracted from the next year's budget for cases of over expenditure. Applying these principles, the resulting amount and percentages for the above annual plans are as given in Table 5.

At present Task Force Members are mainly responsible for annual planning and budgeting system of Kozhikode Corporation. The task force members look into the demands of individual sectors of urban development. There are opportunities built into the planning process to incorporate the views of the beneficiary population, ward committees and municipal council.

As stated earlier, the budgetary allocation under different head is prepared as per the direction of the State Planning Board and the percentage allocation changes from time to time. Table 5 illustrates the allocation in three financial

Table 5 Resulting amount and percentages for the above annual plans

Items of expenditure for annual plan of Kozhikode	In Rs. for 1998–1999	In % for 1998–1999	In Rs. for 1999–2000	In % for 1999–2000
Productive sectors	66,142,026	26.8	74,460,551	22.4
Service sector	60,852,987	24.7	121,494,008	36.6
Infrastructure sector	67,808,676	27.4	78,572,021	23.7
Slum improvement	31,483,241	12.7	12,414,577	3.9
Special programmes			22,264,371	6.7
SC programmes	20,256,587	8.2	22,220,685	6.7
ST programmes	446,110	0.2	406,990	0.0
Total	246,789,627	100	332,333,203	100

Source Computed by author

years. A broad development strategy for each sector is postulated as summarized in Table 1. Based on this broad strategy for each sector and sub-sector project is designed and located in different wards or groups of adjoining wards. Special sector schemes are provided separately and therefore are not included. It amounts to less than 10 % of the total budget. This budget is also included in the municipal budget and monitored.

About 20–30 % budget allocation is given for primary sector such as agriculture, animal husbandry, irrigation and industries. Fisheries could be an important sector in urban areas since Kozhikode is a coastal town. While allocation of budget for industry is understandable in urban areas, it is difficult to rationalize budget for agriculture, irrigation and animal husbandry. In Kozhikode urban area, productive sectors are generally from commercial, service and industrial activities and not from agriculture and animal husbandry. These secondary and tertiary sectors could have higher multiplier effect than agriculture and animal husbandry in the primary sector. No attempt has been made to use these sectors more effectively for urban economic development. Urban land is an important resource of a city and therefore land development is likely to give more returns than agriculture. Budget allocation for various sectors in three consecutive financial years are given in (Table 6).

About 30–35 % of allocation is given to infrastructure and 5 % to slum improvement. These allocations are generally found insufficient and generally more money is spent on these sectors. For example, in the budget for 1998–1999 more money was spent on water supply, transportation and energy.

On the other hand, in many other sectors, the money was not spent. In the budget of 1997–1998, Rs. 19,27,000 was allocated to agriculture and Kozhikode Corporation could spend only Rs. 8,94,005 and as much as Rs. 10,62,995 remained unspent in this sector.

Generally, over 90 % of the budget is spent for development purposes. The balance is utilized in the next year budget. The utilization of balance amount follows

Table 6 Percentage of budget allocation in three annual plans

Sectors	Budget provision (%)		
	97–98	98–99	99–00
1. Productive sector	20	30	20
1.1 Agriculture and irrigation	4.2	24.5	24.8
1.2 Animal husbandry and fisheries	32.17	21.1	9
1.3 Industries	25.7	55.4	66.2
2. Service sector	40	35	40
2.1 Sanitation	29.02	30.6 (include health)	17.9 (include health)
2.2 Water supply	22.56	23.1	6.7
2.3 Health	9.67		
2.4 Housing	6.45	13	50
2.5 Social welfare	12.89	11.9	8.7
2.6 Education	12.89	16.4	6.7
2.7 Sports and cultural activities	6.52	5	10
3 Basic infrastructure	30	30	35
3.1 Energy	22.24	24.4	10.2
3.2 Transport/public work	5.7	59.4	89.8
3.3 Other infrastructure	10.76	16.2	
4. Slum improvement	10	5	5
5. Special sector scheme			
ST/SC schemes			
Total	100	100	100

a certain pattern. The total balance in budget is distributed in such a way that some money is deducted from those sectors where more money has been spent than was allocated and more fund is provided for other sectors where entire amount budgeted is not spent. This effectively neglect some sectors (Table 7).

These adjustments generated a different percentage distribution of expenditure than directed by the State Planning Board. For example, the final budget arrived at for 1999–2000 for Kozhikode had the following percentage distribution of expenditure.

Table 7 Sectoral allocation of budget for Kozhikode Corporation 1999–2000 (as computed by author)

Sectors	Percentage of final budget
Productive sector	22.4
Service sector	36.6
Infrastructure sector	23.7
Slum improvement	3.9
Special sectors	13.4
SC ST development	6.7
Total	100

11.2 Financial Resource Mobilization

Budget gives considerable importance to resource mobilization. The major sources of funds for annual plans are the following:

- State plan allocation
- Balance carried over from last annual plan
- Funds given by State Government for the town
- Funds allocated by Central Government at the town level
- Own funds
- From financial institutions
- From beneficiaries of the schemes
- Voluntary contribution
- Others.

From the experience of the sample 3 years, it is observed that major share of funds come from the State Government for plan allocation. There is considerable scope for augmenting own funds and from financial institutions. Municipal bonds have not been implemented so far.

11.3 Schemes and Projects

The next part of the budgeting approach is to design individual schemes and projects. These are designed through participatory annual planning process discussed. Projects are formulated and identified sector wise (sector number and name) and aims and objective of the project are stated. Budget and share of budget follow this from the plan allocation. Then various sources of funds for the project are also tabulated. Hence the plan document ends up as a big list of projects by ward and by sectors.

Urban Planning being spatial, a GIS-based physical planning support system for annual planning and budgeting is a prime requirement in Kerala. The role of GIS is to generate this supporting system for annual budgeting. For developing a GIS support for annual budgeting, a variety of urban indicators need to be monitored, spatially so that budget can be allocated spatially based on these indicators. Urban indicators and their use in Annual Planning is the subject matter of this R&D project [17, 18, 30, 31].

11.4 GIS in Budgeting System

Introduction of GIS strengthens the process of annual budgeting of Municipal Corporation. Budgeting is a process with backward and forward linkages. GIS integrates various operations of annual budgeting process. This integration is achieved in a computer server, which can be accessed through web. Data input is also possible through web by uploading the data by ward committees and Task forces. The process includes the following:

Surveys: Surveys update the GIS geographic base of the city. The base map is updated with all the projects with location, attributes and progress of work in financial and physical terms every year.
Planning: Planning is conducted using GIS-based approach. It provides the rationale for investment decision which can be visualized in maps.
Budgeting: Budget gives details of allocation of funds by sector and geographic space and sources of funding which is based on State Policies.
Management, monitoring and evaluation: These processes monitor the achievements of annual plans and manage development. The organizational structure of plan management has already been discussed in Fig. 1. This process provides for the impact assessment of annual plan investment and finding reasons for failure, if any. The above cycle is to be repeated every year.

11.5 Mobilization for Management, Monitoring and Evaluation

In order to give an idea of size and composition of the monitoring group, the following organizational structure of Kozhikode Municipal Corporation is presented. At corporation level there were three monitoring groups, one dealing with the productive sector, and another with infrastructure and the third with social welfare groups having 23 members. There are 39 wards and each ward committees have 14 members making a total of 546 members. There is a six-member team for projects in each ward making a total of 234 members. This means about 900 people in Kozhikode will be involved with monitoring and management of annual plans budgeting of Kozhikode Municipal Corporation. Web-based GIS are the most cost effective solution to monitor physical and financial achievement [32]. In this process all the members can have a comparative picture from all other wards and learn from other ward committees.

11.6 Why GIS for Annual Budgeting

As can be seen, Budgeting is an information intensive activity. It is based on annual plan recommendations emanating from several committees and wards. These recommendations are derived based on hard facts at the ground level and participatory deliberation. All these recommendations have a spatial orientation and can be better managed by GIS. It is necessary to retrieve several parts of the information at many stages of plan preparations and decision-making. It can be better achieved by a GIS system than the one is in practice. Since nearly a thousand people are involved in plan preparations, they are all interested in facts, which is the basis of the budget making. A web enabled GIS server can effectively provide for this function than any other alternative. Data can be stored in one server and therefore the participants of the annual planning can access the most up-to-date information any time and from anywhere. Paper work is minimized if we adopt web-based GIS for annual planning and budgeting. Open annual planning and budgeting is only possible with web enabled approach in plan preparation.

A broad framework of budgeting as practiced in Kozhikode has already been discussed. This gave an idea of various users of annual plan budget system in Kozhikode Corporation and types of information generated by these functionaries.

11.7 Preparation of Budget Using GIS

Kozhikode Municipal Corporation consists of an area of 84.282 km^2 divided into 39 revenue wards, 50 electoral wards and 8 villages during the study period which has been expanded by more areas now. Budget is prepared based on a format prescribed by the state planning board and broadly apportions the budget provision under different sectors based on an urban development strategy. In addition, district-planning committee is involved in finalizing the town plan who keeps State Policies in mind with respect to development issues of the district. These processes give the multi-level framework of annual plan budgeting of Kozhikode Corporation. It can be seen that Ward Committees, Municipal Corporation, District planning committee and State Planning Board are intimately connected with the budgeting.

The annual plan of Kozhikode Corporation enumerates about 600 projects subdivided under ten sectoral heads and many sub heads. In addition there were about 200 projects under special sectors such as women development, SC/ST development, etc. It can be safely concluded that around 1,000 projects will be prepared and executed every year in Kozhikode. For the creation of a budget using GIS, the first task is to locate the project within the Kozhikode area map. The second aspect is the generation of attribute database.

11.8 Spatial Units for Budget Analysis

Town Planning Division and Wards are spatial units for analysis. Town level analysis is particularly important for making decision about investment in infrastructure from the state and district level, while Planning Division and ward level analysis is useful to locate planned expenditure spatially below town level. Since facilities and services are provided in a hierarchical manner, ward-level indices may only be pertinent for indicators that can be assessed on the basis of population. For example, budget for transport network such as National or State and District Highways per ward are meaningless.

11.9 Geographic Base Creation of Budgeting GIS

Most of the project undertaken as part of the annual plan occupies space within the municipal corporation. The project generates features that can be represented as point, line, arc or polygon. There are two approaches in locating this feature; one is called fine and the other coarse representation. Fine representation in GIS is relatively costlier than coarse representation. Coarse representation, which is less costly, is possible if the features are related to any landmark or existing geographic features such as road networks or prominent building. If exact coordinates of a project feature are not available, coarse representations may be attempted. Inexpensive GPS can generate these coordinates but it is time consuming. Coarse representation may be to locate a project feature within a revenue or electoral ward or even coarser representation of location in a planning division. National Informatics Centre had developed a methodology for location of infrastructure [33]. This process identifies the water supply mains underground in addition to roads, and point features such as lamppost, etc. This is indeed an expensive process. Alternative such as use of inexpensive GPS with camera with date stamp could be another way of locating the geographical features above ground level in space and time. Pictures can be attached to location in GIS.

11.10 Attribute Base Creation of Budgeting GIS

Attribute base creation involves preparing a database or spreadsheet regarding each projects designed as a part of the peoples planning movement. This activity essentially involves coding the project under the following heads:

- Sector and sub sector number of the project
- Project number in serial order
- Project location by ward (revenue and or electoral) and attaching the data to geographic features
- Project location by planning division
- Project geographic link says location, polygon, photographs, etc.
- Name of the project
- Objectives of the project
- Total Budget estimate

- Plan fund
- Own fund
- State fund
- Central Government fund
- Beneficiary fund
- Bank loans
- Voluntary service/contribution
- Other.

For monitoring the above details of the previous year, annual plan may be prepared in GIS. Here, the following additional details may be included in the spatial database:

- Physical achievement of the project last year
- Financial achievement (expenditure incurred) of the project
- Other remarks.

11.11 Analysis for Budget Allocation

There are several approaches in deciding the allocation of budget by sectors. The most important approach is what is practiced now based on consultation at different area levels of planning. A decision support system based on the exploratory data analysis of relevant urban indicators by wards or planning division can be used for budget allocation. GIS can provide the information support system for the above process. This approach is spatial and uses hard facts. Members of Ward committees can be used to update urban indicators with the help of the local population for such analysis.

About 145 urban indicators have been evolved in this R&D Project [30, 34, 35]. An approach to derive composite index and change index for different urban indicators are also formulated. These indices for ward or planning division can then be subdivided into quartiles of its value. Once the quartiles are represented in a GIS map, the available funds for the particular sector or sub sector can be distributed based on the quartile. The quartile may be below the standards of provision or above and below the standards of provision. Based on this fact budget can be spatially distributed.

GIS-based urban land management has shown that it is possible to develop a net cash flow based on the appropriate land management approach used. This cash flow can be incorporated in the annual planning budget.

There are several issues faced in incorporating this net cash flow in the annual budget. Firstly, there is no head for urban land management in the budget head of the municipal corporation. This is in spite of the fact that land management schemes have been implemented in the past Vinod Kumar (1994). Appraisal of town and country planning practice in Kerala. Institute of Town Planners India, New Delhi (unpublished). That means land as a resource for urban development and augmentation of financial resource is not considered in annual planning of Kozhikode. Since urban land management results in, no loss or no profit scenarios, it should be made a part of productive sector. In one approach this net cash flow can be sub divided under different heads within the approved head. This is an artificial way of incorporating urban land management in the annual plan budget and hence not recommended.

11.12 GIS Tools for Annual Budget Analysis

With the incorporation of GIS [36, 37], in Municipal budget it is possible to display on screen any information related to budget as a map, table or graph any time by anybody. Seagate Crystal Reports are now part of GIS software. This can be effectively used for budget analysis and representations. GIS can be easily used to change or edit the data. It is an ideal tool to keep the inventory of municipal assets, which is dispersed in many wards. Properties can be managed using GIS. In addition to the group involved with annual planning and project formulation, the group dealing with monitoring of the projects will be the greatest user of Budget GIS. Photographs and site maps can be attached to the GIS to give idea of progress of work. Statistical and database analysis can be conducted from GIS of budget. In web GIS [32] these information will be available on demand by any person, anytime and anywhere.

11.13 Spatial Decision Support System (SDSS) for Participatory Municipal Budgeting

Municipal budgeting system needs to ensure equity such that benefit of budget shall go to all wards which are the lowest unit of municipality. In addition to equity there may be many criteria specific to the nature of projects embodying technical efficiency which shall be met in the spatial budgeting. If all aspects are fully known a highly structured multi-criteria analysis can be attempted. There are many tools to help multi criteria analysis, but all such quantified objectives cannot be fully known and structured in a Municipal Budget. In such a situation of semi structured spatial budgeting issues, a SDSS is utilized. It is designed to assist in Ward Conventions, Municipal Development Seminars and District Planning Committee meetings to decide on Budget. As advocated earlier GIS database creation is the prerequisite for SDSS. However GIS fall short of the goals of SDSS for a number of reasons. Analytical modelling capabilities required for a particular type of project often are not part of a GIS. GIS databases have been designed solely for cartographic display of results. SDSS goals require flexibility in the way information is communicated to the user Variables or layers in the database may be insufficient for complex modelling. Data may be at insufficient scale or resolution. A GIS database design is not flexible enough to accommodate variations in either the context or the process of spatial decision-making, but assumes an important role in Municipal Budgeting.

SDSS require a Database management System (DBMS) for budgeting embodying specific analysis procedures in a model base management system (MBMS); a display generator; a report generator and a user interface [38–45]. GIS database management systems are designed to support cartographic display and spatial query. A database of an SDSS must support cartographic display, spatial query and analytical modelling by integrating three types of data namely locational (spatial primitives such as coordinates and chains), topological (attribute-bearing objects, e.g. points, nodes and lines, and relationships between them) and thematic (attributes of the topological objects, including population, elevation, and vegetation). This database must permit the user to construct and exploit complex spatial relations between all three types of data at a variety of scales, degrees of resolution and levels of aggregation. Database management systems found in many GIS use the relational data model.

11.14 Model Base Management System (MBMS) for SDSS

One approach to incorporating analytical models mentioned in last para is incorporating a geoprocessing systems is to develop libraries of analytical sub-routines. This can be easily executed using visual basic and other tools. For example there may be in the library a specific model-based management system which is highly adaptable for public distribution system or similar point-based system. There may be another in the library which is adaptable for various network like water supply, drainage and sewer network. This library approach permits large numbers of models to be made accessible very quickly, because existing programmes can be patched into a system. This will avoid wasteful in terms of replicated code. A second approach, used in business applications of DSS, is to develop a MBMS. This consists of small pieces of code, each of which solves a step in an algorithm. As many of these steps are common to several algorithms, this approach saves large amounts of code. The system developer only has to modify one piece of code to update a step in several algorithms. The MBMS also contains information about how steps are sequenced to execute a given algorithm, using an MBMS facilitates rapid development and testing of new algorithms. Implementation may be achieved simply by adding a new formula to the MBMS. In other cases new code for additional steps also may be added to the model-base.

Graphical and Tabular Report Generators should provide the following capabilities: namely high-resolution cartographic displays; general-purpose statistical graphics, including two and three-dimensional scatter plots and graphs; specialized graphics for depicting the results from analytical models and sophisticated statistical techniques; the full range of tabular reports normally associated with each of the above. The User Interface shall consist of an interface which is easy to use if they are to be effective in decision-making.

11.15 Spatial Budgeting by Ward Conventions in Kerala

The spatial decision problem in municipal budgeting has special features. The budget needs to be allocated over several or all wards. There are many alternatives to spread the budgets among many wards. There are many criteria for locating annual budgets. Some may be qualitative and some quantitative. There is generally more number of decision makers at ward level, municipality level and metropolitan planning committee/district planning committee level. The decision makers may have their own preferences. These decisions are surrounded by uncertainty. In the PB of Kerala, ward conventions are used to identify projects suited for a ward as discussed above. However there are several critical issues in budgeting at ward level. They are:

1. Is the project suited to the wards per the planned strategy for a particular sector?
2. Given the total budget available for the particular project within a sector for the city how many projects can be included in the annual budget for this ward and how much money shall be allocated for the ward.
3. Which are the priority locations where projects can be executed within the ward?

Although question 1 and 3 can be agreed upon by ward convention, Question 2 requires an understanding of the status of all wards as related to this particular project. In other words a complete GIS for the existing status of the similar project in all wards is required. This can be prepared in GIS maps, tabulation and graphs. Simple algorithms are required to allocate funds across many wards. Some examples will be explained below.

12 Demonstration of GIS-Based Spatial Decision Support System (GIS-SDSS) for Calicut

This section presents a demonstration of GIS-based Spatial Decision Support System (GIS-SDSS) that has been devised for Calicut. This demo is limited to a few physical infrastructure projects that are identified from the annual plan of Calicut Corporation. The following are the projects identified:

1. Public Water Taps (provided by the State Water Authority but funded by the Corporation)
2. Pipe Composting (A scheme for household solid waste management).

12.1 Procedure Adopted

The following procedure was adopted for the demonstration of GIS-SDSS.

1. Creation of base data—spatial (on the ward map of Calicut) (Refer Fig. 9).
2. Creation of base data—attribute (Refer Fig. 10).

3. Field survey to record the present status of the project.
4. Data on physical and financial achievement of the project in the previous year.
5. Analysis based on rules set with regard to spatial distribution of the entity and a combination of various attribute data to arrive at an index.
6. Calculation of quartile value of the index and linking the same to spatial data.
7. Allocation of available funds to the respective wards based on the quartile.

12.2 Allocation for Public Water Taps

By following the standard procedure explained in the previous section, the existing water tap locations were identified and a ward map with the water tap locations (as point data) was created. Figure 5 was thus created.

The attribute type database was created with all details as discussed in Sect. 11.10. The database was updated with further details after conducting a field survey on the present status of the water taps. The data on physical and financial achievements of these water taps and their maintenance during the previous year was also collected and incorporated. Also the census data for the wards were collected and incorporated in the wards polygon layer itself.

Number of existing taps coming within each ward was counted through a spatial join operation and added as an attribute to the ward polygon. Using this, number of households served by each tap within a ward was calculated.

$$N_{\frac{hh}{tap}} = \frac{\text{Number of Households in ward}}{\text{Number of taps in ward}} \tag{1}$$

High value of the ratio indicates that the ward requires more number of taps to bring down the ratio. The mean value of the ratio was found to be 413 households per tap, while the range is from 27 to 1,888 for wards with taps, and to infinity for wards without a tap. The wards were grouped into quartiles using this ratio, and highest priority class was assigned to the highest quartile. Lowest priority was assigned to the lowest quartile and the in between quartiles were classified accordingly as given in Fig. 6.

The total budget amount allocated for the public water taps by the Calicut Corporation is 60,00,000 rupees. The method of distribution of fund on the basis of these quartiles is illustrated in Table 8. The council of members decides on what is the sharing among the quartiles. They decide, how much to give to a ward in quartile 1, if we give ≠1 to a ward in quartile 4 (the least priority quartile). This gives a relative sharing, based on the conscience of the council members. This value is illustrated in column C in the table. The amount to be given to a ward in the lowest priority quartile is considered as one share. Number of shares for each quartile is calculated as a product of column C and number of wards in each quartile (Column B). By finding sum of this column, total number of shares required is found. Value of a single share is found by dividing ≠60, 00,000 (the

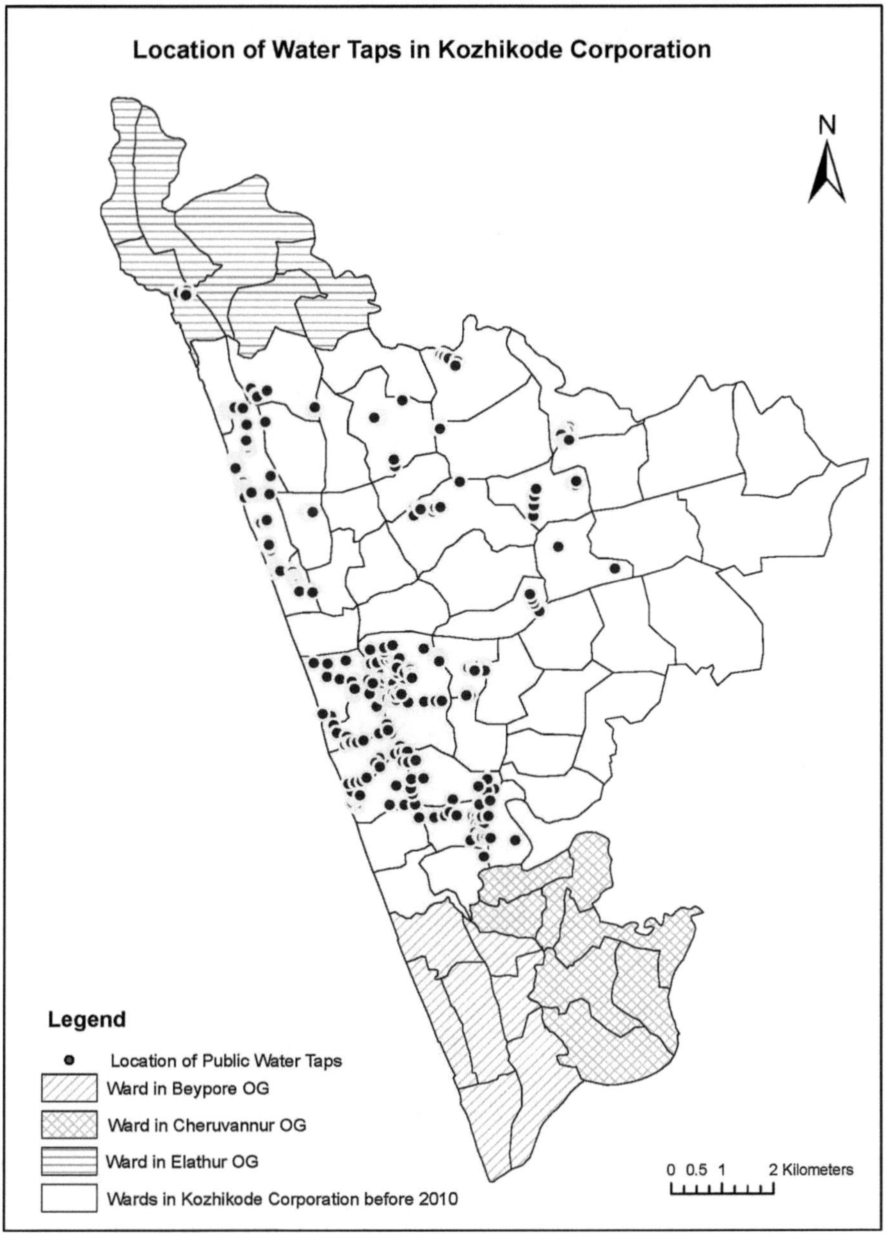

Fig. 5 Calicut corporation ward map with location of public water taps

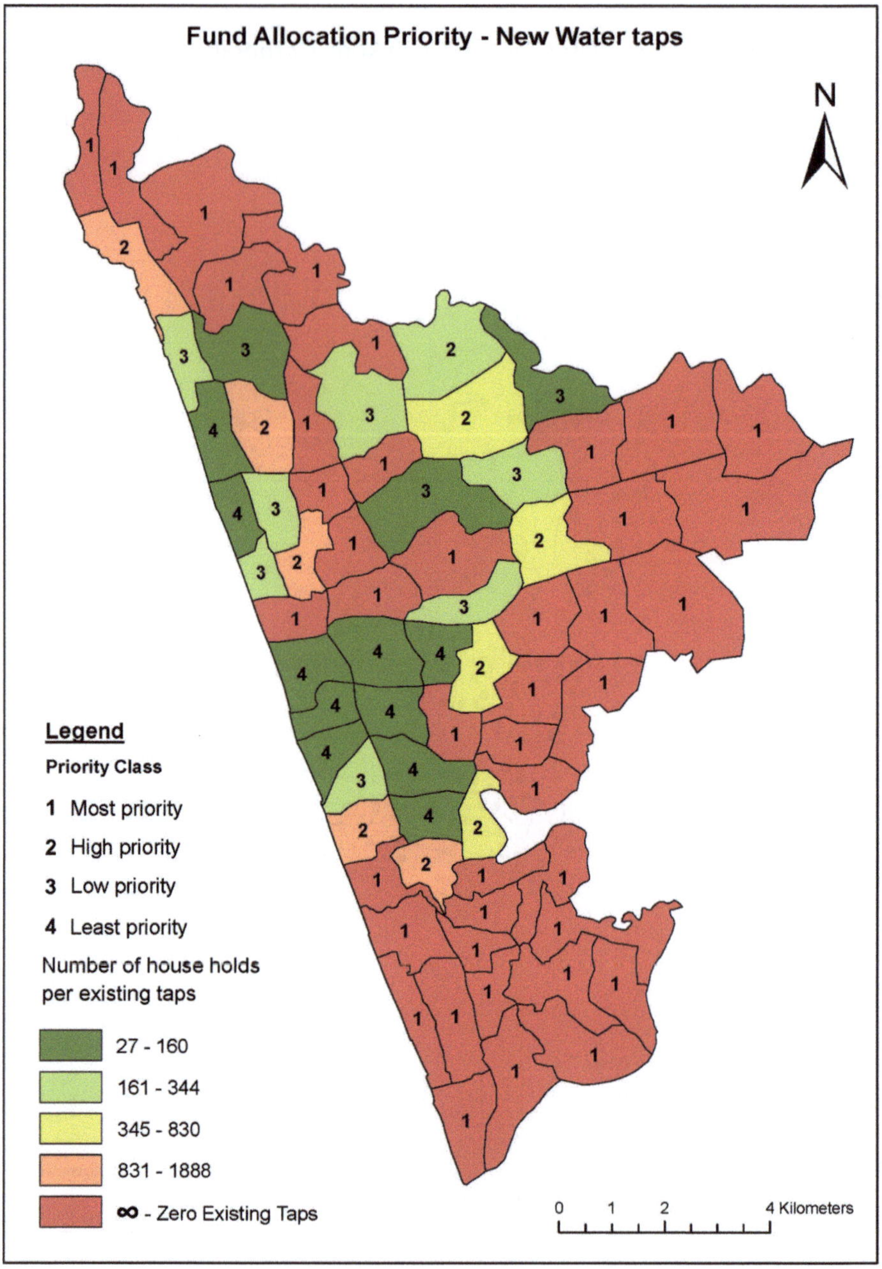

Fig. 6 Fund allocation priority and number of households per tap

Table 8 Allocation of fund to the wards based on the quartile

Allocation of fund to the wards based on the quartile—water taps						
Quartile	Number of wards	Relative sharing among quartiles: yo be decided by the council	Number of shares for each quartile	Value of a share	Amount allocated to each quartile	Amount allocated to each ward in each quartile
A	*B*	*C*	$D = B \times C$	$E =$ Amount/ total number of shares	$F = E \times D$	$G = F/B$
1	41	4	164	Rs. 26,786	Rs. 43,92,857	Rs. 1,07,143
2	10	3	30		Rs. 8,03,571	Rs. 80,357
3	10	2	20		Rs. 5,35,714	Rs. 53,571
4	10	1	10		Rs. 2,67,857	Rs. 26,786
	Total number of shares		224	Total	Rs. 60,00,000	

budget provision for new water taps) by number of shares. Allocation for each quartile is found by multiplying value of a share by number of shares for that quartile. Allocation for each ward in that quartile is found by equally distributing this amount to each ward in that quartile. The amount allocated to each ward is given in column G.

Further analysis was done to identify locations for the proposed taps. A thiessen polygon was created with the existing locations of taps. These polygons subdivide the given area according to the water tap locations. Any point in these polygons will be close to the water tap with in the polygon than any other tap. Hence these polygons give the area served by each tap. Larger the area of the polygon, the more deprived is the place. This map was overlaid with the ward boundaries as the funds are allocated to wards as given in Fig. 7.

It can be seen that many of the thiessen polygons extend to more than one ward. This implies that, a tap coming in one of that polygon will serve the nearby wards as well. Hence the wards can jointly opt for projects using the fund allocated to them to derive maximum utility.

12.3 Allocation for Repairing Surface Damage of Roads

Due to geographical and other reasons, the roads in Kerala require frequent maintenance. Every monsoon season leaves the roads in bad condition, and local governing bodies need to repair these roads every year. Calicut Municipal Corporation has National Highways (NH) and State Highways (SH) passing through and a number of other roads. Municipal Corporation needs to take care of all roads except NH and SH as they are maintained by National Highway authority and State public works department.

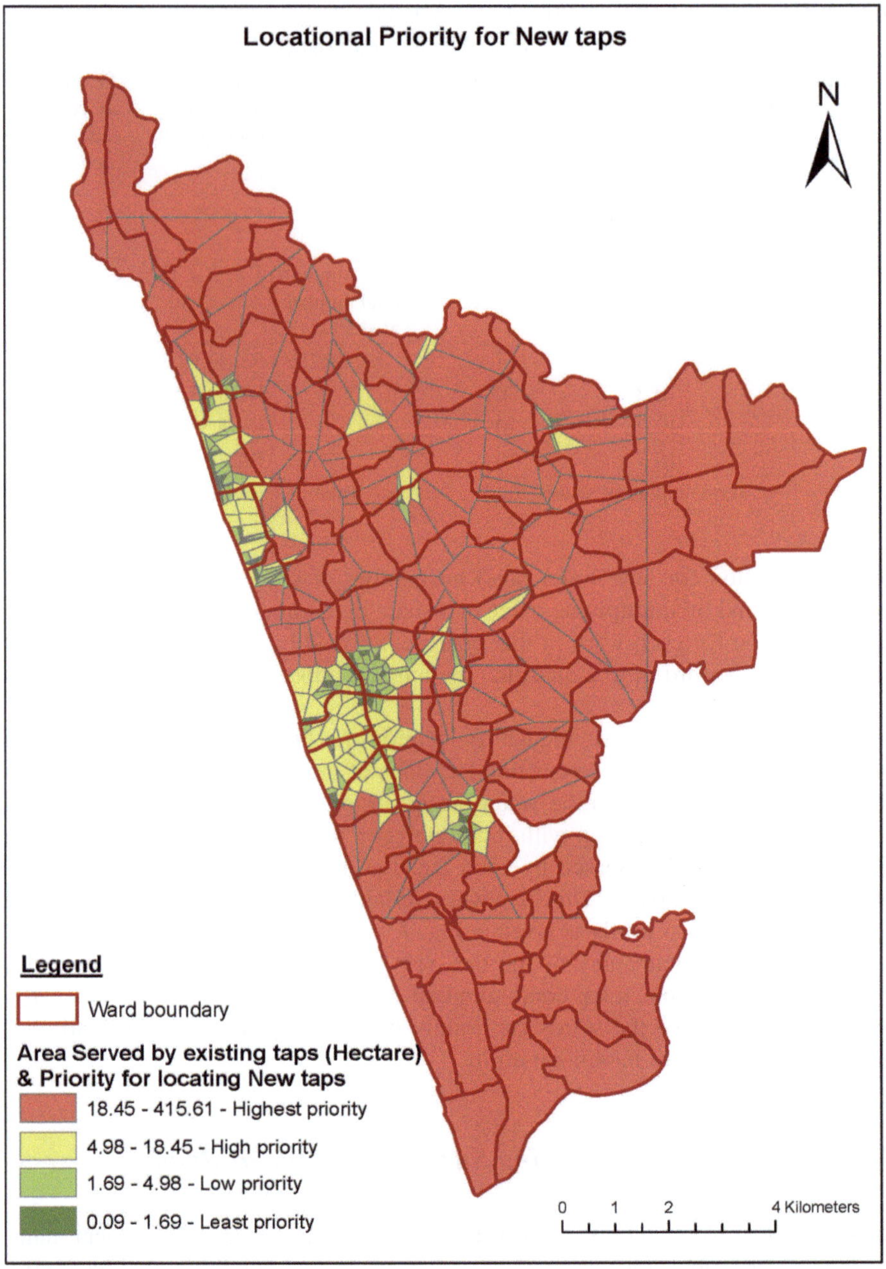

Fig. 7 Location priority for proposed taps

The allocated fund is used for repairing the existing damages, and to meet expenditure towards regular maintenance of these roads in the year ahead. Hence the fund allocation mechanism should take care of two things,

(a) Existing level of surface damage of roads in each ward,
(b) Expected expenditure for maintenance of roads in each ward.

The overall methodology to derive the fund allocation is given as a flowchart in Fig. 8.

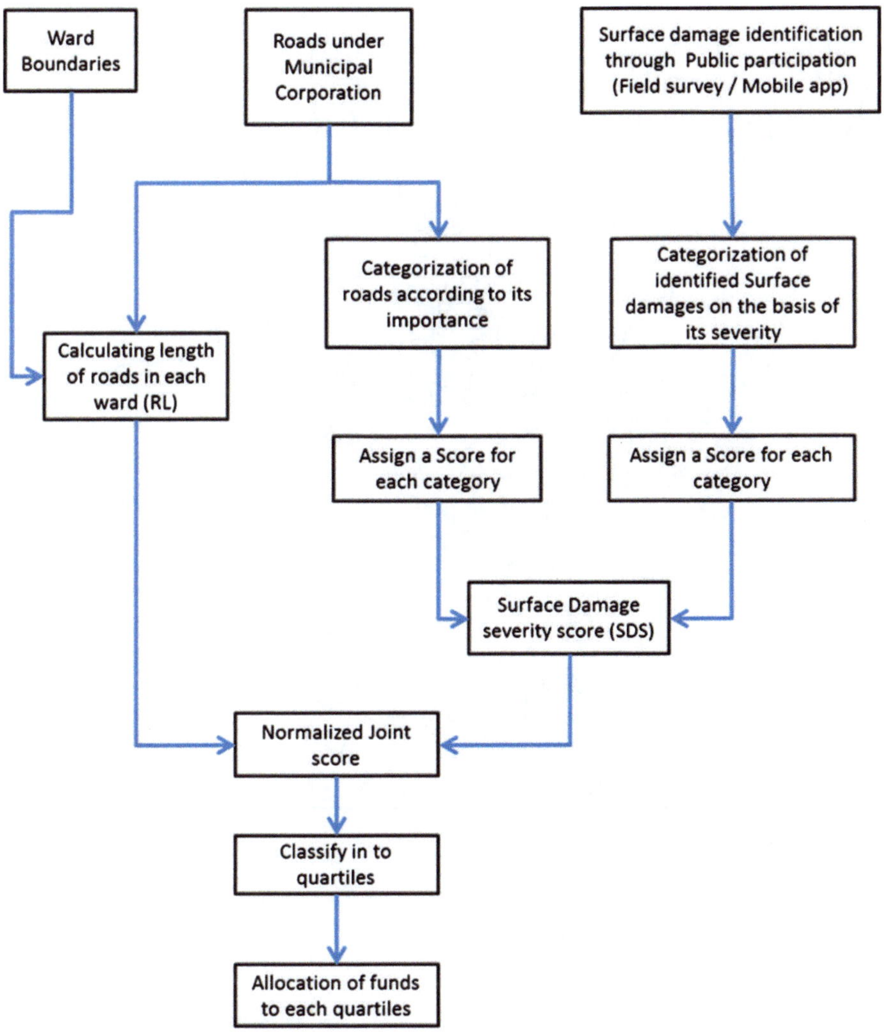

Fig. 8 Flow chart—procedure for allocating fund for road surface damage to wards

Existing level of surface damage is assessed through physical inspection of roads and identifying pot holes on each road in wards. Alternatively this can be done through public participation using a mobile application, which can take a photograph, identify the geographical location of pot holes and upload to a central GIS database. This information is used in GIS analysis as a point feature as given in Fig. 9. Such points are grouped based on their geographical location; this is required as many people would have uploaded images of the same damage. So this operation will merge all the entries to a set of unique locations of surface damage. These point features are inspected by looking at the photographs and onsite inspection whenever required and they are classified into three categories

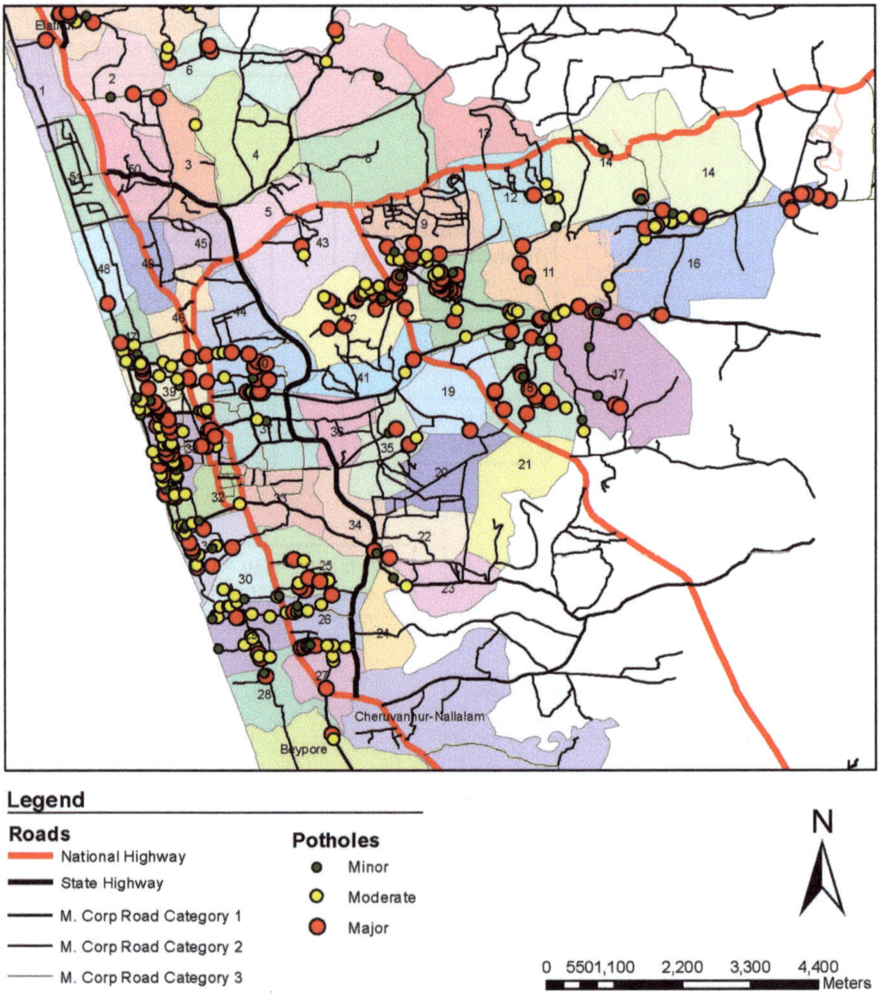

Fig. 9 Spots of surface damages identified on roads

Table 9 Score for surface damage categories

Category	Score
Minor	1
Medium	3
Major	6

Table 10 Score for different categories of roads

Category	Score
Local roads	1
Minor—collector roads	3
Major—collector roads	6

according to the severity of damage. A score is assigned to each point feature according o the category as given in Table 9.

Roads are also classified into three classes according to their importance in the traffic scenario, and a scoring is assigned as given in Table 10.

These two scores are summed up to get a single score—surface damage severity score (SDS)—which reflects the severity of the damage and its significance. This scoring can be used to rank wards on the basis of the existing damage. It is also required to maintain the roads throughout the financial year, hence a factor which reflect extend of roads in each ward is also required. Total length of roads in each ward is calculated for this purpose using intersection operation between wards polygon layer and roads polyline layer, which splits up continuous polylines at the ward boundaries. Outcome of this operation is grouped and summarized for each ward. This gives the length of roads (RL) in each ward.

Further SDS and RL are combined into a single score using relative weighting. Hence the corporation can decide the relative importance that need to be given for repairing existing damage and regular maintenance. In a year, when the roads were not mutilated badly, they may give more importance to regular maintenance and the wards can be ranked accordingly. On the contrary, when the roads are badly damaged, the funding preference can be for maintenance and the wards may be ranked accordingly. The scoring is done according to Eq. 2. Where W_{SDS} is the weight given for surface damage and W_{RL} is the weight given for regular maintenance.

$$JS = \frac{W_{\mathrm{SDS}} \,.\, SDS + W_{\mathrm{RL}} \,.\, RL}{W_{\mathrm{SDS}} + W_{\mathrm{RL}}} \tag{2}$$

In this example, W_{SDS} and W_{RL} were given weights 1 and 2 respectively. The score was calculated for each ward and later normalized with the maximum score to get a consistent range between exercises. In this particular case, two wards were found to be outliers, with a value of 1 and 0.5 and all other values range from 0 to 0.34. Hence these two wards were separated as outliers and the normalization was done with the remaining wards. This score was used to categorize wards into quartiles and fund was allocated accordingly. The final result is given in Fig. 10.

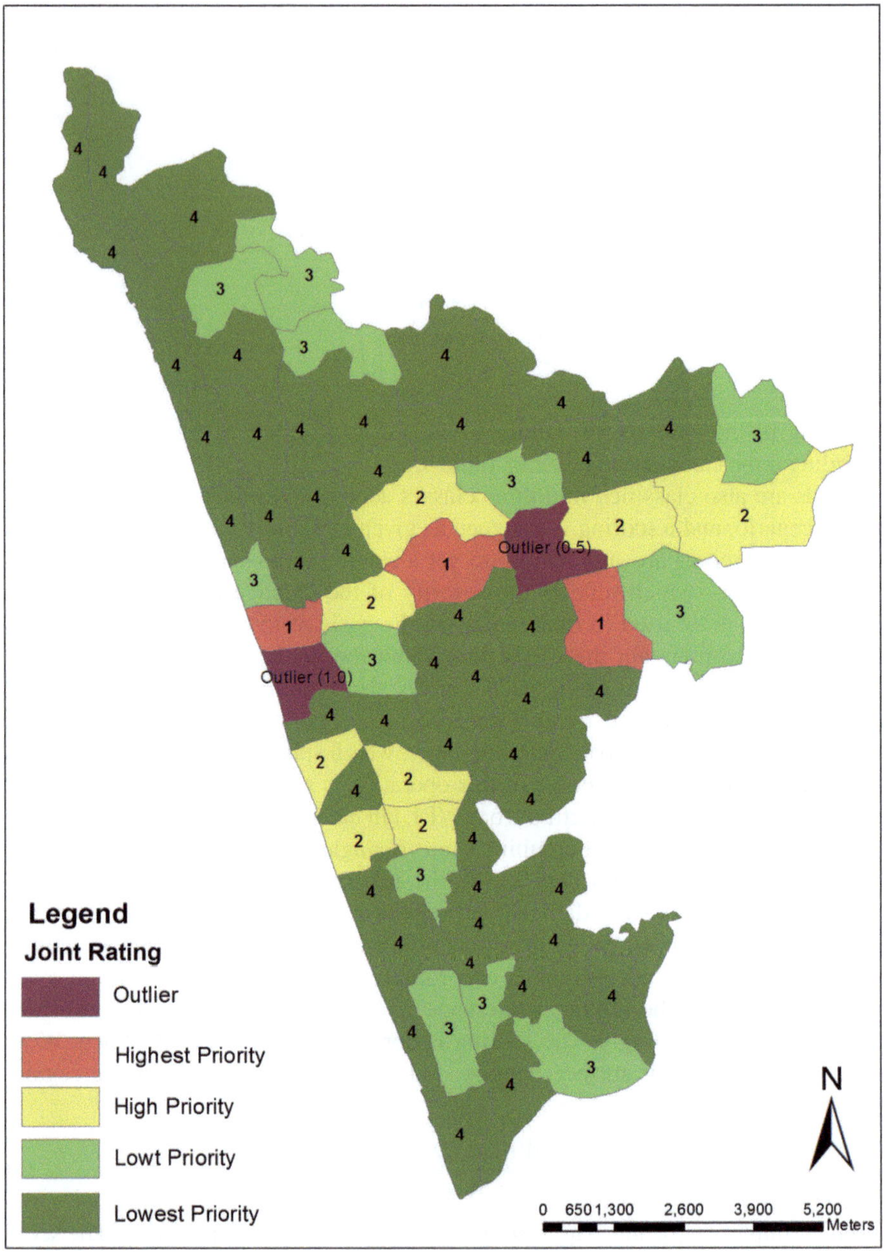

Fig. 10 Quartiles for fund allocation based on joint score

The total budget outlay for repairing roads in Calicut Corporation is 63, 47,000 rupees. Two wards that are separated as outliers are compared with quartiles to understand the case as given in Table 11. It can be seen that, joint score of

Table 11 Comparison of outliers with other quartiles

Quartile	Number of wards	Average number of pot holes	Average joint score	Average length of road	Average SDS
1	3	23.33	94.6	9.06	171.0
2	8	14.88	63.6	7.40	112.4
3	12	8.33	34.6	7.89	53.5
4	46	1.43	10.0	5.45	9.0
Outliers					
Ward No. 38	1	60	258.3	14.80	487.0
Ward No. 10	1	38	131.8	10.32	243.0

Table 12 Fund sharing among quartiles and wards—repairing roads

Quartile	Number of wards	Relative sharing among quartiles: to be decided by the council	Number of shares for each quartile	Value of a share	Amount allocated to each quartile	Amount allocated to each ward in each quartile
A	*B*	*C*	$D = B \times C$	$E =$ Amount/ total number of shares	$F = E \times D$	$G = F/B$
Outlier	2	7	14	Rs. 52,892	Rs. 7,40,483	Rs. 3,70,242
1	3	4	12		Rs. 6,34,700	Rs. 2,11,567
2	8	3	24		Rs. 12,69,400	Rs. 1,58,675
3	12	2	24		Rs. 12,69,400	Rs. 1,05,783
4	46	1	46		Rs. 24,33,017	Rs. 52,892
	Total number of shares		120	Total	Rs. 63,47,000	

ward 38 is 258.3 compared to the average joint score of highest quartile is 94.6. This is due to the wide difference in the number of pot holes in these wards as seen in the table. Ward no 38 has 60 pot holes where as the highest quartile has average 23.3 potholes per ward. Not only ward no 38 has more road length but also its roads has a higher weightage as it lies in the core the city. Some of these roads may require resurfacing since they have a large number of pot holes. Hence a larger share compared to other wards is to be allocated for these outlier wards.

This amount was distributed among wards by a procedure similar to the distribution in case of water taps. Only difference being outliers were considered separately and its relative share was decided by the council members along with the quartiles. The allocation of funds is given in Table 12.

13 Conclusion

The Municipal Budgeting System of Kozhikode has changed beyond recognition. Prior to Janaki Asuthranam, the peoples planning movement, Municipal Commissioner and his clerk used to make the annual budget. There was no need to consult any one, especially the citizen. There was no need for field surveys for assessing requirements. The budget prepared was a revision of last year budget to fit in the allocation of money this year as grant. The money received as grant was meagre. With the advent of State and Central Financial Commission, the percentage of money transfer to Municipal Budget as a percentage of total state budgets has increased. Further, centrally funded projects also create a flow of money to Municipal budget. Multi-level budgeting system as part of Janakiya Asuthranam, require that many thousands of citizens are participating in the budget making system in Kozhikode. There are review committees, task forces at multi-level to make decisions and find out anomalies. In the absence of most up-to-date information and sound rationale for allocation of funds it became difficult to allocate the increasing budget provisions. Often it has been found sound rationales and felt needs of ward committees are not truly represented in the final adopted budget. These points towards introduction of GISs and its use for giving a SDSS for annual budgeting. In the month of June 2014, one Municipal Councillor attempted to commit suicide in the office of the Mayor of Kozhikode Corporation for issues connected with budget allocation in her constituency. At least this will not happen if budgeting is based on SDSS and is used in multilevel decision-making. Although Kozhikode Municipal Corporation received a copy of Arc GIS as part of UNDP-Ministry of Science and Technology project (Mission for Appropriate Technology for Urban renewal), it is not used for decision-making or creating GIS-based SDSS.

Acknowledgments The authors gratefully acknowledge the contribution made by Febin Siby and Rubeena Moideen Kutty, the final year (2014) students of Architecture at National Institute of Technology Calicut for the help rendered in collecting required data for the preparation and demonstration of GIS-based Spatial Decision Support System for Calicut Corporation for this chapter.

References

1. Eva-Maria Verfürth, Giovanni Allegretti (2013) More generous than you might think. http://www.dandc.eu
2. http://www.psupress.org/books/titles/978-0-271-03252-8.html
3. http://siteresources.worldbank.org/INTEMPOWERMENT/Resources/14657_Partic-Budg-Brazil-web.pdf
4. Lewit D. Porto Alegre's budget of, by, and for the people
5. http://www.participatorybudgeting.org/about-participatory-budgeting/where-has-it-worked/
6. http://www.cfe.org.uk/uploaded/files/Participatory%20Budgeting.pdf
7. http://www.watsonblogs.org/participatorybudgeting/historypb.html
8. http://www.buergerhaushalt-europa.de/documents/Volumen_zwei_Endbericht_Buergerhaushalt_Europa5.pdf
9. http://www.torontohousing.ca/participatory_budgeting
10. http://pbhamont.ca/ http://books.google.com/books?hl=en&lr=&id=qk6YyKRRIk0C&oi=fnd&pg=PT3&dq=arvind+kejriwal+swaraj&ots=7RX3EZkh1b&sig=y63_H4Ud8uBksDdSFkGzMUCeG_k#v=onepage&q=arvind%20kejriwal%20swaraj&f=false
11. http://www.aamaadmiparty.org/page/goal-of-swaraj

12. http://www.participedia.net/wiki/Poitou-Charentes%27_participatory_budget_in_the_highschools
13. http://permaculture.org.au/2010/05/27/chicagos-1-3-million-experiment-in-democracy/
14. YES Magazine (2010, April 10) Chicago's $1.3 Million Experiment in Democracy; For the First Time in the U.S., the City's 49th Ward Lets Taxpayers Directly Decide How Public Money is Spent,
15. http://www.yesmagazine.org/people-power/chicagos-1.3-million-experime
16. Saad-Sulonen J (2012) The role of the creation and sharing of digital media content in participatory e-planning. Int J e-Plann Res 1:1–22
17. Vinod Kumar TM (2004) Emerging role of professional planners under the new governance and GIS, PSS and SDSS SDR spatio-economic record 11(6)
18. Vinod Kumar TM (2001) GIS for municipal budgeting, spatio-economic development record, vol 8, no 6. November–December 2001, New Delhi, pp 4–15
19. Government of Kerala, Department of town planning and Calicut development authority (1994) Development plan for Calicut urban area 2001, Calicut
20. Government of Kerala (1967) Department of town planning and Calicut development authority. Interim development plan for Calicut urban complex
21. Government of Kerala (1995) Finance Commission Report, Thiruvanathapuram
22. NIUA (1989) Upgrading Municipal Services: Norms and Financial Implications. Research Study Services No. 38, NIUA, New Delhi
23. Armstrong MP, Densham PJ (1990) Database organization strategies for spatial decision support systems. Int J Geogr Inf Syst 4(1):3–20
24. Antenucci JC et al (1991) Geographic information systems: a guide to the technology. Van Nostrand Reinhold, New York (ISBN 0-442-00756-6)
25. Burrough PA (198) Principles of geographical information systems for land resources assessment. Oxford University Press, Oxford (ISBN 0-19-854563-0); ISBN 0-19-854592-4, paperback
26. Elmasri R, Navathe SB (1989) Fundamentals of database systems. The Benjamin/Cummings Publishing Company Inc, Redwood City
27. Aronoff S (1989) Geographic information systems: a management perspective. WDL Publications, Ottawa (ISBN 0-921804-00-8)
28. Korte GB (1992) A practioner's guide: the GIS book. OnWord Press, Sante Fe (ISBM 0-934605-73-4)
29. Laurini R, Derek T. Fundamentals of spatial information systems. Academic Press Limited, London (ISBN: 0-12-438380-7)
30. Pangotra Prem P (1998) City monitor 1996–1997 Ahmedabad-Bangalore-Hyderabad-Pune. Ahmedabad Management Association, Ahmedabad
31. Ravindra A (1998) Creating well governed and managed city, case study. Bangalore (Mimeo), Bangalore
32. Kenneth E, Anthony PK (1997) WebGIS, NCGIA Core Curriculum in GIScience
33. Chandra M, Kandwal AK, Bhargava A (2000) Trenchless technology role of GIS based utility mapping. Paper presented in the workshop on trenchless technology, IIT Delhi, New Delhi, Jan 2000
34. Indo-US Financial Institutions Reform and Expansion Project—Debt Market Component FIRE (D) (1999) Project notes urban performance indicators systems a comparative approach to monitoring urban performance and application in Tamil Nadu, New Delhi, August 1999
35. Kirloskar Consultants Limited, Chennai (1998) Comparative performance assessment and ranking of urban local bodies in Maharashtra, Chennai
36. Huxhold WE (1991) An introduction to urban geographic information systems. Oxford University Press, Oxford, ISBN 0-19-506534-4
37. Fischer MM, Nijkamp P (1993) Geographic information system, spatial modeling, and policy evaluation. Springer, Berlin, New York, p 42
38. Armstrong MP, Densham PJ (1990) Database organization alternatives for spatial decision support systems. Int J Geogr Inf Syst 3(1). (Describes the advantages of the extended network model for network-based problems)

39. Bonczek RH, Holsapple CW, Whinston AB (1981) Foundations of decision support systems. Academic Press, New York Basic text on DSS
40. Densham PJ, Rushton G (1988) Decision support systems for locational planning. In:Golledge R, Timmermans H (eds) Behavioural modelling in geography and planning. Croom-Helm, London, pp 56–90
41. Geoffrion AM (1983) Can OR/MS evolve fast enough? Interfaces 13:10. (Source for six essential characteristics of DSS)
42. Hopkins L (1984) Evaluation of methods for exploring ill-defined problems. Environ Plann B 11:339–348
43. House WC (1983) Decision support systems. Petrocelli, New York. (Basic DSS text)
44. Sprague RH (1980) A framework for the development of decision support systems. Manag Inf Sci Q 4:1–26. (Source for DSS development model)
45. Sprague RH, Carlson ED (1982) Building effective decision support systems. Prentice-Hall, Englewood Cliffs, NJ (Basic DSS text)

Part IV
E-Smart City Governance-Futures

Smart City E-Governance: Issues and Future

T.M. Vinod Kumar

Abstract Issues for developing Smart City E-Governance in India are unique. Two basic issues are the levels of E-Governance Infrastructure in position, namely information and communication technology (ICT) and relative ability of population subgroups to access the E-Governance infrastructure. There has not been uniform social development of urban and rural population and also population subgroups such as male and female, scheduled caste (SC), scheduled tribes (ST), Muslims and non-Muslims in India. Therefore uniform environment of developing Smart City E-Governance does not exist. These target groups require differing policy set to make their access to Smart E-Governance possible. This partitioning of subgroups into different levels of accessibility to Smart E-Governance is the major issues of Smart E-Governance in India. There has been rapid development in ICT such as telephones, mobiles and Internet in India in absolute terms but when it is converted to percentage figures it gives very negative impression doubting the possibility of Smart City E-Governance in India. Also, E-Government survey conducted by United Nations in 2012 places India in the lowest quartile. However recent development of E-Commerce in India gives considerable hopes for E-Governance. E-Commerce and E-Governance shares the same ecosystem but E-Commerce has shown very rapid development and GDP generated can even surpass health and education sector in India. These developments are mainly generated by potential smart cities in India. This gives positive assurance that carefully selected about 31 mega cities, million pilus cities in India and can be developed for Smart City E-Governance. Other towns from about 8,000 census towns have to wait to overcome their population-related access to Smart City E-Governance. This concluding chapter in the last part highlights the further reaches of E-Governance discussed in all chapters of the book as part of its future.

T.M. Vinod Kumar (✉)
School of Planning and Architecture, New Delhi (SPA-D), Calicut, Kerala, India
e-mail: tmvinod@gmail.com

T.M. Vinod Kumar (ed.), *E-Governance for Smart Cities*, Advances in 21st Century Human Settlements, DOI 10.1007/978-981-287-287-6_11

Keywords Barriers to developing smart E-Governance in India · Relative access of population subgroups to E-Governance · E-Governance and E-Commerce in potential smart cities · ICT infrastructure for E-Governance · Telecommunicati on and Internet · Further reaches of smart city E-Governance from all chapters of this book

1 Introduction

The National Democratic Alliance (NDA) Government in India has an electoral promise to be fulfilled of creating 100 Smart Cities to accommodate the growing urban population in India, who they consider as largely the new and old middle classes worthy of settling in smart cities. Implied in this policy is a feeling that new urban inhabitant shall live in Smart Cities to reap maximum socio-economic and spatial dividends. This can accelerate the GDP growth rates in the future. Undoubtedly this logic can be questioned. Details about these cities are still unknown and are being specified and blue prints are being prepared behind the scene. Foreign capital investors are interested to give soft loans for this purpose. The fact remains that Rs 7,000 plus crores have been allotted for this scheme in the first budget of NDA Government this year.

These cities can be Greenfield projects along the freight and industrial corridors being implemented/planned or brownfield projects as part of metropolitan urban agglomeration, mega cities, and twin cities, ring towns, counter magnet, new Industrial Township, new capital cities or satellite towns as the case may be. There has been no discussion on equitable distribution of 100 smart cities among states and Union territories, which is a good sign since spatio-economic development largely depends on many factors including the geographic location. Whether it is Greenfield or Brownfield smart cities the intention of smart cities is to provide India with rapid economic development, which essentially involves wealth creation by innovative and large-scale urban production and services and rapid rate of employment generation for the growing population of India. A broad goal in urban planning and smart infrastructure is expressed in the manifesto of the Bharatiya Janata Party (BJP) which vaguely expresses the six systems that constitutes the Smart city, which is discussed in Chapter E-Governance for Smart Cities. There is no mention of Smart E-Governance of these cities in the manifesto of the party ruling India but there is a mention of implementation of E-Governance or simplify improved Governance. This book is an attempt to formulate E-Governance for Smart Cities. As detailed in Chapter E-Governance for Smart Cities, Smart City E-Governance is the fifth phase of development of E-Governance and one needs to build up from the fifth phase to many more unknown phases. E-Governance development is also cumulative such that the fifth stage also includes stages one to four.

The objective of this chapter is to study the major issues faced in India for instituting a Smart City E-Governance. It also explores the futures. Futures are studied in several chapters of the book with limited number of City Case Studies,

Domain Studies and specific study of selected Tools and issues. The Smart City Governance essentially involves regulation that protects the common interest of Smart Cities in the state and national perspectives and citizen and stakeholders' active involvement and participation in implementing what is good for the Smart City through practice of E-Democracy. This is achieved by using a user friendly system of E-Governance which works 24 h and 7 days a week, in a most efficient manner. Regulation can be legislated policy made into state level laws and the procedure of administration of laws. Regulation can encompass all sectors relevant to a Smart City and can be considered domain-specific laws and regulations or municipal governance-specific. Regulations as laws and administrative procedure can be made more effective by the active participation of citizen and their responses and feedbacks. This chapter elaborates on the major issues Smart City E-Governance and investigates what potential exists for progress? This is discussed from Sects. 2–5. The second aspect highlights the Smart City E-Governance futures, which is discussed briefly, based on contribution of this book by many authors in enhancing further reaches of E-Governance for Smart Cities. This will be discussed from Sect. 6 onwards briefly, chapter wise.

For effective governance of mega cities and million plus cities or a group of small selected settlements widely dispersed and not easily accessible like island groups of Lakshadweep or Andaman's or mountain states of Himachal Pradesh or North Eastern India, a fully information and communication technology (ICT) enabled Smart City E-Governance is required. Smart governance is all about proactive role of electorate in all aspects of governance in a continuous all-encompassing fashion using all available channels of communication provided by ICT. In other words, Smart city E-Governance is empowering Smart people for the development of Smart cities. Here the duality of electorate and elected gets blurred in practice and vanished de facto but not de-jure, which this book defines as Smart E-Governance. It is a game changer and more powerful than the so-called participatory governance practiced in India today. ICT creates a virtual face-to-face community, which is the foundation of E-Government built on the firm foundation of E-Democracy. The use of ICT in Governance is important since Smart E-Governance is not a type of governance in which the electorate only casts vote once in 5 years and passively judges the performance of the elected to caste vote in the next election. Under Smart E-Governance, the electorate is actively engaged in all aspects of governance, starting from city development issues identification (local or citywide), formulation of appropriate policies, SWOT analysis and strategic plans, participatory zonal plans, formulation of development codes and audits its implementation at multiple levels of a smart city. The Indian Constitution endorses people participation in Governance, but what is practiced is unsuited and obsolete for Smart cities. Smart city E-Governance goes much beyond the existing ritual called 'participation' being practiced. This is only namesake and insincere implementation of participation in Governance. The first factor that helps E-Governance is the deployment of city wide ICT, which includes cheap and universally available broadband. The second aspect is to ensure easy population access to use ICT in participatory governance. It has been statistically proved that there exist barriers for access to ICT for different social groups, with differing income, educational attainment, caste and class which needs to be overcome.

There are 7,925 Census Towns in India in 2011. All of them cannot be converted to Smart Cities even if loans and grants are available to achieve this objective but the present government limits it to 100 in the next 5 years. In reality, the potential for smart city development exists only to a very small number of cities in India, but its impact for national economy and creating a model of smart living for the rest of the cities can be appreciable and can inspire other cities to follow the model by creating a set of assets that enable smart cities. Therefore, it is important to identify these small number of potential towns for smart city development, in a systematic way using empirical data. No way can it be arbitrary sharing models between states for Central Government Project funding. Other cities shall march towards Smart City potential, based on well-defined policy goals and plan objectives laid down by the union government and federal states. Capital requirement for such an ambitious project can be enormous and impractical to meet if mere political and unscientific judgement is used in selection of cities. As noted earlier, the Annual Union of India budget has devoted Indian Rupees of about 7,000 Crores for initiating work on 100 smart city project, which can only be considered as initial seed money. Therefore, the political strategy of distributing certain number of Smart City projects by States and Union territories is not realistic, although they may look reasonable in terms of equity in a federal state like India and autonomy of the states. Such political action if at all attempted, at the most can create technology demonstrator of smart city which cannot be called smart city in the absence of six components of smart cities such as a rudiment of smart economy, potential of smart people, enabling smart environment, effective smart mobility, demonstrated smart living and smart government discussed in Chapter E-Governance for Smart Cities, present in the selected cities for smart city up gradation. New Smart Cities can run a risk of not finding these six systems in place even after the city is constructed. Equity-based political approach, will create a smart city not well functioning resulting in loss of investment creating a non-functional smart city. Some elements of smart city characters' shall exist in the potential smart cities so as to build over it to a successful smart city. Selection of potential smart cities based on empirical data may help in identifying potential candidate smart cities that are feasible. It is also not important that there shall be large number of Smart Cities in India so that it propels Smart economic development. For example Mumbai, single handed contribute a lion share to GDP of India and that benefits national tax revenues and economic and regional development of Western India as well as India as a whole but it is not counted as one of the three smart cities that exists in India and 20 that exists in Asia. A small number of smart cities properly developed based on their potential can accelerate economic development of India than unsuccessful and failed attempt of arbitrary selection of smart city for development and investment based on equity among states. In this process of selection, the other cities not eligible are not totally left out, but can also follow the path of Smart City development when their potential qualifies for Smart City by directing appropriate efforts. These cities need to concentrate on the fundamental of Smart City potential such as millennium development goals, Graduate Enrolment Ratio (GER) and Human Development Index (HDI), which by no means are unimportant considerations in national development.

It may be re-emphasized that Smart city is a system with six components. They are Smart People, Smart Economy, Smart Mobility, Smart Environment, Smart Governance and Smart Living. There shall be strong indications in the selected city that these six elements exist in some form or other and there is great hope for further augmentation of these six systems. Among the six components, Smart People is the most important component of the six systems that creates Smart City which is the prime mover of smart city. It is the smart people which create smart communities and smart economy. In many Smart cities, Municipal website itself is the Smart Government where all Government services can be accessed 24 h and 7 days a week from mobile phones, home laptop, tablets or nearby computer centre such as "Akshaya" in Kerala. The first question is potential Smart City have Smart people which leads to Population Characteristics of Smart City. This is discussed in Sect. 2. In order to actively participate in E-Governance, GER of Smart Cities shall be increased rapidly than the existing low rate in India. The issues are presented in Sect. 3. Finally if one considers a set of variables related to potentials of some important components of Smart Cities, what shall be the potential smart cities in India is mapped below.

2 Population Characteristics for Smart City E-Governance

The population characteristics determine the potential for Smart City E-Governance. The state where smart city is to be located shall have higher than average HDI. HDI of India is 0.55 in two decimal points and states and union territories are sorted out from descending order of HDI to those just above the Indian average [1]. It gives 15 entries with Kerala on top with 0.92 and Tamil Nadu at bottom at sixteenth rank with 0.61. It is a contradiction in Tamil Nadu that HDI is low (0.61) but GER is one of highest (32.9 as against the National figure of 19.4) largely due to migration of students from other states to study there in private sector colleges and deemed to be universities. They are tabulated in Table 1. It can be seen that big states like Uttar Pradesh and Bihar do not appears in this list. Delhi which will be the second largest city in the world after Tokyo in 2050 is the eighth rank and the most industrialized Tamil Nadu is in the bottom sixteenth position in HDI ranking. Generally it is seen that higher the percent urban population higher is the HDI of the State. Percent literate and GERs do not show one to one relationship with HDI. GER is an important population characteristic for Smart Cities and Smart City E-Governance.

After HDI, the second factor that shall be considered is Percent of urban population in a state. Since smart city is an urban economy the state that envelops the city shall have a higher percent of urban population. When it was tabulated along with HDI, it is found that barring four states Himachal Pradesh (10.04), Nagaland (28.97), Sikkim (24.97) and Manipur (30.21) all others have more than average

Table 1 HDI, Urban %, Literate %, 2011 and graduate enrolment ratio 2011

S. No.	State/Union Territories	HDI	Urban (%)	Literate (%) 2011	GER 2012–2013
I	Kerala	0.92	47.72	94.00	21.90
II	Chandigarh	0.89	97.25	86.10	41.40
III	Lakshadweep	0.80	78.08	91.85	0.34
IV	Mizoram	0.79	51.51	91.30	21.60
V	Nagaland	0.77	28.97	79.60	21.50
VI	Andaman and Nicobar Islands	0.77	35.67	86.60	11.40
VII	Daman and Diu	0.75	75.16	87.10	3.50
VIII	Delhi	0.75	97.50	86.20	32.50
IX	Puducherry	0.75	68.31	85.90	31.20
X	Manipur	0.71	30.21	79.20	35.90
XI	Maharashtra	0.69	45.23	82.30	27.60
XII	Sikkim	0.68	24.97	81.40	24.20
XIII	Himachal Pradesh	0.65	10.04	82.80	26.00
XIV	Goa	0.62	62.17	88.70	33.20
XV	Punjab	0.61	37.49	75.80	19.40
XVI	Tamil Nadu	0.61	48.45	80.10	32.90
	India	0.55	31.16	74.00	19.40

Source References [1, 6]
Note 1 HDI Human development index. Ranking I, II, III, IV, refers to ranking of HDI of Indian states and Union Territories
2 GER Graduate enrolment ratio

percent urban population of India (31.16 %). There seems to be a correlation with percent urban and HDI [2, 3]. The third factor is in fact is e-literacy for all. Statistics on e-literacy is not available. Barring Kerala, there is no significant achievement of creating full e-literacy in any state in India. In the absence of e-literacy statistics, percent literate is also considered. It is found that all the states selected based on HDI also satisfies with above average India percentage literacy which is 74 % in 2011. The fourth important factor is GER in percent, [4, 5]. The GER is computed on Graduate enrolled in University and colleges divided by college going age population of 18–22 group of the State. In the year 2012–2013, the country average India GER was 19.4 % and all states in Table 1 barring Lakshadweep (9.4) and Damn Diu (3.5) are having above 19.4 % GER. This low figure may be due to the fact that college going age group may be using facility in Kerala for Lakshadweep and Mumbai for Damn and Diu instead local colleges which may not have threshold to sustain large universities for all their needs. These four indicators are surrogates for Smart people in India. It is assumed that higher GER and percent literate may lead to higher e-literacy. Table 1 tabulates these indicators for 16 states and union territories based on descending order of HDI and Table 2 tabulates the existence of mega cities and million plus cities based on Table 1.

Table 2 Potential smart metropolises (mega cities and million plus agglomeration) 2011

S. No.	State/Union Territories	Mega cities 2011	Million plus urban agglomeration 2011(descending order of population per State and Union Territories)
I	Kerala	None	Kochi, Kozhikode, Thrissur, Malappuram, Thiruvananthapuram, Kannur, Kollam
II	Chandigarh	None	Chandigarh
III	Lakshadweep	None	None
IV	Mizoram	None	None
V	Nagaland	None	None
VI	Andaman and Nicobar Islands	None	None
VII	Daman and Diu	None	None
VIII	Delhi	Delhi	None
IX	Pondicherry	None	Puducherry
X	Manipur	None	None
XI	Maharashtra	Greater Mumbai	Pune, Nagpur, Nashik, Vasai Vihar City, Aurangabad
XII	Sikkim	None	None
XIII	Himachal Pradesh	None	None
XIV	Goa	None	None
XV	Punjab	None	Ludhiana, Amritsar,
XVI	Tamil Nadu	Chennai	Coimbatore, Madurai, Thiruchirapalli

Source Table x.1, Census of 2011
Note 1 Mega cities are five million and above
2 Ranking in descending order: I, II, III, IV, … refers to ranking of HDI of Indian states and Union Territories

Column three is the name of selected mega cities (five million and above population) in 2011 namely Greater Mumbai, Delhi and Chennai. Kolkata do not figure here. The fourth columns are million plus agglomerations in these 16 states and union territories. There are 19 million plus agglomeration selected for smart city up gradation and remaining 9 cities are not million plus or mega cities. Hence there is scope for developing 31 settlements out of 7,925 census towns to be upgraded to Smart cities in India with differing city size. The others form 100 can be Greenfield Smart Cities in the industrial and freight corridors or cities around new industrial estates and SEZ areas. Alternatively some of the million plus agglomerations can have more than one smart cities centred on a dominant activity like Cyberabad in Hyderabad or Smart City in Kochi.

Figure 1 plots three mega cities and 19 million plus agglomeration in India map. These exclude all other nine settlements below one million populations.

It can be seen that potential smart cities are mostly located in southern India and western India. They are the developed part of India as of today. For balanced Smart City development in the east, Smart new towns with probable population

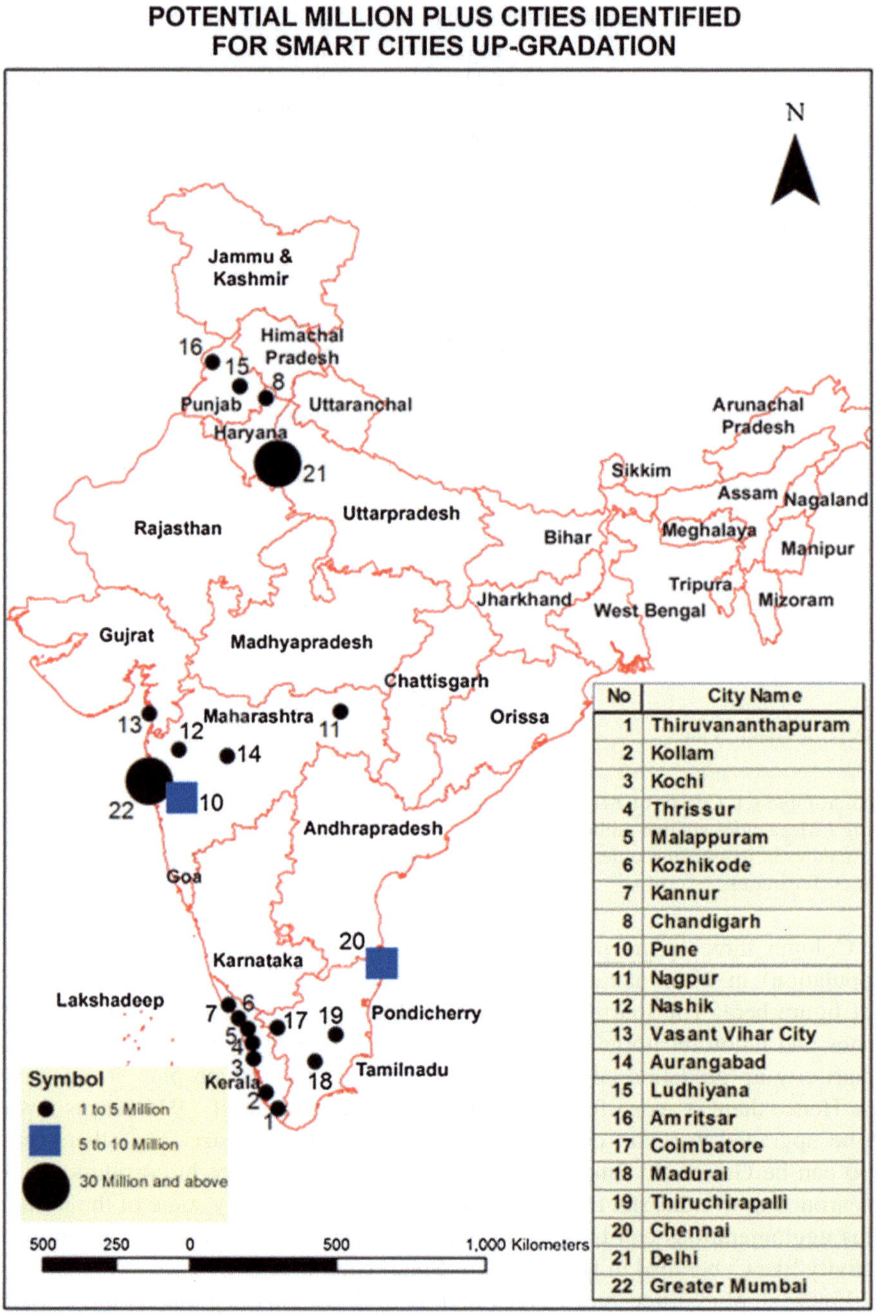

Fig. 1 Potential million plus cities identified for smart cities up-gradation

base from outside can be developed in Eastern parts of India similar to Delhi Mumbai Industrial Corridor or a system of Smart New Towns in the proposed Asian highway in the north east that connects to South East Asia. Therefore, building new smart city say in some industrial corridors suites better in Eastern Region of India than up-gradation of existing towns to smart cities.

3 Graduate Enrolment Ratio

GER is important for Smart City and Smart E-Governance for effective functioning. As indicated, GER is generally expressed in percentage and it is number enrolled in college divided by college going age group. Here PhD group above 22 is not included. Based on National Sample Survey (NSS) 61st round corresponding to year 2004–2005 and NSS 64th round corresponding to year 2007–2006 GER % has been tabulated in Table 3 for scheduled caste (SC), scheduled tribe (ST), other backward caste (OBC), Muslims and Non-Muslims. It can be seen that SC, ST, OBC and Muslims have GER much less than half in comparison with other higher castes. The position of ST and SC are only one-third of the higher caste. Unless there is a policy to increase GER of SC, ST and Muslim population of India they cannot participate effectively in the E-Democracy the fundamental basis of E-Smart City Governance.

Table 3 summarizes GER as related to NSS 61st and 64th rounds. It can be seen that total GER increased from 14.19 to 17.21 % during these four years periods which is a reflection of more investment in college education by India during this period by establishing more centres of higher education. Non-Muslim GER was much higher (15 and 19 %) than Muslim (9 and 10 %) and also GER for Other Backward Class OBC (11 and 15 %). GER was much lower for ST (8 and

Table 3 Enrolments and GER (18–22 years)

	NSS 61st round (2004–2005)		NSS 64th round (2007–2008)	
	Enrolment (000)	GER (%)	Enrolment (000)	GER (%)
(a) *General and reserved categories*				
Schedule Case (Sc)	1,898.5	8.72	2,485.5	11.54
Schedule Tribe (ST)	767	8.44	652	7.67
Other backward castes (OBC)	5,027.4	11.48	6,599.6	14.72
Others	7,787.2	22.52	8,886.6	26.64
Total	15,480.1	14.19	18,623.7	17.21
(b) *General and minorities*				
Muslims	1,308.8	8.5	1,521.4	9.51
Non-Muslims	14,170.9	15.1	17,102.4	18.54
Total	15,479.7	14.19	18,623.8	17.21

Source NSS 61st and 64th rounds

Table 4 GER rural and urban areas
Source NSS 61st and 64th rounds

	Rural	Urban
NSS 61st round (2004–2005)	8.42	16.18
NSSA 64th round (2007–2008)	11.06	19.03

8 % stagnant) and was more or less higher for Schedule Caste SC (9 and 12 %) in comparison with ST. Since progress in the two NSS period is not satisfactory, it calls for target oriented GER development policies for such groups.

Table 4 computes GER urban and rural for the period 2004–2005 and 2007–2008. On the NSS 61st round corresponding to 2004–2005, Urban GER was 16 % while rural was only 8 %. There was increase in these values in 2007–2008 and rural became 11 % and urban 19 %. These figures shows considerable effort is required to increase this percentage to higher.

It can be seen that urban figure was 16.18 and 19.03 % in these two periods. Hence it can be safely concluded that smart cities selected will have higher than those figures and there still be problems of SC, ST, OBC and Muslims getting lower scores in smart cities. Extra policy measures are required to bring up GER of these groups if at all they can be part of Smart City R-Governance.

Figure 2 shows growth of GER from 1950 to 2013 in India. There has been greater growth during 2011–2012 than in the past. It was a dismal figure during 1950–1951 soon after independence of India. It was between 10 and 15 during 2006–2012. It became nearly 20 in 2012–2013. However what India has achieved is small in comparison with other countries which are shown in Fig. 3.

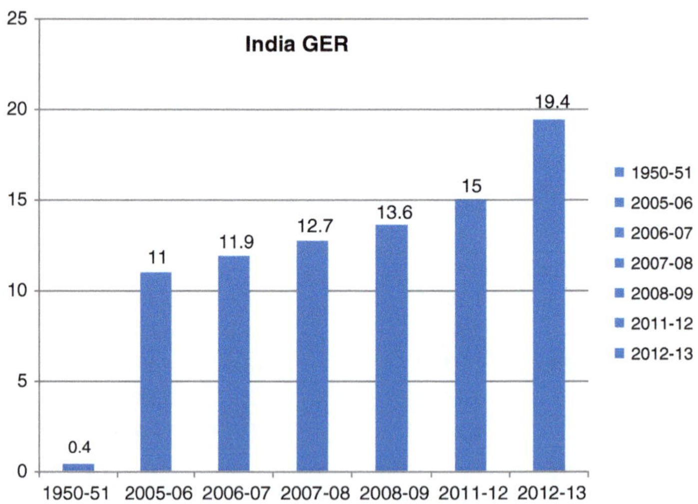

Fig. 2 Graduate enrolment ratio 1950–2013. *Source* Government of India and Confederation of Indian Industries. "ASHE 2013 annual status of higher education of states and UTs in India", Deloittee November 2013

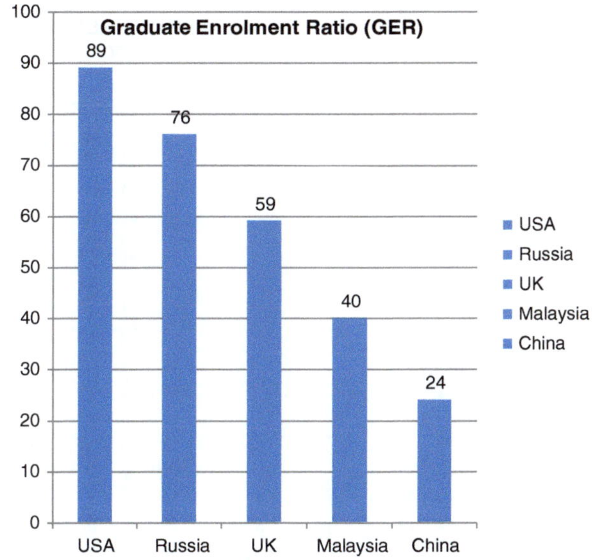

Fig. 3 Graduate enrolment ratios of some countries. *Source* Government of India and Confederation of Indian Industries "ASHE 2013 annual status of higher education of states and UTs in India", Deloittee November 2013

Figure 3 shows GER of USA (89 %), Russia (76 %), UK (around 59 %), Malaysia (40 %) and China (24 %) and can be seen that they are much higher than India (less than 20 %) [6].

Figure 4 shows disaggregated data of GER by different social groups in India for the year 2009–2010. This shows advancement in values from 2004 to 2005 and 2007 to 2008, presented in Table 4. However, disparity of urban females with respect to urban male is more. It is also seen that economically and socially backward SC, ST, OBC has not seen much advancement in GER during these two periods. For example total urban GER is about 35 %. It can be seen that GER urban total is two times higher than Total GER of all India in 2009–2010. Other social groups have also increased than 2004 and 2007 figures given above. It can now be safely concluded that GER shall not be a limiting factor for 22 selected Smart cities and nine small settlements even in the absence of city specific data.

Linear Regression relationship between GDP and GER is captured in Fig. 5 as scatter map shows relatively lower position of India. It is lowest among countries considered and there is distinct cluster emerging between developed and developing countries. It also shows that higher the GER higher shall be the GDP or vice versa.

While GER is only one variable that augments Smart city E-Government of a population, the real availability of ICT infrastructure and access to Internet in India need to be studied.

For participatory Smart City E-Governance, ICT infrastructure shall be developed and easily available. This is discussed in Sect. 4, constraints in Sect. 5, outlook in Sect. 6.

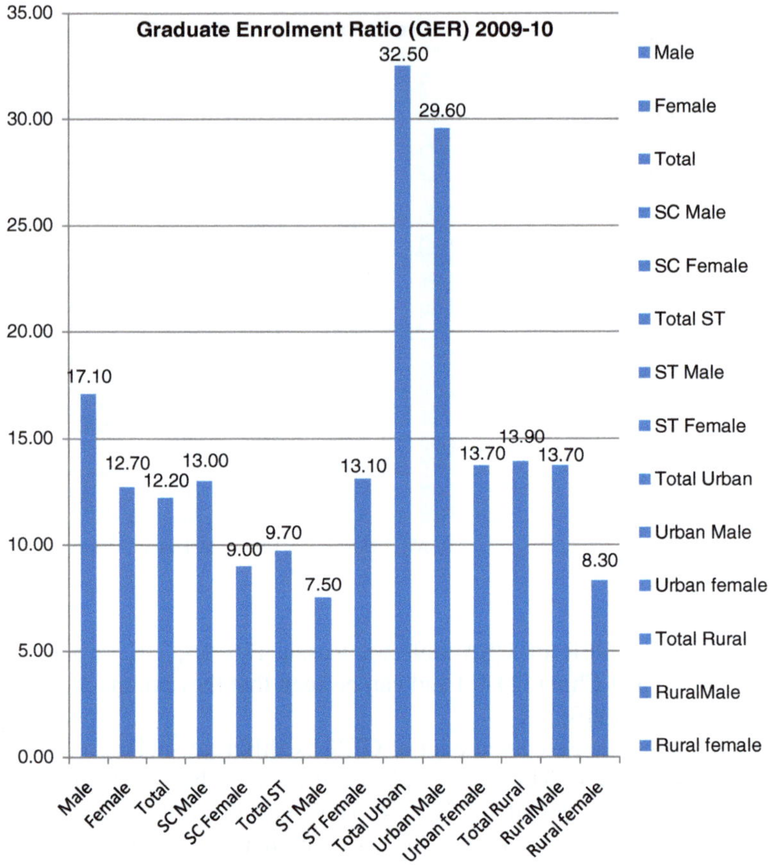

Fig. 4 GER 2009–2010 by Classes. *Source* Government of India and Confederation of Indian Industries "ASHE 2013 annual status of higher education of states and UTs in India", Deloittee November 2013

4 Status of Telecom and Internet in India

Status of telecom and Internet subscribers in India as on 31-3-2013 is given in Table 5 [7].

It can be seen that urban tele-density achieved as on 31-3-2013 in India is as high as 146.96 for a combination of wireless and wirelines. This is very encouraging. This means endowment of ICT in Smart City is feasible to all selected potential smart cities. All India and rural figures of tele-density are as low as 73 and 41 % respectively. Tele density has been achieved largely by private sector in India who contributes to as high as 86 % share of the telephone and wireless communication market. Private sector is responsible for developing competitiveness in tele market and progressively reduced the cost of service to consumers. In no case it can be achieved by public sector.

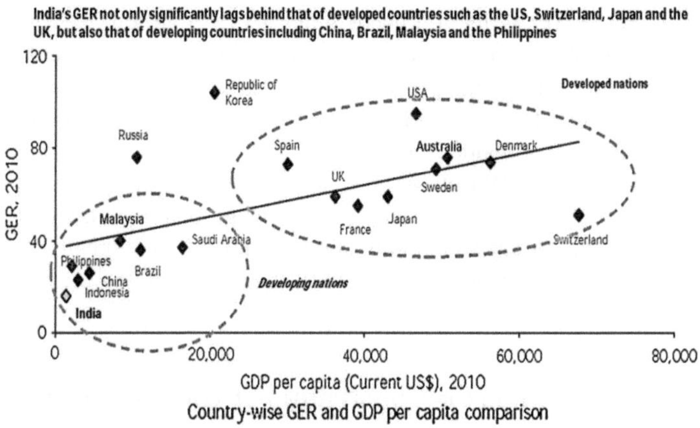

Fig. 5 Country wise GDP and GER relationship

Table 5 Status of telecommunication and Internet in India as on 31-3-2013

Telecom subscribers (wireless + wireline) 31-3-2013	
Total subscribers	898.02 million
% change over the previous quarter	0.28 %
Urban subscribers	548.80 million
Rural subscribers	349.22 million
Market share of private operators	85.51 %
Market share of PSU operators	14.49 %
Tele-density	73.32
Urban tele-density	146.96
Rural tele-density	41.02
Internet/broadband subscribers 31-3-2013	
Total Internet subscribers	164.81 million
Narrowband subscribers (excluding subscribers who accessed Internet through wireless phones)	6.56 million
Broadband subscribers (excluding subscribers who accessed Internet through wireless phones)	15.05 million
Subscribers who accessed Internet through wireless phones	143.20 million

Source Reference [7]

The Internet and broadband subscribers as on 31-3-2013 is given in Table 5. Of a total of 165 million broadband subscribers, there are only 15 % broadband subscribers and 143 % wireless subscribers. Hence there is high potential for "M-Government" or "T (Tablet)-Government" than broadband Internet-based government using PC, Linux or Mac by consumers in India. However overall Internet connectivity is not encouraging and needs much more attention. E-Government in India is more desktop/laptop oriented today than Mobile phone oriented and this need to be corrected by making apps oriented to small Mobile phone screens. It is also important to move towards 4 G systems and consumer cost for Internet services need to be lowered considerably [8].

However it is likely that 19 million plus cities and three mega cities identified for Smart City Up gradation, may have higher percentage of tele-density and Internet connectivity more than urban and country figures and full city specific data is unavailable. This may not be the case of other 9 less than one million cities selected.

5 McKinsey's Findings on Constraints of Internet in India

The impact of the Internet in India is constrained by obstacles and gaps in the Internet ecosystem compared with many peer countries [9]. On most indicators of the strength of the Internet's foundation (i.e., the preconditions for future growth), India ranks in the bottom quartile. The exceptions are human capital, where India has a large pool of technically trained workers, and the reasonably well-developed access to finance. On most other dimensions, India's Internet infrastructure and e-engagement levels are limited, and its Internet foundations have significant scope to improve. India's low levels of user adoption and engagement are due to the following obstacles:

Limited availability of Internet infrastructure: India has low rank on Internet infrastructure and environment. For example, India has only approximately 6 % of the number of secure Internet servers per capita than Brazil or South Africa has. Average bandwidth per capita in India is significantly lower than in many other aspiring countries. The penetration of PCs is only 47 per 1,000 people, which is much lower than in Argentina, Mexico, the Philippines, or Vietnam. Internet penetration among India's large rural population is just one-twelfth that of the urban population. Low availability of basic infrastructure, such as reliable electricity supply, is a key bottleneck in rural areas.

High cost of access and usage: At $61 per Mbps (on a PPP basis), India has one of the highest median costs of broadband access among comparable aspiring countries—more than four times that of China, Brazil and Argentina, and 20–30 % higher than that of Vietnam and Malaysia.

Lack of awareness and low digital literacy: Only 35 % of businesses in India offered online services such as Web presence, compared with an average of 56 % in aspiring countries. In an online survey of India's SMEs in the organized sector, they cited the lack of education on using the Internet as among the top three reasons that prevent consumers from using the Internet.

Narrow range of applications and services: Internet applications are yet to scale up in a wide range of areas that impact society, such as agriculture, education, health care, and citizen services. Access to online government services across the country is low, with a large quantity of government data, such as land or health records, yet to be digitized and large flagship Internet infrastructure projects such as the National Optical Fibre Network yet to become fully operational. India is ranked in the bottom quartile on government e-participation index, which measures the breadth and usage of online services offered by the government.

6 McKinsey's Findings on Internet Outlook for India

McKinsey's offers seven key findings concerning the impact of and outlook for the Internet in India [9]. However, major conclusions for E-Governance of India below are that of the author.

1. India's base of about 137 million Internet users in 2013 is the third-largest in the world in absolute terms. The Internet's role in communication, social networking, and informing and influencing India's consumers in categories such as apparel, books, financial services, and travel is already comparable with that of developed countries. However in comparison with other countries Internet penetration in India is very low. Figure 6 gives the comparative picture.
   India's users spend less time online per capita than users in developed countries, a reflection of high cost per use or even quality of service including poor bandwidth. Table 6 gives usage statistics of Ernst and Young which may be considered as of lower side [10].

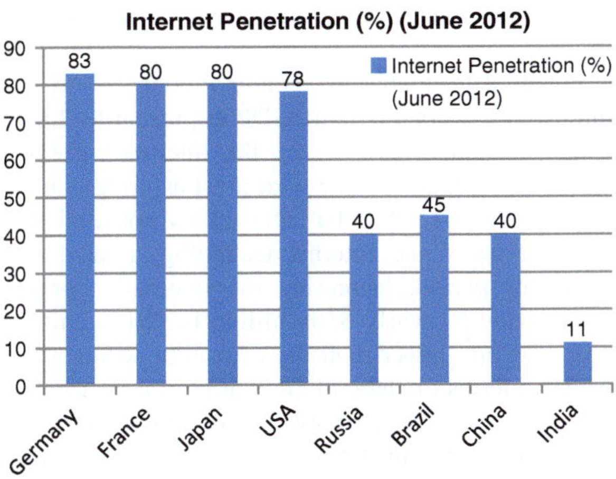

Fig. 6 Internet penetration by selected countries (2012). *Source* Reference [8]

Table 6 Average hours per month Internet use by person by year

Year	2006	2011	2012	2013	2014	2015
Hours per month/person	12.6	17.4	18.9	19.2F	20.4F	21.1F

Source Earnest and Young (2012) "Rebirth of E-Commerce in India": New Delhi
Note F Forecast

Table 7 User mode of access of Internet (%) by year

Year	2007	2008	2009	2010	2011	2012	2013	2014	2015
Work	34	38	40	29	32	14	10F	8F	6F
Home	30	26	23	37	32	24	19F	16F	14F
Cybercafe						19	13F	9F	7F
Mobiles	36	36	37	34	27	43	58F	67F	73F
Total	100	100	100	100	100	100	100	100	100

Source India Brand Equity Foundation (2012) The rise and rise of E-Commerce in India: New Delhi quoted from World Bank, IAMA and ARNACA Research
Note F = Forecast

2. India is likely to have the second-largest user base in the world, and the largest in terms of incremental growth, with 330–370 million Internet users in 2015. Table 7 gives the percentage access by mode for Internet in India from 2007 to 2015 and it can be seen that use of mobile to access Internet is increasing.
   Given current downward trends in the costs of Internet access and mobile devices, India is on the verge of an Internet boom. In an evolution pattern unique to India, users who access the Internet only through a mobile or tablet device will constitute around 73–75 % of new users and 55 %of the aggregate user base in 2015, leading to increasing demand for content that is optimized for a small screen. E-Governance portals in India however are not optimized for small screen mobile phones and tablets. Banking sites which normally get interlinked with E-Governance portal however is moving towards small screen mobile screen.
3. India has the potential to double its economic contribution from the Internet in the next three years, from 1.6 % of GDP at present to 2.8–3.3 % by 2015. Despite the large current base of users, the Internet currently contributes a modest 1.6 % to India's GDP, in line with most aspiring countries. This could grow to 2.8–3.3 % by 2015 if India achieves its potential for growth in the number of Internet users and Internet technology-related consumption and investment over this period, increasing the Internet's contribution to GDP from $30 billion today to nearly $100 billion in 2015. This would make the Internet-related economy larger than the education sector and as large as the healthcare sector, in terms of share of GDP at present. Currently, India's information and communication technology (ICT) exports are the most significant component of the Internet's impact on GDP. But private consumption, private investment and public investment have greater potential to grow in future. Telecommunication and Internet investment in India is largely (85 %) private sector investment but E-Governance portal is largely sustained by public sector. There is a need to open up this activity for private sector in tune with the E-Commerce ecosystem. Advancement in E-Governance can be another factor that increases the percent GDP contribution by Internet.
4. The impact of the Internet in India is constrained by current gaps and obstacles in the Internet ecosystem. This gap what is faced by E-Commerce is also felt by E-Governance. While India scores well on the availability of human and financial

capital, but it rates poorly on Internet infrastructure, Internet engagement, the E-Commerce/E-Governance platform, the ease for Internet entrepreneurship that can impact on E-Governance.

5. Although the Internet ecosystem is becoming more vibrant, the benefits have been relatively concentrated. India's Internet start-ups are scaling up through creative adaptations to overcome infrastructural and systemic bottlenecks. Yet, while large enterprises have gained from their early adoption of the Internet, there is scope among individual consumers, SMEs and the government sector to significantly increase engagement. Today, India's measurable consumer surplus from the Internet is estimated at $9 per user per month, at the low end of the range for aspiring countries ($9–$26) and well below the range for developed countries ($18–$28). Even by 2015, with overall Internet penetration likely to reach 28 %, rural penetration is likely to be just 9 %.
6. India can achieve broad-based Internet impact by aiming for the digital inclusion of nearly 40 % of its population, to reach a user base of 500 million by 2015, rather than the likely target of 330–370 million. What we have seen in terms of GER % for different social group seems to be repeated to Internet access to different social groups. Most of the additional 150 million to 160 million users would be individuals and small businesses in semi-urban and rural parts of the country. Extending Internet access to these segments of the population, and promoting the usage of many more online services, would enable India to derive much more of the intended benefits from government programmes of inclusive growth in employment, education, health care, nutrition, and financial services. A proactive policy approach like free gift of mobile phones and monthly payments for mobile use by Government to those below poverty level and SC, ST and OBC and free laptop or tablets to all students will help a great deal in this direction. Some of these are already happening but evaluation of the impact is not studied.
7. Concerted actions by policy makers and businesses can help India achieve an inclusive Internet transformation; reduce the cost of Internet access across devices, content and applications; increase access to low-cost, high-speed connectivity in rural and semi-urban India beyond the top cities; promote widespread digital literacy through the introduction of devices and content tailored to the local context; devise Internet applications in new areas such as agriculture, health care, education, energy, utilities, and public information; and create a more favourable business environment for Internet entrepreneurs to support rapid innovation.

Next Variables that constitutes E-Governance index is presented in Sect. 7.

7 E-Government Index

Figure 7 shows the set of variables that makes an E-Government Index which is computed for member nations by United Nations periodically [11, 12]. E-Government development index is a composite measurement of the capacity

E-government development index is a composite measurement of the capacity and willingness of countries to use e-government for ICT-led development

E- government development index

Based on a comprehensive survey of the online presence of all 193 UN member states, which assesses the technical features of national Web sites as well as e-government policies and strategies applied in general and by specific sectors for delivery of essential services

It is calculated as a weighted average of three normalized scores of the most important dimensions of e-government: scope and quality of online services, development status of telecommunications infrastructure, and inherent human capital

Online services

- How extensive the online presence is
- How good/bad the Web sites are (design and user friendliness, extent of content offered)
- Score is equal to the actual total score less the lowest score divided by range of total score values for all countries

Telecommunications infrastructure

- Internet users
- Number of fixed telephone lines
- Number of mobile subscribers
- Number of fixed Internet subscriptions
- Number of fixed broadband facilities

Human capital

- Adult literacy rate (⅔ weight)
- Combined primary, secondary, and tertiary gross enrolment ratio (⅓ weight)

SOURCE: ITU; UNESCO; UNDP Human Development Report, World Bank; UNCTAD; IMF; UN; McKinsey analysis

Fig. 7 Measuring E-Government development index

and willingness of countries to use E-Government for ICT-led development. Essentially three groups of variables are used. They are on line services, telecommunication Infrastructure and Human Capital. Further details are as given in Fig. 7. United Nations brings about such statistics for 193 UN member states periodically.

The basic criteria for computing E-Governance Index by United Nations are detailed out in Fig. 7.

Table 8 shows United Nations E-Government Index of Southern Asian Countries in 2012 and 2010. In 2010 India ranked at 119 at the bottom and deteriorated further to 125 out of 193 in 2012. Countries like Maldives, Iran and Sri Lanka have better ranking than India. The worst index as expected was that of Afghanistan. Considerable effort is required by Government of India and State Governments to achieve better E-Governance Index. This country specific Index need not deter us in developing Smart city E-Governance in some limited number of cities.

E-Commerce penetration and E-Government status seems to be interconnected for potential smart cities in India, since ecosystem of both E-Governance and E-Commerce are one and the same. Wherever E-Commerce is thriving, there can also be great progress in E-Government since same IT technology is used in these two cases. E-Commerce statistics 2013 is given in Table 9 [13].

Table 8 E-Government index of Southern Asia: E-Government development in Southern Asia

Country	E-Government development index		World E-Government development ranking	
	2012	2010	2012	2010
Maldives	0.4994	0.4392	95	92
Iran	0.4876	0.4234	100	102
Sri Lanka	0.4357	0.3995	115	111
India	0.3829	0.3567	125	119
Bangladesh	0.2991	0.3028	150	134
Bhutan	0.2942	0.2589	152	152
Pakistan	0.2823	0.2755	156	146
Nepal	0.2664	0.2568	164	153
Afghanistan	0.1701	0.2098	184	168
Sub regional average	0.3464	0.3248		
World average	0.4882	0.4406		

Source United Nations Department of Social and Economic Affairs (2012) E-Government survey 2012, E-Government for the people, New York

Table 9 E-Commerce statistics

	India	USA	China	Australia	Brazil	Sri Lanka	Pakistan
Internet users (millions)	137	245	538	20	79	3.2	29
Penetration (%)	11	78	40	89	40	15	15
Online buyers(million)	25	156	270	11	27	2	NA
Online buyers as % of Internet users	18	64	50	55	34	63	NA
Consumer E-Commerce	13	224	210	30	19	2	4

Source Internet World Statistics 2013: KPMG and Internet and mobile association of India report

Here development of E-Commerce points towards the future possibility of E-Governance. It can be seen that India ranks much lower for per cent of Internet users (137 million) online buyers (18 %) and Consumer E-Commerce (13 %) in comparison with China (50, 210 %), US (64, 224 %), Brazil (34, 19 %) and Sri Lanka (63, 2 %). Based on current status of Telecommunication and Internet usage in India, it is important to extrapolate the potential of India based on expert views. The view of expert is more oriented towards E-Commerce but it is also valid for E-Governance.

Participation of Women in E-Commerce in India is noteworthy and since the ecosystem of E-Commerce and E-Governance are same, there is scope for more participation of women in E-Governance which is discussed in Sect. 8. To elicit more women participation there shall be separate section for women in the E-Governance web portal highlighting their governance issues.

8 Contribution of Women in E-Commerce in India

Times of India (TOI) dated 27-3-2014 reports that Indian women fuelled online shopping worth over half-a-billion dollars last calendar year and that figure is galloping five-fold to $3 billion in the next three years. Women-influenced sales would be 35 % of Indian E-Commerce market estimated at $8.5 billion by 2016, as reported to TOI by venture capital firm Accel Partners, one of the prolific backers of start-ups. These projections come in the backdrop of a frenetic growth in Internet penetration through smartphones and women lapping up the convenience of shopping online. Indian E-Commerce industry totalled $2 billion last calendar, with women-influenced sales at $511 million, or at 26 % of overall market. The huge contribution by women to the E-Commerce market—excluding travel and ticketing—contrasts the brick-and-mortar retailers, who derive most sales from men even in categories like fashion and apparel.

In the year 2013, fashion E-Commerce sales doubled to $559 million compared to the previous year and will potentially go up about $2.8 billion by 2016 rivalling electronics and mobiles which have traditionally been the biggest draw online. "Women are emerging as a strong customer base for us, having grown from contributing 20 % to our overall sales two years back to now clocking 40 % of sales. The offline fashion and apparel industry has been dominated by offerings for men. We expect women will contribute 50–55 % sales in the next two years as we increase our private-label presence," said Mukesh Bansal, co-founder of Myntra. Facebook investor Accel said working women clicking online increased 43 % and formed roughly 10 % of the active Internet user base in India.

Prashanth Prakash, partner at Accel Partners, said Indian women are embracing the concept of fast fashion online which they did not do so far offline. "Apparel, jewellery, home decor and baby products are some of the retail categories which have seen exponential growth in Indian E-Commerce in last two years," he said.

Domestic E-Commerce will witness a compounded annual growth of 63 % in the next three years considering the huge headroom for growth. Consider this: online sales are just 1 % of the $43 billion fashion, accessories and footwear market; 0.2 % of the $45 billion jewellery market; 7 % of the book sales and 4 % of mobile phone and tablet shipments.

"Private devices like smartphones and tablets are taking more women to E-Commerce. This is more relevant to private purchase categories like lingerie, which is shifting online in a big way. These devices also provide them an option to indulge in recreational, relaxed shopping," said Richa Kar, founder of Zivame, an online lingerie store which raised $10 million in funding.

About 9 % of domestic E-Commerce traffic came through mobile devices last year, reporting an eight-fold jump. Mobile-driven shopping stood at 10 % of China's $106-billion E-Commerce industry. The rapid adoption of mobiles for online shopping will open up newer forms of payment options. Use of mobile wallets, popular in China, is likely to gain momentum here too, the report said. India's Internet users will double to 400 m in 3 years 70 % new users are added through mobile devices Core online shoppers would be roughly 10 % of the Internet base by 2016 their average purchase value to double from 1,860 in 3 years.

Any achievement in E-Commerce in India will also influence positively E-Governance since they share the same ecosystem. It is likely women's role in E-Commerce will reflect in their active role in E-Governance.

Although, there is a difference between E-Governance and E-Commerce in India, E-commerce technology can be used profitably in E-Governance. A good example is big data management. This synergy can be mutually beneficial for E-commerce and E-Governance, if E-Commerce firms are given role of mentor in E-Governance of Smart cities on a profit sharing basis based on well worked out policy frame. Tablet and smartphones are Indian accelerators for E-Commerce and this shall be accepted also for E-Governance. The present practice of desktop oriented E-Governance shall be replaced with mobile phone or tablet-based M-Governance keeping both systems available.

After analysing potential and constraints of E-Government development in India and its future, it is important to summarize the contribution of this book in exploring further reaches of E-Governance practice in India.

Some aspects of Smart City E-Governance futures presented in various chapters is presented in Sect. 9.

9 E-Governance for Smart Cities

T.M. Vinod Kumar in the first chapter "E-Governance for Smart Cities" considers E-Smart City Government as a web page with capability to transact all municipal services with Citizen (C2G), Business (B2G and B2C) and Government (G2G, G2C and G2B). All complex transaction which may take many days and many visits to different offices related to Municipal Government can be undertaken using laptop, tablet or mobile phones and the website without any unnecessary travel and all governmental services can be accomplished within a short time. This service is available 24 h and 7 days a week for citizen for any aspect of services related to birth, marriage or death or more. Such services exist in many emerging smart cities, but do not encompass all possibilities. Hence in India we do not have the required web-based E-Government for Smart Cities. Figure 8 gives all such possibilities in the evolution of this website in many stages of development.

In Fig. 8, Stage 1 Municipality merely uses its website for information publishing. Information can be Master Plan, Zonal Plan, Urban development codes, Urban Land Management Schemes, Form-Based Codes, Municipal Orders raising house tax by locality, Municipal ordinances or Municipal laws and amendments. It can also be email id, phone numbers and office address of Mayor and all other political and administrative functionaries. It may be possible that all such information is not published but only few selected information. Such website is available in most of the municipalities in India in some form or other in a less comprehensive way. There are reports of non-maintenance and updating of these Municipal websites. This is

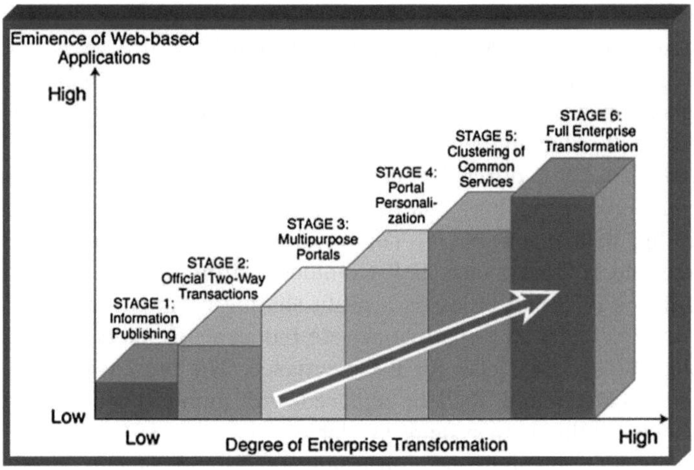

Fig. 8 E-Governance web-based applications

because there is no reliable structure to support and update information in the website. Here, the eminence of web application is very low since no interaction and transaction of municipal business is possible using website. In the second stage, two way transactions are made possible in a simple and rudimentary form. For example an application to get licence to establish an advertisement panel in the road side can be executed here by filling forms and paying fees electronically. House tax, water bill and electricity bills can be paid this way. The website redirects, the citizen to designated banks secure website to pay fees. This stage can be called official two way transaction with security of transaction. In the third stage a multipurpose portal is made available by municipality which can perform a large number of municipal transactions involving many different government agencies and departments in combination without moving from one website to another. Possession certificate for a newly constructed home is a good example for this stage where certificate from many government agencies such as electricity department, water supply department and municipality are required to occupy the house. In the fourth stage Portal Personalization by citizen is possible. Here citizen easily construct a municipal website for his specific use by incorporating only services on a page, he is interested to use periodically by eliminating all others which he may not use at all. He can add other municipal services without much difficulty later. In stage five citizens can cluster services to two or many. So he can have a cluster for all interaction with municipality as a citizen and another as shop keeper which is his main occupation. Municipality can also cluster services based on some logic with no input from users. In the stage 6, full enterprise transformation of web take place incorporating all line agencies and all municipal functions seamlessly on website to serve citizen, business and government. Here citizen is taken to as many domains he wants to use as by this predesigned clustering based on logic and intelligence. He can make a structured query to get into that part of municipality enterprise website he wants to access seamlessly. The aim of the Municipal Web is to serve citizen from his birth to death and even beyond from the same web portal. For example if a parent wants to enrol their daughter to nursery schools, he may be guided by the Municipal enterprise website from all steps starting with identification, determining distance to school, availability of transport, quality of education, fee structures and so on and enrolling his daughter and pay fees without moving outside their home. Stage 6 does not exist in India and it has to be achieved by all smart cities.

The author recommends four alternative models to achieve Smart City E-Governance in India. In Model-1, smart governance is just the governance of 'an aspiring smart city' like any other type of city with no transformation in existing practice of Governance is envisaged. Here there is only a marginal extension of existing E-Governance practice which may not be very much visible and unique. Model 2-advocates low level of transformation which is not even apparent to accommodate capacity for innovative decision making and implementation process. It is not a complete structural change in Governance at all but marginal. It is supported by multi-level training programmes by institutions such as Kerala Institute of Local Administration (KILA) which trains political and administrative executives in different levels. Since city is a spatial entity, location-based decision assumes importance. Geographic Information System (GIS) allows for developing Spatial Decision Support System (SDSS) which can be used for variety of tasks in Governance. This requires periodic training.

SCADA (supervisory control and data acquisition) is a system operating with coded signals over communication channels so as to provide control of remote equipment (using typically one communication channel per remote station). The supervisory system may be combined with a data acquisition system by adding the use of coded signals over communication channels to acquire information about the status of the remote equipment for display or for recording functions. It is a type of industrial control system (ICS). Industrial control systems are computer-based systems that monitor and control industrial processes that exist in the physical world. SCADA systems historically distinguish themselves from other ICS systems by being large-scale processes that can include multiple sites, and large distances. These processes include industrial, infrastructure, and facility-based processes, as described below:

- Industrial processes include those of manufacturing, production, power generation, fabrication, and refining, and may run in continuous, batch, repetitive, or discrete modes.
- Infrastructure processes may be public or private, and include water treatment and distribution, wastewater collection and treatment, oil and gas pipelines, electrical power transmission and distribution, wind farms, civil defense siren systems, and large communication systems.
- Facility processes occur both in public facilities and private ones, including buildings, airports, ships, and space stations. They monitor and control heating, ventilation, and air conditioning systems (HVAC), access, and energy consumption.

SCADA can be used for governing power supply or water supply or even traffic management.

The third model believes that smart governance is all about creating smart administration. Smart Administrator shall be well versed in Indian Constitution and Nitti gritty of innumerable legislations of Union and State Government. He should be in a position to work with people, get participation of people, elected political executives and local business, using his social engineering skills. He shall be an expert in E-Governance, big data processing and data security. He should be having capability for speedy decision making and speedy implementation using modern E-Governance practices and, project management tools like Primavera, SDSS and real time responsive systems like SCADA and SDSS. Hence this model incorporates Model 2 and of course Model 1. He should have very high professional capability in ICT with academic attainments and facility shall be there to renew his ICT capabilities by annual training programmes. Big data, spatial data processing and data security shall be his concerns. He should have supporting staff in creation of database which can be spatial or non-spatial. He should have capability to collect revenues as per law and spend budget in a timely fashion using E-Governance tools. He will automate all rule-based administrative decision using SDSS and SCADA through E-Governance leaving only little important decision to be made by him.

The fourth Model of Smart Governance advocates that smart governance is all about rearranging the position of government within the urban system as dictated by the urban system itself. This involves the highest level of transformation since it is not only about the transformation of the internal organization by smart governance (vide Model 3), but also of the external organization that connects Smart city with other cities inside or outside the State or Nation. In this model all economic activity are community driven transcending national boundaries and state boundaries. Autonomy to articulate this community action transcending national and state boundaries are required and calls for amending constitution by introducing concurrent list incorporating smart cities. This establishes a highly responsive urban system of smart city with legal interacting capabilities with external cities even if it is not part of the same nation. In this model, one can question, how Smart City Bangalore, Hyderabad or Jaipur be part of a State for Governance on the basis of linguistic division of State.

10 E-Governance State of the Art Survey: Stuttgart, Germany

Satyendra Singh surveys the state of the art of E-Governance in Stuttgart. Stuttgart is a Smart City in Germany. While Stuttgart city is over half a million in population size, Stuttgart metropolitan region is more than 5 million. A deliberate promotion of E-Governance in all walks of life was intimated by European Union. Stuttgart went ahead adopting E-Governance for Smart Living appropriate to a Smart City. E-Governance has to be participatory and transparent. For participation E-Democracy is to be strengthened. E-Governance of Stuttgart aimed at higher participation than exists today and total transparency and used E-Democracy technologies like E-voting and many others to strengthen E-democracy. Within the scope of a one chapter in a book, author very briefly covers salient aspect of E-Governance practiced and showed the security elements involved. Health smart card and E-Governance for Smart mobility was highlighted in his chapter. The European Innovation Partnership for Smart Cities and Smart Communities combines ICT, energy management and transport management to come up with innovative solutions to the major environmental, societal and health challenges facing European cities today. Thereby, E-Governance covers most aspect of Smart city. Germany's current E-Government 2.0 programme framework which is conceptualized keeping in mind the overall modernization of the Public Administration at federal level has also laid down framework to be implemented at local (municipality) level. The idea is to create a fully integrated E-Government landscape in Germany which connected federal government, federal-state governments, and municipal administrations. This survey is an effort to explore the state of the art of E-Government services in Stuttgart metropolitan region and also examine the role of E-Participation to promote inclusion and empowerment. An appraisal of select best practice i.e. Sustainable Mobility management in Stuttgart is examined as how it fits to the dimension of emerging Smart cities and what challenges and issues it poses on planning a future oriented city. The inferences of this investigation are going to benefit development of emerging cities in developing world where efforts are being made to develop new smart cities. The basis of this survey is designed by providing answer to the following basic questions: How the functioning of administration made more transparent, efficient, innovative and responsive? How various administrative procedures are made available to citizens? How management of all these processes has been made understandable to all? How the information is disseminated to various departments and how the participation of citizens is ensured?

11 E-Governance an Effective Tool of Municipal Governance Cases from Hyderabad and Ahmadabad

Vinita Yadav discusses the E-Governance practice of Hyderabad in Telengana and Ahmadabad in Gujarat in India. Hyderabad is considered as smart city. With many E-Governance innovations introduced in Gujarat last few years, Gandhi Nagar may be considered as an emerging Smart City in 2014. This is rare in India since no New Towns in India had reached the status of Smart City which also includes Chandigarh the state capital of Haryana and Punjab and New Delhi the national Capital. While identifying smart cities for an award for the period 2007–2011 in the Asia region, five factors were taken into consideration: broadband connectivity; a knowledge-based workforce; digital inclusion; innovation; and marketing and advocacy. Three Indian cities, namely Bangalore, Hyderabad and Jaipur, figured in a list of 20 cities from the Asian region; four in China, four in Korea and three in Japan made it to the list. Hyderabad is one of the three declared Smart Cities of India, other being Bangalore in Karnataka and Jaipur in Rajasthan in India.

Ahmadabad Urban Agglomeration (UA) is a million plus UA with a population of 63, 52,254 in 2011. It is the seventh largest UA in terms of population. The city embarked on the path of ICT in 2002. One decade of progress made it into a path to Smart City. ICT is used for public service delivery through six civic centres located in five zones. Each of the e-city centres is having 10 nodes. It has increased to forty-six in 2013. E-Governance includes payment of dues such as property tax, vehicle tax, professional tax and other dues. The registration of marriages and issuing birth and death certificate has become easier after E-Governance. The issuing of licenses for shop, establishments, health, hawkers and granting building permission online has increased transparency. The development of grievance handling mechanism has not only ensured recording of complaints but also their redressed in the shortest possible timeframe.

Hyderabad with a population of 68, 09,970 is fifth largest metropolitan city. It is highly urbanized city and initiated a lot of steps to initiate good governance in Municipal Corporation. Smart city is the one, which utilizes the devices to increase efficiency. In Hyderabad, services provided by E-Governance are provision of birth and death certificates, trade licenses, property tax, advertisement fee, rent payment of commercial properties and payment of contractors. The real time records about street lighting, town planning, municipal park facilities and entomology are collected and analysed. There are multiple platforms i.e. Citizen Service (E Sava) Centres, Municipal Corporation offices, banks and web portal available 24 $\times$ 7 to provide such services. E-Sava is not only specific to Greater Hyderabad Municipal Corporation (GYM) but is implemented across Andhra Pradesh and Telangion. Mobile phones have been used for reporting disaster, co-ordinate political protest and finding the rates of a product to increase profitability but camera was not used for public domain (NIUA 2012). Offsite Real-time Monitoring System (OSRT) provides information about the distant events in real time. OSRT is a tool to ensure accountability with internal, upward and downward dimensions in public domain. The camera of mobile devices is used efficiently to register the workers attendance as well as status of cleaning the bins within the Greater Hyderabad Municipal Corporation (GHMC) spread over 625 km^2 area with 7.40 million population since 2011. The supervisors check the attendance of 14,000 workers on the basis of pre-scheduled jobs. The images of bin, their coordinates, bin and its surrounding area are captured and analysed to study the effectiveness of its implementation.

12 E-Governance and Its Role in Infrastructure Services of UAE, Case Study—Dubai

Ashmita Karmarker provides with the details of E-Governance of Dubai and how this city has transformed itself to adopt various means in order to establish itself as a smart city. Similar to other Emirates, Dubai had a Traditional Government. Traditional governments were always small, both in size and scope. However, this governance valued consensus as well as participation, and the traditional form of such participation would exist within the context of a *majlis* or council. In this framework, issues relevant to the community were discussed and debated. Opinions were expressed and the sheikh would take these opinions into consideration prior to taking a decision. Traditionally, the ruler of an emirate—the Sheikh—was the leader of the most powerful tribe, while each tribe, and often its sub-sections, also had a chief or sheikh. The Sheikhs maintained their authority only as long as they were able to retain the support of their people. This, in essence, was a form of direct democracy. Part of that process was the unwritten, but strong, principle that the people should have open access to their ruler, and that he should hold a frequent and open *majlis*, in which his fellow citizens could voice their opinions. It is now evident that it is these elements of governance that have served as a solid foundation in maintaining the unique identity of the country against a backdrop of rapid economic and social changes. This went on well as long as Dubai settlement was small in size. When Dubai became big in population, to replicate traditional governance it embraced ICT in all forms. Higher Internet penetration and mobile penetration was further augmented by city wide wireless Internet access. The E-Government of Dubai has implemented various ways to interact with its residents/visitors and provide them with easy access to government-related information, power to participate in government-related activities and also in day to day use of services. This chapter has highlighted especially on the smart infrastructure services which forms a major link between the residents and the E-Government of Dubai. Dubai is investing heavily in adopting and implementing ICT in its government and private sectors. The Global Information Technology Report 2010–2011 indicates that the UAE leads the MENA region in leveraging ICT for increased economic diversification and competitiveness. The E-Government programme in the UAE is a key initiative of the UAE-Government Strategy 2011–2013 that lays the foundation to achieve UAE Vision 2021.

The smart city Dubai has made a commendable progress in the field of security. The city records the identity of each person crossing the border through eye scanning and starts maintaining the database thereafter. And once the person re-enters the database is checked at the security point against the person's criminal records, outstanding fines and immigration details registered within in any emirates becomes accessible at the touch of a button. The UAE national identity management infrastructure is a strategic initiative to enhance homeland security and develop a federated identity management system enabling secure E-Government transactions. A federated identity is the means of linking a person's electronic identity and attributes, stored across multiple distinct identity management systems. Such systems would allow individuals to use the same user name, password or other personal identification as a part of the programme. The UAE issues smart identity cards for all of its population. The UAE national identity card is one of the world's most advanced and secure smart cards. The card is provided with identification parameters stored securely in the smart chip. It thus enables establishing a person's identity on-site (physically) and remotely (virtually), enabling secure and trusted transactions. The multi factor authentication which provides both match-on-card4 and match-off-card5 features, facilitates validation, verification and authentication of any given identity. The cardholder can then access all identity-based services sign into the networks of more than one enterprise in order to conduct transactions. Since technology drives the very core of a smart city, the usage of smart devices, the smart card technology has enabled the residents of Dubai to carry a single card to reach to any destination through public transportation service. These services include the metro railway, Dubai taxi, Dubai water transport and buses. The card is available at any transportation node, supermarkets, retail shops and railway stations.

Dubai electricity and water department (DEWA) is taking major steps in saving the resources by involving the residents in various awareness programmes. The launch of smart grid technology, smart meters for the consumption of the resources, various types of bill generations showing the consumption rate for each property and various award ceremonies for residents saving the resources are the steps taken by the E-Governments. The application which can be downloaded on mobile and Internet for the payment of the water and electricity and the services charges for a property, checking the history has been introduced by DEWA. This application can also be used for various enquiries and suggestions from the public side to the E-Government.

There are various measures taken by the Road and transport authority to control the flow of traffic within the city. The measures include the installation of toll gates, speed checking cameras along with the design measures for the roads such as access points, road diversions on and off street parking provisions as well. The devices used to maintain the traffic are controlled by the authority and are made connected to all the commuters through Internet and mobile phone apps. The apps help to check the vehicle registration details, traffic fines, payment of toll taxes, fines and invitation of management improvement suggestions from the users.

The payment for parking has been made easy and the commuter has been given the choice of payment

- At the parking spot by paying cash at parking meters at hourly basis
- Buy a parking card from the municipality centres and use the provision at monthly basis
- Use their mobile phones to pay the parking fees via m parking app as described below.

Dubai Police has launched its official mobile application as part of the M-government initiative announced by His Highness Sheikh Mohammed bin Rashid Al Maktoum, Vice President and Prime Minister of UAE and ruler of Dubai earlier this year. The Dubai Police app will provide access to various services it provides.

13 E-Governance for Public Realm

T.M.Vinod Kumar and Bimal in their chapter on E-Governance for Public Realm; structure a well organized participatory planning and design process to create form-based development codes for E-Governance of public realm in Panniyankara, Kozhikode. Although *Janakiya Asuthranam* or peoples planning exists in Kerala since mid-1990s, it can only articulate the felt needs of the ward. It has not graduated to detailed local area planning, urban design or generator of urban development codes. This gap is fulfilled by this chapter using participatory web-based interactive interface for planning and design of local areas by participants located around the study area or any part of the world. It may be noted that Master Plan in the scale that is prepared now in Kozhikode, cannot see the status and generate interventions on public realm locally unless Kerala accepts Zonal Plans integrated with Master Plans which is not practiced in Kerala. Hence public realm remains the most neglected aspect of spatial planning in Kerala. Open spaces and public realm appears to disappears in the last two Master Planning exercise of Calicut (Kozhikode) and now as the third Master Plan is prepared it is now leaving below than standard public space for healthy living of the people of Kerala and Kozhikode in particular. What is allocated at city level is much less than that is given in guidelines prepared for making public realm in Indian cities. The situation of public realm becomes worse now in Kerala. People of Kerala are agitating for their self-interest against public interest for lowering the standards of right of way for all roads including National and State Highways standards of Indian Road Congress.

The impact of higher mobility in Panniyanakra by the introduction of mono rail and location of station is the stimulus for this chapter. To preserve open spaces and subject it to peoples planning and design, a public realm website is hosted which open to interaction to all, including local people, outside people, government planners and other government functionaries to intervene in planning and design of public realm. The planning process follows the following sequence namely field surveys and analysis, SWOT analysis, strategic intervention, participatory visioning of the area, zonal planning, regulation plans, and formulate a series of standards for public realm such as public space standards, built form standards, landscape standards, architectural standards, signage standards and environment standards. Administrative process to regulate the area is also arrived at and way in which building permits are given based on form-based codes are also developed for the website. The website incorporates one alternate design and allows participating people to upload and suggest other feasible alternatives. Prior to implementation all inputs to the web is studied by an appointed high-level committee and implement it from time to time. This website is expected to be part of the Municipal website and can be utilized for granting building permits based on Form-Based Codes and regulation plans conforming to Zonal Plans and Master Plans. Urban development monitoring, and maintenance involving multiple agencies of Government can be undertaken seamlessly. It also provides for generation of periodic database for annual plans and municipal budget.

14 Use of ICT, Smart Systems and Smart Grids for Improving Governance in Electricity Supply in India

Veena Aggarwal and Parimita Mohanty examine electricity supply in the context of smart cities in India. The chapter describes with some examples how information, communication and digital technology is being used in the Indian electricity sector by distribution companies (discoms) to improve consumer's services, reduce technical and commercial losses and improve operational efficiencies. E-Governance is a means of enhancing quality of public service and public life through the use of ICT. The 'Smart' concept originally comes from the integration of ICT in products and systems with the proclaimed aim of optimising their efficiency. A smart efficient and sustainable energy infrastructure is one of the most important building blocks of a smart city. Smart cities of the future will overall consume lesser energy, use more renewable energy and will have low carbon footprint. This chapter describes some of the initiatives of electricity regulator of India which are supported by Information technology and is helping the regulator reach out to the public and thereby improve governance in the sector. The Electricity Sector in India has seen a series of reforms beginning with measures to increase private sector participation in capacity addition in the early 1990s to the reforms in distribution in early 2000s with the enactment of the Electricity Act, 2003 and setting up independent regulatory commissions. The establishment of independent electricity regulatory commissions was seen as a major move in the Indian power sector towards better accountability and transparency. It was assumed that crucial issues of pricing, competition, and consumer interest could be best taken by a statutory authority that would not be dependent on the state government. Institutionally, electricity distribution is not arranged at a city level but at a state level with commonly three to four utilities per state and each utility serving multiple districts comprising both urban and rural consumers. Hence a state utility will serve many cities coming under several separate Municipal jurisdictions. Overall Planning for electricity is undertaken Vis-a Vis the area served and may not necessarily be in tandem with administrative boundaries of cities. For electricity utilities, Municipal bodies are large consumers of electricity and offer prospects for trying out new initiatives on energy efficiency (for instance in water pumping) and renewable energy (for instance solar energy-based traffic lights and street lights). However from a planning perspective, there is not enough interaction between the two. The city level electricity supply companies mentioned in the earlier paragraph are an exception and could be interesting case-studies for understanding how cities can integrate Municipal services with electricity supply services. This chapter also discusses the support being provided by Government of India through the Restructured Accelerated Power Development and Reform Programme (R-APDRP,) and the recent developments on smart electricity grids. It tries to understand how smart systems in electricity converge with the overall initiatives on smart cities. It also discusses some of the challenges in the use of smart technology in electricity and also the institutional impediments in integrating electricity supply services with other Municipal services, as is one of the ultimate aims of smart cities. Notwithstanding these challenges, if the level of automation, integration and resource efficiency, as envisaged under smart cities does become a reality, urban dwellers will see a tremendous improvement in quality of life. Finally authors sketches the role of E-Democracy the major decision maker in power supply from generation to consumption. Smart grid initiatives which generally become part of Smart Cities are still in the very initial and preliminary stage in Delhi with not much of information in public domain, but the chapter uses a case study of a mini Smart Grid established in TERI campus to discuss the details of issues. The chapter first discusses some of the major challenges in the electricity supply in India. The Regulatory Information Management System is the major policy and programmes that are driving ICT in the electricity sector. The Regulatory Information Management System (RIMS) have not shown much progress in Delhi. Subsequently, the chapter focuses on some of the information technology driven initiatives that are helping utilities in improving their interface with consumers; and regulators in reaching out to consumers for improved transparency and accountability in the sector. The chapter examines some of the recent technology driven initiatives which are helping utilities improve their efficiency, reduce outages and losses. India's smart grid programme and its implementation are also discussed with a case-study of TERI's renewable-based smart mini-grid. The chapter examines the challenges in implementation of smart technologies in India, particularly concerns of costs vs benefits, issues of security of assets, data privacy issues, capacity constraints. It also examines the institutional impediments in integration of all basic urban services, including electricity, as envisaged in a smart city. Finally the chapter suggests that policy makers and regulators need to be careful in ensuring that the drive towards 'smartness' is cognisant of ground realities in India and that smart initiatives are implemented to facilitate and not inhibit universal service delivery.

15 E-Governance for Photo Voltaic Power Grid: Solar City Gandhi Nagar, Gujarat, India

Jignesh Bhatt and Omkar Jani discuss their experience in Gandhi Nagar, Solar City based on Smart Hybrid solar photovoltaic Grid. Electrical energy generation, distribution and consumption include different personnel at various hierarchical levels such as Energy Production Engineers, Power Transmission and Distribution Network Maintenance Team members, Billing and Corporate Agency persons as well as the End User or Consumer. Enabling simple, user-friendly, reliable, efficient, timely and seamless access to critical and vital information to all such personnel remains the most desirable feature for a Smart City. Reliable availability of critical data on timely basis for different decision making authorities plays key role for successful E-Governance of energy distribution systems. Emerging approaches based on hybrid communication technologies possess capability to transform the present ones into smart, intelligent and adaptive sites and can become basis of exploring more options including anytime-anywhere monitoring in future.

Solar Photovoltaic is the proven method of electricity generation with significant sustainable future potential. Grid-tied photovoltaic rooftop system is gaining wide acceptance as one of the most widely accepted models for wide-scale deployment due to their simplicity, ease of installation, operation and maintenance. E-Governance of power grid implies the delivery of electrical power assisted or supported by communication devices providing efficient and effective communication between Government to Government (G2G), Business to

Employee (B2E), Business to Government (B2G), and Business to Citizen (B2C) and finally, Government to Citizen (G2C) with responsibility and accountability.

Gandhinagar in Gujarat has emerged as city in India taking leadership in application of advanced hybrid communication technologies and thereby transforming itself into Smart City. Earlier Gandhinagar has been facing frequent power shortages as well as demand-supply fluctuations causing not only inconvenience, but also resulting in large work and production losses as well as chaotic situations. For sustainable growth and enhanced lifestyle of citizens, reliable and affordable power is mandatory requirement and Gandhinagar city administration has been doing its best to approach every citizen through hybrid communication network especially for power distribution and thereby resulting in a strong trustworthy relationship of citizens with the government. Today Gandhinagar is considered to be one of the most advanced and world-class self-reliant city having most resilient power infrastructure in the India. Well planned, systematically designed, and customer oriented power distribution network encouraging participation of citizens has played key role in making Gandhinagar more liveable and a place to stay today. Power supply in Gandhinagar has been distributed into separate zones and interested citizens are invited to express their willingness to participate in the programmes and after completion of formalities including registration, document verification, scope and size of solar PV installations are fixed up. Commercial aspects such as subsidy, revenue sharing and billing details are worked out on actual basis. Using seamless data connectivity using Global Positioning System (GPS), GIS and Virtual Instrumentation GUIs, real-time data of generation and consumption has been availed to the end user for determination and implementation of his choice to run important equipment's, if any.

Remote Energy Parameter Monitoring System using Hybrid Networking Technologies for Photovoltaic Power grid of Gandhinagar Solar City has been presented with necessary details. Due to relatively reasonable installation cost, new installations can be planned with presented system and existing ones can be expanded in terms of coverage. Thus obtained savings in installation investment can further encourage inclusion of more sensors and hence increase overall scope to acquire more parameters. It also enhances spatial resolution that results in more fine-grained measurements and better monitoring of critical regions.

Emerging approaches based on hybrid communication technologies are expected to transform the present ones into smart, intelligent and adaptive sites and can become basis of exploring more options including anytime-anywhere monitoring in future.

16 Can Smart Cities Be Inclusive Cities? Spatial Targeting (ST) and Spatial Data Infrastructure (SDI)

Smart City is often considered as an all-inclusive city integrating various aspects of Quality of Life of the people (QOL). There is also a belief that Smart cities excel in exclusion of lesser smart people from main stream development. In this context,

Sridharan asks an important question; can Smart Cities be inclusive cities? He finds his answers in Spatial targeting (ST) and spatial data Infrastructure (SDI). Spatial Data Infrastructure is all about facilitation and coordination of the exchange and sharing of spatial data between various stakeholders from different territorial levels. Unless spatial inclusion is planned and managed through Spatial Data Infrastructure some of its socio-economic and spatial issues can be left behind in the development process. Use of SDI in city governance reduces efforts of citizen in terms of time lost in dealing with the public services. However, little attention is paid to how this knowledge could be used in strategic urban governance. Through a case study of Hubli-Dharwad City in Karnataka, his chapter shows how the poverty hot spots are left behind though many innovative programmes are carried out in the city through E-Governance. It points out the need to introduce SDI in Cities as an essential part of Smart City building process especially in E-Governance of Smart Cities. Lack of recognition of Spatial Data as an Infrastructure, and including Spatial Data Infrastructure as one of the infrastructure in the planning process has resulted in spatial, infrastructure and governance exclusion in the city development. Though at the national level, there is National Spatial Data Infrastructure (NSDI) as a pioneering organization, its role and relevance has been subdued and its backing through State Level Acts and City level implementation will solve the problems associated with spatial exclusions. Smart city functions on the basis of data, and its backbone is Spatial Data. Truly smart cities are long way off. We have seen SDI creates 'digitally inclusive regeneration platforms' (E-Topia) where various stakeholders such as users of different infrastructure, decision analysts (technocrats) and decision makers (bearcats and politicians) can converge. However, in the absence of this convergence the decisions on city governance and budget allocation become exclusionary. Added to this, 'regulatory fractures' that are widespread in Indian cities, which further alienates the marginalized groups in terms of infrastructure, spatial and governance access. He hope that the 100 Smart Cities proposed by the present Government, will bring in SDI as an important component of city governance system.

17 M-Governance: Smartphone Applications for Smarter Cities: Tapping GPS and NFC Technologies

Ummer Sahib discusses role of M-Governance in providing easy and convenient access to government services when and where required in a Smart City. M-Governance is the delivery of services via mobile communication devices. It is a tool and method that facilitates citizen to citizen, citizen to government, government to citizen, Government to employees and contractor to government interactions with agility and ubiquity. With one in five people having smartphones in the world today and this figure being double in cities, the delivery of services on smartphones is fast becoming a norm. However, the smartphone have-nots will have to seek middle tire enablers to avail these services, as has been the case with E-Governance services. The objective of this chapter is to describe Global Positioning System (GPS) and

Near Field Communication (NFC) technologies that are built-in smartphones and to explore their potential applications in M-Governance.

Embedding GPS technology on smartphones has made "Location" easy to know and measure, just how wrist watch has made knowing and measuring "time" easy. Today, GPS technology is used by a common man without even knowing its definition and thanks to web mapping technologies that gives context to a GPS location. The GPS technology section of the chapter explores Location centric M-Governance applications that are currently used and explore new applications.

NFC Technology is a close sibling of GPS where by "Proof of Presence" Is used to manage mobile work forces of government in offering government services and to monitor contractors adhering to service provisions. The NFC technology section of the chapter focuses on various NFC applications on Government to Employees and contractors. The author presents a global use of GPOS and NFS technologies in M-Governance.

18 E-Budgeting of Kozhikode Municipal Corporation

Naseer, Bimal and T.M.Vinod Kumar surveys the existing participatory budgeting procedures internationally briefly and in details what is practiced in Kozhikode Municipal Corporation using the most recent experiences. This budgeting process is based *on Janakiya Asuthranam* (Peoples Planning) developed since 1990s. Here the annual planning and budgeting become a people's movement than a clerical work in the Municipality. Here the policy postulated by State Planning Board of Kerala from time to time is integrated with the people's perception of how money should be spent spatially by wards and municipality based on current issues. Since it is a celebrated model of participatory budgeting practiced in Kerala for many decades, it is presented in all details. This chapter attempts, e-budgeting innovation in allocation of annual budget spatially using Spatial Decision Support System (SDSS). From the real example of budget allocated for a project such as road maintenance and distribution of public water taps, in several municipal wards, the use of SDSS in GIS is demonstrated for Kozhikode Municipal Corporation. Here attempts are made to use Geographic Information System in municipal budgeting and budget allocation by providing Spatial Decision Support System Annual budgeting in Kozhikode is rule based with definite logic of spatially distributing allocation by wards and other administrative divisions within Municipality. E-Budgeting is a spatial data intensive exercise and Geographic Information System is the best tool available to update necessary spatial information for helping people decide in the e-budgeting process as an ongoing exercise.

19 Smart City E-Governance Futures

No nation can have 100 % of their cities as Smart Cities barring a few city states like Singapore. So far India have only three smart cities in three states namely Bangalore in Karnataka, Hyderabad in Telengana and Jaipur in Rajasthan, with two from south India and one from North India but have nearly 8,000 towns. Asia has about twenty smart cities. There are attempts to design smart new towns in Industrial corridors in India which is not yet complete. While Indian constitution is the basis of City Governance, the design of E-Governance need not be one and the same for all cities and all potential smart cities in all states and union territories. This is true world over. The experience and culture of smart living in a smart city by smart people shall shape the design of smart E-Governance. Also smart cities can learn the innovation practiced in another smart city. In smart cities intensity of interaction of practice of regulation with e-democracy may be intense, but it may not be of the same intensity in non-smart cities. While it is easy to advocate a minimum set of E-Governance system for all Municipalities in India, it is not possible to spell out E-Governance systems for all Smart Cities. The smart city has to decide, design and implement E-Governance System as per their needs which can be different and showing greater amount of innovation. As per Indian constitution, it is the State Government that legislate municipal constitution by Municipal Legislation and municipality design and adopt City E-Governance for their use. The Union Government can at the most induce design and development of E-Governance system by grants and aids. The aim of Smart City E-Governance is to simplify the implementation of existing regulatory governance by discarding those that has no relevance today and further simplifying in terms of time and money spent in regulatory regime to the most efficient and less time consuming procedures for citizen. This can reap great benefits in City Governance.

It is not necessary that all smart cities in India shall be million plus cities or mega cities which is already documented earlier in this chapter. There can be smart cities having less population and spread over many islands as in the case of Lakshadweep and there can be some mega cities and some million plus cities as indicated in the earlier part of the chapter. It is also true that size of city alone do not make the city for candidate for up gradation to the smart city. Since smart city is a complex system consisting of smart people, smart economy, smart environment, smart mobility, smart governance, and smart living, it is not possible that every city shall show the emergence of this complex system. The absence of this six system means possibility of emergence of smart city there is minimal today but can be there tomorrow. Hence it is impossible to practice equity in distribution of equal number of smart cities in each states and union territories like distribution Indian Institute of Technology, Indian Institute of Management and All India Institute of Medical Sciences by each state.

20 Smart City E-Governance and Continuing Education

In non-smart cities Governance is an activity of the empowered Government officers. The officers write examinations and attend training courses from time to time to attain proficiency in their work. They generally conduct their work with no participation of people and no transparency. On the other hand in smart cities, there is full participation of people in Governance and all activities related to governance are transparent. Policies of regulation shall be fully convincing to people and they jointly regulate in this governance system. The first requirement is smart city population shall be smart population and there shall be ample scope for continuous learning on all aspects of city life including participatory governance. In such a situation government servants shall access superior expertise in information technology use, management and related governance discipline than that exists today. In other words we require smart government servants with considerable expertise. Policies and plans being prepared by the elected Members of Parliament and legislative assembly shall be fully known to all citizens before legislation is passed in Parliament and Assemblies and there shall be scope for discussion and modification by informed smart citizen using ICT. Annual budget shall be totally participatory using E-Governance system and Spatial Decision Support System and Geographic Information System. Smart citizen shall have full grasp of geospatial technologies that help them decide their budget. This calls for provision of continuous education for all using E-Education system. They should be part of collecting and collating geospatial information in GIS by field surveys annually for their use in the next budget.

21 Smart City E-Governance and ICT

There shall be a policy, plan and budget for total percolation of ICT in all urban land uses and all smart city activities. Latest broadband technologies like WiMAX and LTE shall be cheaply available to all. Last mile connectivity shall be executed in the most cost-effective manner. There shall be free Wi-Fi in all public spaces such as airport, railway station, planes, trains, ship, market places, schools, factories and so on. The status of ICT in India has been discussed in this chapter.

22 Smart City E-Governance and E-Commerce

E-Commerce and E-Governance shares the same eco system. E-Commerce has given us technologies like big data processing and cost- and time-effective logistical system. They are also in the path of continuous innovation and adaptation to local endowments and conditions. If a Government policy encourages E-Commerce and builds bridges between E-Governance and E-Commerce for effective utilization of information that can be in public domain, it is possible to develop these two differing activities in the same eco system most efficiently. On the basis of experience in mobile phones and E-Commerce, these sectors can generate their own funds for their expansion since it brings about more profits and consumer benefits from low price of the commodity and gets more choice, which may not be available in the market.

23 Smart City E-Governance and Smart Infrastructure

Smart infrastructure is intelligent infrastructure which is fully aware of its functioning and is capable of managing its functioning with a control system like SCADA. Human interventions waste considerable response time and is therefore replaced by smart infrastructure with non-human intervention dictated by control system. While smart electric grid is being established in many parts of India, smart water supply systems are slow in its growth in India which is the main reason this book do not have a chapter on E-City Governance for Water Supply . Even smart grid connections are in early stages in India and are so early to conduct an evaluation study. Delhi metro uses miniSCADA for its management which is still un-documented. Delhi Metro Corporation uses a mini SCADA and a water supply system being commissioned in Kozhikode also uses SCADA. It is hoped more such smart infrastructure with high capacity Governance system inbuilt may be used in Smart Cities. The main challenge is to find out appropriate technology for India and have Research and Development conducted in IITs and NITs and fast deployment in smart cities and even in non-smart cities. One should be careful that smart city technologies with Governance modules in built are having small shelf life and depending on other countries is not a good policy. The policy India follows on defence hardware like joint production need to be encouraged for Smart City technologies and foreign direct investment.

24 Smart City E-Governance and Smart Cards

India is in the process of implementation of Unique Identity system called Aadhar and also compiling population register work as part of census 2011 which seems to be complete barring final processing and issue of card. It covers iris mapping and finger printing and collects all relevant information of address, Bank account, Driving licence, LPG cylinder registration and so on. It is capable of linking with many other databases in a microchip in the smart card. Now it is important to use this information embedded in the smart card in the most efficient manner. For example a person using metro need not stand in a queue and buy ticket but smart card automatically using sensors record the station where he got into the metro and where he got out of metro and a charge is automatically deducted from his bank account. The same functions can be extrapolated to all Government services. Here not a second is wasted by the person for this. Also if a minor traffic offence is committed by a car, the database linking smart card and bank of the owner of the car can be fined automatically and instantly and money withdrawn by transport authority from the bank. Not a single second is wasted here. The same thing can be done when gas cylinder is supplied by sensing the smart card, purchase at retail trade and also many such uses such as crossing a toll payment station on a road. These are all smart governance systems using smart card, which is totally absent in India.

25 Smart City E-Governance and Smartphones

Most operations that can be done by smart card and laptop can be conducted with smartphones. With rapid rise of per cent of mobile using population in India, it is possible to use smartphones for E-Governance. Bank transaction using electronic payment can be effected by smartphone and it is also possible to pay utility bills and a host of other government services. Here also mobile application development and its deployment is lagging in India.

References

1. Human Development Report 2010, UNDP National Council of Applied Economic Research, Planning Commission Government of India 2012, India Human Development Report 2011 Towards Social Inclusion
2. FICCI, Ernst & Young (2009) Making the Indian higher education system future ready. New Delhi
3. Ministry of Human Resources Development GOI and Confederation of Indian Industry (2013) ASHE-2013 Annual status of higher education of states and UTs. Deloittee Nov 2013
4. FICCI, Ernst & Young, Planning Commission (2012) Higher education in India twelfth five year plan (2012–2017) and beyond. New Delhi
5. Planning Commission and Confederation of Indian Industry (2012) ASHE 2012 annual status of higher education of States and UTs in India 2012 (Summary report). New Delhi
6. Government of India and Confederation of Indian Industries (2013) ASHE 2013 annual status of higher education of States and UTs in India. Deloittee Nov 2013
7. Telecom Regulatory Authority of India (2013) The Indian telecom service performance indicators Jan–Mar 2013. Mahangar Doorsanchar Bhavan, New Delhi
8. Earnst & Young (2012) Rebirth of e-commerce in India. New Delhi
9. McKinsey & Company (2013) Online and upcoming: the internet's impact on India. New Delhi
10. India Brand Equity Foundation (2012) The rise and rise of e-commerce in India, New Delhi quoted from World Bank, IAMA and ARNACA Research
11. World Bank (2011) World development Indicators World Bank UNESCO Global Education Digest 2011
12. United Nations Department of Social and Economic Affairs (2012) E-Government survey 2012. E-Government for the People, New York
13. Internet World Statistics (2013) KPMG and internet and mobile association of India report

MIX
Papier aus verantwortungsvollen Quellen
Paper from responsible sources
FSC® C105338

If you have any concerns about our products,
you can contact us on
ProductSafety@springernature.com

In case Publisher is established outside the EU,
the EU authorized representative is:
Springer Nature Customer Service Center GmbH
Europaplatz 3, 69115 Heidelberg, Germany

Printed by Libri Plureos GmbH
in Hamburg, Germany